TRANSITION TO HYDROGEN

Pathways Toward Clean Transportation

This book is a comprehensive and objective guide to understanding hydrogen as a transportation fuel. The effects that pursuing different vehicle technology development paths will have on the economy, the environment, public safety, and human health are presented with implications for policy makers, industrial stakeholders, and researchers alike. Using hydrogen as a fuel offers a possible solution to satisfying global mobility needs, including sustainability of supply and the potential reduction of greenhouse gas emissions. This book presents research on issues that are at the intersection of hydrogen and transportation, since studies of vehicles and energy carriers are inseparable. It concentrates on light-duty vehicles (cars and light trucks), set in the context of other competing technologies, the larger energy sector, and the overall economy. The book is invaluable for researchers and policy makers in transportation policy, energy economics, systems dynamics, vehicle powertrain modelling and simulation, environmental science, and environmental engineering.

ALEXANDER WOKAUN is Head of the General Energy Research Department at the Paul Scherrer Institute, Villigen, Switzerland, and is a Professor at ETH Zurich. He received his PhD from the Swiss Federal Institute of Technology (ETH) in Zurich, and subsequently worked at IBM and the Bell Laboratories. From 1982 to 1985, he directed a programme on surface energy transfer processes at ETH Zurich. In 1986, he became Full Professor of Physical Chemistry at the University of Bayreuth, with a research programme comprising studies of materials surfaces, catalytic reaction mechanisms, and laser surface modification. Since 1994, he has led the General Energy Research Department of the Paul Scherrer Institute, and has taught in the Chemistry Department at ETH Zurich. Development of a sustainable energy system is the focus of his research programme, which comprises technologies for the use of renewable energies, energy storage, clean and efficient energy converters, and assessment of the environmental consequences of energy use. Since 2006, he has been a member of the Advisory Group on Energy of the European Union. He chairs the Steering Committee of the Competence Center Energy and Mobility, which facilitates large-scale collaborative projects of ETH domain institutions. In July 2007, he was appointed Vice Director of the Paul Scherrer Institute for the area of energy research.

ERIK WILHELM earned his Doctorate of Science from ETH Zurich in the Laboratory for Energy Systems Analysis at the Paul Scherrer Institute. His research focus is on powertrain simulation, hybrid control optimisation, and heuristic vehicle design. He has an Honours Bachelor degree and a Masters in Chemical Engineering from the University of Waterloo, Canada.

TRANSITION TO HYDROGEN

Pathways Toward Clean Transportation

Edited by

ALEXANDER WOKAUN
ERIK WILHELM
Paul Scherrer Institute, Switzerland

CAMBRIDGE
UNIVERSITY PRESS

CAMBRIDGE UNIVERSITY PRESS
Cambridge, New York, Melbourne, Madrid, Cape Town,
Singapore, São Paulo, Delhi, Tokyo, Mexico City

Cambridge University Press
The Edinburgh Building, Cambridge CB2 8RU, UK

Published in the United States of America by Cambridge University Press, New York

www.cambridge.org
Information on this title: www.cambridge.org/9780521192880

First published 2011

Printed in the United Kingdom at the University Press, Cambridge

A catalogue record for this publication is available from the British Library

Library of Congress Cataloguing in Publication data
Transition to hydrogen : pathways toward clean transportation / [edited by]
Alexander Wokaun, Erik Wilhelm.
p. cm.
Includes bibliographical references and index.
ISBN 978-0-521-19288-0 (hardback)
1. Hydrogen as fuel. 2. Fuel cell cars. 3. Alternative fuel vehicles. 4. Transportation–Energy conservation. 5. Environmental responsibility. 6. Sustainable living.
I. Wokaun, Alexander. II. Wilhelm, Erik. III. Title.
TP359.H8T73 2011
629.22′9–dc22 2011008035

ISBN 978-0-521-19288-0 Hardback

Contents

The colour plates will be found between pages 76 and 77.

Contributors

Robert Alvarez
Empa – Swiss Federal Laboratories for Materials Science and Technology, Laboratory for Internal Combustion Engines, Ueberlandstrasse 129, CH-8600 Duebendorf, Switzerland

Christian Bauer
Laboratory for Energy Systems Analysis, Paul Scherrer Institute, CH-5232 Villigen PSI, Switzerland

Steven W. Bond
Empa – Swiss Federal Laboratories for Materials Science and Technology, Laboratory for Air Pollution/Environmental Technology, Ueberlandstrasse 129, CH-8600 Duebendorf, Switzerland

Mathias Bosshardt
General Energy Research Department, Paul Scherrer Institute, CH-5232 Villigen PSI, Switzerland

Brigitte Buchmann
Empa – Swiss Federal Laboratories for Materials Science and Technology, Laboratory for Air Pollution/Environmental Technology, Ueberlandstrasse 129, CH-8600 Duebendorf, Switzerland

Timur Gül
Energy Economics Group, Paul Scherrer Institute, CH-5232 Villigen PSI, Switzerland

Stefan Reimann
Empa – Swiss Federal Laboratories for Materials Science and Technology, Laboratory for Air Pollution/Environmental Technology, Ueberlandstrasse 129, CH-8600 Duebendorf, Switzerland

Warren W. Schenler
Laboratory for Energy Systems Analysis, Paul Scherrer Institute, CH-5232, Villigen PSI, Switzerland

Andrew Simons
Laboratory for Energy Systems Analysis, Paul Scherrer Institute, CH-5232 Villigen PSI, Switzerland

Hal Turton
Energy Economics Group, Paul Scherrer Institute, CH-5232 Villigen PSI, Switzerland

Silvia Ulli-Beer
Dynamics of Innovation Systems Group, Paul Scherrer Institute, CH-5232 Villigen PSI, Switzerland

Martin Weilenmann
Empa – Swiss Federal Laboratories for Materials Science and Technology, Laboratory for Internal Combustion Engines, Ueberlandstrasse 129, CH-8600 Duebendorf, Switzerland

Erik Wilhelm
Laboratory for Energy Systems Analysis, Paul Scherrer Institute, CH-5232, Villigen PSI, Switzerland

Alexander Wokaun
Head of the General Energy Research Department, Paul Scherrer Institute, CH-5232, Villigen PSI, Switzerland

Foreword: on the transition to hydrogen

The transition to a sustainable energy supply is one of the major challenges that humans will face during the twenty-first century. This transition is inevitable, but there are many scenarios discussed for how and when this will happen. Humans have relied on renewable energy for most of history, and will do so again, as affordable supplies of fossil fuels decline in the coming decades (or centuries). However, there is wide agreement that the world should not give up the benefits of modern technology. The transition to a 'green' energy supply has already begun, and technologies for collecting and converting energy from the environment, new means of energy storage, and increased energy efficiency have progressed greatly. That is why there is legitimate hope that the coming change will be possible without a major reduction in the quality of life.

Individual mobility and long-distance travel have become vital elements of human existence. This mobility and the freedom it enables are taken for granted by most, and even considered a fundamental right. Most of this mobility is provided by the more than 900 million motor vehicles that now populate roads around the globe. An enormous number – and expected to grow to more than 1.1 billion in less than a decade. If all the motor vehicles on the globe were put bumper to bumper, the resulting giant traffic jam would be 4.5 million kilometres long, or wrap around the globe more than 100 times.

The vast majority of these vehicles are fuelled by gasoline and diesel – road transport is more than 96% reliant on fossil petroleum. It is obvious that sooner or later alternatives will be required. This will require a major transformation for our present society. From a historical perspective, the age of oil is a very brief and recent interlude that will disappear again soon. So the question is not if we are able to make this transition, but how we will manage it. Road transport is not the largest end-use for fossil energy, but technologically it is very challenging to change. Liquid hydrocarbons have a very high energy density and can be easily stored under ambient conditions onboard vehicles. Combustion engines are robust and reliable under different climates, and mass production has made them relatively affordable. Increasing the efficiency of conventional motor vehicles and their internal combustion engines is therefore the easiest and most efficient way to conserve energy and extend fossil energy reserves.

The energy sector will increasingly produce electricity from renewable resources; in Europe, this is mainly from wind power for economic reasons. The electrification of the automobile is therefore seen by many as an answer to the question of sustainable transport with energy stored onboard vehicles in the form of electrolytic hydrogen or in lithium batteries. Battery electric vehicles have a high tank-to-wheel efficiency. For well-defined driving patterns, and if recharging is possible while parked, a battery electric vehicle may offer the simplest substitute for a conventional car. Although a hydrogen system at first glance might appear less efficient and more complex, it also offers great potential to store large amounts of energy, which can be filled during off-peak hours where generation capacity exceeds demand. Thus, hydrogen may be available for vehicles as a valuable 'by-product' of the future energy sector. Hydrogen vehicles also profit from the high gravimetric energy density that hydrogen offers, i.e. hydrogen vehicles offer relatively large range and fast refuelling, with no limitation on vehicle size. The amount of on-board energy also allows for safety and comfort features that we are used to, such as heating and air conditioning, without sacrificing too much cruising range. A vehicle-use profile similar to that enjoyed with today's conventional cars will thus be possible with fuel cell cars.

Fuel cells, hydrogen technology, batteries, and electric propulsion systems – the technology elements that are needed for a transition to sustainable road transport– seem to be ready. A development path exists for these technologies, and cost figures and efficiencies are described in a sufficient detail. The question, however, remains how the shift from the current technology will take place. The new technologies are disruptive in the sense that they will change the automotive business in many ways. Electricity and hydrogen as energy carriers for transportation will require a completely different supply and distribution infrastructure compared to oil. Car manufacturers will also likely need to change their business model and their supply chain. Although it may require years or even decades until the change appears, it is likely that, once triggered, the transition will occur rather rapidly. Thus, it makes sense to prepare by developing technology and by playing through possible future scenarios.

Most of the scientific literature on the use of hydrogen and fuel cells for mobility focuses on the technology of vehicles and their components. To evaluate the ecological benefits of zero-emission vehicles propelled by electricity from hydrogen or batteries, it is also important to consider the path of hydrogen (or electricity) production, distribution, and supply. Only a true well-to-wheel calculation can provide the basis for comparing transport systems. Several studies have addressed well-to-wheel pathways, and this becomes a fairly complex task, especially if fluctuating sources like wind power must be taken into account. For example, analysing system efficiency becomes irrelevant if a battery cannot absorb all the charging energy available, or if charging energy is needed when renewable energy is scarce or unavailable.

The effects considered in a well-to-wheel analysis should include not only energy efficiency and CO_2 emissions, but also evaluate and compare second-order factors like particulate and SO_2 emissions and potential hydrogen losses. Furthermore, a fair comparison should incorporate life-cycle analysis of all relevant components, i.e. vehicles and infrastructure. Only the future will tell if the assumptions and calculations in this book are comprehensive and correct, or whether the scenarios proposed will materialise. However, the scenarios should indicate at least a basic direction in dependence between the assumed variables.

This book presents such a broad approach, providing a summary of well-to-wheel studies that incorporate a comprehensive set of parameters. As expected, there are no simple conclusions on the 'best' technology path, but the results of the studies presented may help us to understand the options and their implications for our ecology and economy. The results also show clearly that after the oil age has been left, it is unlikely a single well-to-wheel energy path will ever again account for 96% of the energy used for road transport.

Dr Rittmar von Helmholt

Preface

When the project that forms the basis for this book was initiated in 2006, hydrogen was enjoying immense popularity as a possible alternative to fossil fuels as an energy storage vector. Record-breaking attendance at fuel cell exhibitions was common, and the news media was full of the ambitious market introduction plans of fuel cell company representatives and the grand hopes of politicians. It is now 2011, and the euphoria surrounding hydrogen has shifted to batteries and all-electric vehicles; the 'electron economy' is being advertised in much the same way. Hype surrounding transportation technology is a common phenomenon, and history provides us with many examples of announcements of 'major breakthroughs'. Some, like fuel injection, caught on; others, like the 'Nucleon' nuclear powertrain, did not. Whether hype hampers or helps society to find and to implement the best solutions for our pressing social and environmental problems is debatable. What is certain is that a thorough analysis of technological alternatives, and perseverance in execution once the best solution (or set of solutions) has been identified, should never be superseded by action plans driven by media attention.

The Alliance for Global Sustainability is the framework that facilitated collaboration between John Heywood's group at the Massachusetts Institute of Technology and Alexander Wokaun's team at the Paul Scherrer Institute in investigating vehicle technology alternatives 'Before the Transition to Hydrogen'. With generous financial support from the Swiss Competence Center for Energy and Mobility (CCEM), these two working groups met biannually with sponsors and partners from industry and professional associations, to share ideas and to engage in exciting multidisciplinary research. The results of the Swiss team's research are contained in this book. Even though, occasionally, the opinions of the partners from opposite sides of the Atlantic regarding the potential of the various transportation energy technologies investigated did not converge, the collaboration was highly successful because of the high value that both research teams placed on objective and unbiased analysis. In particular, the MIT group's strength in combustion engine technology and policy research proved to be highly complementary to the PSI group's technical know-how regarding fuel cells, global energy system modelling, and scenario analysis.

This book was facilitated to a large degree by the professionalism of the very capable chapter authors who contributed content, peer-reviewed each other's work, and participated in lively discussions enabling effective quality control. The challenge inherent in pulling together five PhD dissertations that focused on very different aspects of hydrogen in transportation was never overwhelming, and the editors gratefully acknowledge the support of all 15 authors. We would like to specifically thank Dr Philipp Dietrich for contributing his technical support throughout the project and for providing his invaluable assistance with this book as part of the group of authors. We are deeply indebted to Dr John Heywood for his unwavering engagement in the 'Before the Transition to Hydrogen' project together with the members of his analysis team.

That this book so quickly found a publisher, and that our sponsors continue to be interested in our results and supportive of our efforts, testifies to the fact that, although hype can support magazine sales, only objective analysis can support good ideas.

Executive summary

Increasing the security of the supply of energy for transportation and reducing environmental impacts from personal vehicles are prominent among the great challenges of the twenty-first century. To investigate pathways toward achieving these goals, a wide variety of analytic approaches are used in this book including life cycle assessment, heuristic vehicle design and drivetrain simulation, evaluation of hydrogen emissions from industry and transportation, modelling of technology innovation and market penetration, as well as energetic and economic modelling of future transportation scenarios. This coordinated effort has analysed a broad portfolio of technology options and implementation strategies with an emphasis on hydrogen as a transportation fuel. A summary of the main conclusions follows.

Reducing energy use – Reducing overall energy use depends on the combination of vehicles, drivetrains, energy carriers, and energy resources. Hybrid technologies can reduce non-renewable vehicle energy use by 10–60% (from mild hybrid to plug-in series hybrid architecture). Reducing vehicle weight using the lightweighting technologies considered can reduce energy use by 0.2–5% (keeping vehicle size and powertrain constant). Switching from current fuels to alternate fuels or electricity can reduce vehicle energy use by 5–75%, depending on the fuel or electricity mix considered. Electric vehicles shift conversion efficiency losses from the car to the power plant, so the maximum primary energy saving is not 75%, but rather about 40–60%, depending upon the efficiency of generation.

Reducing CO_2 – Hydrogen only represents an attractive transportation fuel for meeting stringent CO_2 reduction targets if it is produced using primary energy resources with zero or low CO_2 contents. Life cycle assessment of fuel production and consumption pathways reveals that there are several alternatives to conventional fuels that offer slight reductions in CO_2 emissions, but only a few that reduce

CO_2 by more than 50%. The highest reduction is achieved by using electricity from renewable or nuclear sources in electric vehicles. The next most effective is using hydrogen produced by electrolysis, followed by hydrogen produced from fossil fuels with carbon capture and storage.

Hydrogen production – Starting with the surplus capacity of current industrial hydrogen production reduces the barriers to the development of hydrogen production and distribution. A range of concepts is available for both centralised and decentralised hydrogen production from low-CO_2 primary energy. The associated costs are significant, but comparable with the anticipated costs of adaptation to a changing global climate.

Hydrogen emissions and levels – A future energy system using hydrogen for 10% of the total final energy demand by the end of the twenty-first century will likely have anthropogenic emissions of molecular hydrogen to the atmosphere that remain in the same range as today's levels of hydrogen emissions. The clear advantages of hydrogen-based transportation in reducing greenhouse gases and other pollutant emissions are therefore unlikely to be accompanied by any relevant negative effects due to the hydrogen leakage.

Increasing security – The insecurity of supply represented by an oil-based transportation system is based on both the geographic unevenness of the resource distribution and the rising cost of the remaining supplies. As the transportation sector shifts to a different mix of primary energy resources and energy carriers, we can expect that the primary energy resources (gas, coal, nuclear, and renewables, in particular, biomass and solar energy) will have both increasingly long time horizons and generally broader geographic distributions. The shift in energy carriers (from conventional fuels to synfuels, hydrogen, and electricity) will be increasingly regional. Both of these changes tend to increase the security of supply.

Restructuring the transportation sector – Achieving stringent climate protection goals in order to restrict a rise in global temperature will require a thorough restructuring of the global energy system at a significant cost. Radical changes in the transportation system will be integral to this restructuring, with dynamic shifts in the market shares of competing vehicle technologies. Self-reinforcing interactions between citizens, car manufacturers, and industry, as well as regulators and governments will be required. In addition to financial incentives, the inherent attractiveness of the environmentally friendly technologies will be of key importance in driving these changes.

Regional differences – Regional studies and fleet simulations show that the emerging solutions will differ between continents and regions. Regional availability of fossil fuels will favour the continued use of conventional and hybrid technologies, while the market entry of alternative technologies will depend on the availability of biofuels, on the one hand, and on cost reductions for battery electric and fuel cell vehicles, on the other.

Policy measures – Policy measures are needed to support the transition to a clean transportation system in both the short and long term. Global scenarios clearly show that, in the absence of climate policy, only low-cost measures, such as lightweighting and the introduction of hybrid technologies, are likely to be implemented. Only strict greenhouse gas targets would encourage biofuels and transportation based on renewable and/or nuclear energy. Hydrogen fuel cell vehicles represent an attractive option for a climate policy that limits atmospheric CO_2 concentrations to 450 ppm.

In summary, no single solution exists to facilitate the transition to a clean and secure future transportation system. Advanced conventional and hybrid vehicles running on conventional, synthetic, or biomass-derived fuels are expected to co-exist for decades with electric vehicles based on rechargeable batteries and fuel cells. Low CO_2 electricity is expected to make an important contribution to lowering the greenhouse gas footprint of transportation, both by direct use in battery electric vehicles for shorter driving ranges and for decentralised hydrogen production.

While change can only come gradually, the changes required to deal with the global transportation energy problem will require a paradigm shift in perceptions and attitudes that must be led by individual countries, communities, and stakeholder groups. The developed world's technology and economic resource advantages mean that it will necessarily lead the supply side of implementing such shifts in the transportation sector, while the developing world's greater growth rates mean that it can play a proportional role in redesigning transportation demand.

Abbreviations

ATR	autothermal reformer
BAU	business-as-usual
C	carbon
CADC	Common Artemis Driving Cycle
CC	combined cycle
CCS	carbon capture and storage
CED	Cumulative Energy Demand
CGR	coal gasification and reforming
CH	Switzerland
CH_4	methane
CHP	combined heat and power
CO	carbon monoxide
CO_2	carbon dioxide
DDP	deterministic dynamic programming
EC	European Commission
ECU	engine control unit
EEFSU	Eastern Europe and Former Soviet Union
E_{H2}	hydrogen emission factor
E_{CO}	carbon monoxide emission factor
Empa	Swiss Federal Laboratories for Materials Science and Technology
EPR	European pressurised reactor
FAO	Food and Agriculture Organization of the United Nations
FCV	fuel cell vehicle
GDP	gross domestic product
GHG	greenhouse gas
GMM	global multi-regional MARKAL (model)
ΔH	enthalpy change
H_2	hydrogen

H_2/CO	molar ratio of hydrogen to carbon monoxide
H_2O	water
HDA	heuristic design algorithm
HEV	hybrid-electric vehicle
HFCV	hydrogen fuel cell vehicle
HHV	higher heating value
HVDC	high-voltage direct current
ICE	internal combustion engine
ICEV	internal combustion engine vehicle
IEA	International Energy Agency
IGCC	integrated gasification combined cycle
IPCC	Intergovernmental Panel on Climate Change
ISO	International Organization for Standardization
J	joule(s)
Kph	Kilometres per hour
l	litre(s)
LAFM	Latin America, Africa, and the Middle East
LCA	life cycle assessment
LCI	life cycle inventory analysis
LCIA	life cycle impact assessment
LDV	light-duty vehicle
LHV	lower heating value
LPGV	liquefied petroleum gas vehicle
m	metre(s)
MAGICC	Model for the Assessment of Greenhouse-Gas-Induced Climate Change
MARKAL	market allocation (model)
MCDA	multi-criteria decision analysis
MOX	mixed oxide
NAM	North America
NEDC	New European Driving Cycle
NGV	natural gas vehicle
NMVOC	non-methane volatile organic compounds
NO	nitric oxide
NO_2	nitrogen dioxide
NO_x	nitrogen oxides
O_2	oxygen
OECD	Organisation for Economic Cooperation and Development
OEMS	original equipment manufacturers
OOECD	Other Organisation for Economic Cooperation and Development
OH	hydroxyl radical

PE	primary energy
PM	particulate matter
ppmv	parts per million volume
PSA	pressure swing adsorption
PSI	Paul Scherrer Institute
PV	photovoltaics
PWR	pressurised water reactor
RD&D	research, development, and demonstration
RES	reference energy system
rpm	revolutions per minute
SCR	solar carbo-thermic reduction of Zn/ZnO
SMR	steam methane reforming
SNG	synthetic natural gas
SRES	Special Report on Emissions Scenarios (of the IPCC)
STD	solar thermal dissociation of Zn/ZnO
t	metric tonne(s)
TWC	three-way catalytic converter
UCTE	Union for the Co-ordination of Transmission of Electricity
VKT	vehicle kilometres travelled
VSM	vehicle substitution model
W	watt
WEUR	Western Europe
WGR	wood gasification and reforming
Wh	watt-hour
WHO	World Health Organization
λ	lambda

Abbreviations for tonne (t), watt (W), watt-hour (Wh), and metre (m) are combined with the following prefixes to indicate larger amounts:

E	exa–	10^{18}
G	giga–	10^{9}
k	kilo–	10^{3}
mg	milligrams	10^{-3}
mg/km	milligrams per kilometre	10^{-3}
ppb	parts per billion	10^{-9}
ppm	parts per million	10^{-6}
Tg	teragrams	10^{12}
Tg/year	teragrams per year	10^{12}

1

Introduction

WARREN W. SCHENLER

1.1 Study motivation

Future transportation systems face a broad range of requirements. They must be economic, clean, low-carbon, efficient, and reliable (to name just a few). These characteristics must be based not only on the vehicles, but also upon their fuel supply chains. Using hydrogen as a fuel offers a possible solution to satisfying global mobility needs, including sustainability of supply and the potential for significant reduction of greenhouse gas (GHG) emissions.

Because hydrogen is an energy carrier that must be produced from primary energy resources, sustainability also implies that the hydrogen must be cleanly and economically produced, whether this involves carbon capture and storage (CCS), renewable energy resources, or long-term nuclear solutions.

This does not mean that a sustainable future must be static, and the evolving vision of sustainability is increasingly coming to incorporate concepts of flexibility and resilience.

This book is about research issues that are at the intersection of hydrogen and transportation, because the study of vehicles and energy carriers is inseparable. This book presents analysis of light-duty vehicles (LDVs; i.e. cars and light trucks, not heavy trucks), set in the context of other competing technologies, the broader energy sector and the overall economy.

1.1.1 Size and complexity of the problem

The size and complexity of the transportation system are undeniable. Transportation represented 29% of US energy use in 2009 (EIA, 2010b), and the share is similar globally (EIA, 2010a). The historic availability, affordable (though volatile) price, and excellent physical and chemical properties of oil have long made it the

Transition to Hydrogen: Pathways Toward Clean Transportation, ed. E. Wilhelm and A. Wokaun. Published by Cambridge University Press.

dominant transport fuel, with 95% of US transportation energy demand, including ship, rail, and air use, being satisfied by petroleum (EIA, 2010c). The size of the transportation infrastructure is vast and permeates the very fabric of our human environment on both the physical and social levels. On the supply side, this includes oil production, transport, refining, and distribution; on the demand side, it includes not only the stock of vehicles and roads produced, but also demographic patterns and a society built on division of labour and cheap transportation. Both the scale and complexity of this infrastructure mean that changes must be gradual.

The analytic and policy complexities of the transport sector are significant. Both the costs and benefits of our current system are great, and they affect a large number of stakeholders in different ways. Some of these costs and benefits are unremarkable; for example, the costs of traffic accidents and the benefits of personal freedom are largely uncontroversial, if difficult to quantify. The dominant problems of dependence upon oil are the increasing cost and decreasing reliability of future supplies, and the environmental consequences of our ever-increasing demand, due primarily to the costs associated with predicted future climate change. Both of these problems have costs and benefits that are unevenly distributed around the globe, with implications for social equity, global development, and international security.

The role of carbon dioxide (CO_2) as a GHG deserves, perhaps, special mention in this discussion of problem complexity. While the relative contribution of ever-increasing CO_2 levels compared to other factors affecting climate change (e.g. solar wind, oceanic oscillations, and cloud/water vapour feedback) is still a subject of some debate, the consensus of the scientific community is that further increases in CO_2 levels are likely to lead to long-term costs due to climate change, increased storm intensity, changes in rainfall patterns, and rising sea levels. There are also arguments about the economic priorities of avoiding CO_2 emissions versus adapting to climate change while addressing other global problems (Lomborg, 2009). But it is clear that reducing CO_2 emissions is a very significant global concern.

1.1.2 Energy security

Predictions of future gaps between dwindling oil supplies and growing demand will not be fulfilled as long as market prices can function to incentivise new production and allocate production to the most valuable uses. But that does not mean that the production of cheap crude will not decline, or that those people unable to afford future prices will not suffer. The expectation is not the 'end of oil' but the end of cheap, easy, and reliable oil. Not 'peak oil', but a transition to resources that are ever more difficult to find, extract, and afford. This is a present continuation of the historical shifts from wood to coal to oil, and ongoing shifts to gas and presumably nuclear (interrupted, but now possibly resuming). The depletion of current reserves, and the increasing marginal cost of extracting oil from new reserves or existing reserves of lower quality is the driving force behind the development of competing energy resources. As Sheik Ahmed Zaki Yamani said, 'The Stone Age didn't end for lack of stone, and the oil age will end long before the world runs out of oil' (Maass, 2005). We will leave oil behind when it is too expensive to compete.

For the present though, it is not just cost and global availability that are a concern for energy security. Geopolitical concentration in supply and the international competition for supply as global demand patterns shift are not new in the history of oil. The fungibility and easy transport of oil mean that the market is global, but prices are still volatile. This volatility leads to economic disturbances that are painful in and of themselves; and also serves to deter the development of competing energy resources.

1.1.3 Climate change mitigation

The current transportation sector is a major source of CO_2, contributing 23% of total emissions worldwide in 2006 (IEA, 2008b), and the difficulty of reducing emissions is out of proportion even to its contribution. Vehicle emissions are the hardest source of CO_2 to eliminate because they are numerous, small, mobile, and distributed. In contrast, fixed site emitters like power plants are larger and much more amenable to efficiency gains, fuel switching, or CCS (a technology that is still in the demonstration phase). Even small, fixed site emissions, such as from residential heating, are much more amenable to fixes like better insulation or replacement by biomass (wood) or heat pumps. One of the advantages of using hydrogen as a transportation fuel is that emissions are moved from the tailpipe back up the supply chain to where the hydrogen is produced. Depending upon the source of the hydrogen, the CO_2 emissions per vehicle kilometre can be increased (e.g. hydrogen from coal), reduced (from natural gas), or virtually eliminated (renewable energy or nuclear).

These major problems of energy security and climate change provide great impetus to transition beyond oil to a different future energy and transportation system, but this does not guarantee that future transportation systems will be based on hydrogen. Hydrogen has a long history of being called the ultimate fuel, and the latest wave of enthusiasm and inevitability has only recently begun to ebb again. There is significant need for objective, unbiased analysis of the potential that hydrogen holds in transportation, and where its use can most benefit the economy in general.

The dominant expectation is that hydrogen in vehicles will be used by fuel cells rather than internal combustion engines, due to the high efficiency that exceeds the thermodynamic limits of combustion. This is compatible with and benefits from the current trend to increased electrification in vehicle drivetrains found in hybrids and electric vehicles. The chief disadvantages of hydrogen vehicles are the still high costs of fuel cells and the relative balance that must be struck between tank size and limited driving range (currently as much as 400–600 km), due to the low volumetric energy density of current hydrogen storage methods. These problems have meant that commercial demonstration of hydrogen-fuelled vehicles is still lacking. Battery-powered vehicles currently have a more limited range (100–200 km), but with the additional advantage that it is much easier to transport and deliver electrons (electric power) than protons (hydrogen), as the distribution infrastructure for electricity is already in place. In contrast, a hydrogen vehicle fleet will likely involve building a new distribution infrastructure of pipelines and tankers that will not only prove more expensive, but also form a barrier to market penetration.

1.1.4 No guarantees

The potential benefits of hydrogen are very great, but its competition is far from standing still in both the areas of drivetrains and energy carriers. Improvements to the internal combustion engine (ICE) continue through electromechanical controls (e.g. variable valve timing and compression, and mixed multi-fuel metering), multi-fuel engines, and many other schemes. Hybrids and electrical vehicles are at or near the cusp of commercialisation, and battery energy densities are continuing to climb. On the fuel side, there are many alternatives to produce alternative fuels with volumetric energy densities that are higher than hydrogen, including natural gas, syngas, a range of synthesised liquid fuels like alcohols, bio-oils, Fischer-Tropsch, etc. These gases and liquids can stem from a wide range of primary energy resources, including coal, biomass, algae, and direct microbiological sources. Thus, the advantage that hydrogen may be produced from many different energy resources is in fact shared by other fuels, each with their own unique advantages. Hydrogen is therefore by no means guaranteed to be the fuel of the future, and only a careful and detailed analysis of its costs and benefits in comparison with other drivetrains and energy carriers will reveal its strengths and weaknesses. It is indeed likely that no one energy carrier will continue to dominate all economic, environmental, and social (or consumer) criteria, so individual choices in the marketplace will have a key role in determining the vehicle of the future.

1.2 Primary objectives

The primary objectives of the research reported in this book were to perform a comprehensive and holistic analysis of the impacts of implementing hydrogen as a fuel into future LDV fleets while considering the effects on other energy sectors. Important issues have included centralised versus decentralised production, competing distribution methods, suitable vehicle drivetrains and onboard storage, full energy chain cost, environmental and safety concerns, the feasibility of transition to a hydrogen infrastructure, and the effect of different policy options on implementing such a transition.

1.2.1 Fuel use and emissions identities

As a background to discussing the structure of the research programme that forms the basis for this book, it is worthwhile to recall the basic identity equations that form the basis for total vehicle fuel use and emissions, namely:

$$Fuel\ use = (population) \times (cars/person) \times (km/car) \times (fuel\ use/km) \quad (1.1)$$

and adding an additional factor for emissions:

$$Emissions = (fuel\ use) \times (emissions/fuel\ use) \quad (1.2)$$

This chain of factors illustrates that, if the goal is to reduce the total amount of fuel used or emissions, it is not sufficient to look only at the individual vehicle. The equations must be summed over the entire vehicle fleet to understand total fuel use and emissions, and it must be understood how the fleet is projected to change over time in order to forecast total fuel use and emissions into the future.

1.2.2 Trends in fuel and emissions factors

Based on the factors in Equations (1.1) and (1.2), it is worth mentioning how trends in the relatively recent past have influenced each factor in the chain.

Population growth

Population growth, and policy options to control it, is beyond the scope of this book, but it is worth noting that differential, national population growth rates form a key driver for transportation demand, and resource demand in general.

Fleet growth

The per capita ownership of light vehicles is dependent upon per capita income, and hence economic growth over time. This factor includes shifts between vehicle types (e.g. from scooters to cars to SUVs), and thus fleet composition. Fleet penetration depends on competing vehicle characteristics (price, mileage, safety, etc.). Factors relevant to fleet penetration studies include vehicle fuel type and availability, various policy options (e.g. taxes, subsidies, feebates, etc.) and social acceptance.

Annual vehicle use

Travel per vehicle is primarily determined by demographic patterns (e.g. commute distances) and per capita income. Increasing income can also shift mileage between vehicle types.

Fuel use

Engine efficiency gains over the last decades have been significant, while the vehicle fuel use per kilometre has stagnated. The reason for this has primarily been the consumer choosing vehicle performance over economy, particularly in vehicle classes exempt from mileage standards (Bandivadekar *et al.*, 2008a). Reallocating engine efficiency gains to fuel economy is important for reducing total fuel use and carbon dioxide emissions. Other key ways to reduce fuel use and emissions include vehicle downsizing and lightweighting.

Emissions

The emissions per unit of fuel used can be divided between local emissions (NO_x, VOC, particulates, etc.) that are determined by engine and exhaust treatment technology, and global emissions (i.e. CO_2) that cannot practically be scrubbed from vehicle emissions and depend upon the carbon content of the fuel.

1.3 Scope of work and approach taken

Based on the factors and trends discussed above, the scope of the research and the approach taken followed the basic steps below, that is, to:

- Analyse individual technology impacts (for both energy carriers and drivetrain technologies)
- Analyse vehicle trade-offs for a large number of vehicle designs
- Analyse changes in fleet composition, and the resulting changes in total fleet burdens over time
- Analyse technology choice strategies using least-cost optimisation

More concretely, this scope of work was translated to the following separate research tasks that are shown in Figure 1.1 below, which shows the structure of the research programme that generated the results described in this book.

- Well-to-tank analysis of hydrogen production from multiple primary energy resources (including renewables), in addition to competing fuels (Chapter 2).
- Heuristic, rule-based generation of vehicle designs, including drivetrain simulation to produce multi-attribute characterisation on a vehicle and kilometre basis (Chapter 3).
- Assessment of atmospheric hydrogen concentrations related to fossil-fuelled vehicles and estimates of the effects of the introduction of hydrogen vehicles (Chapter 4).

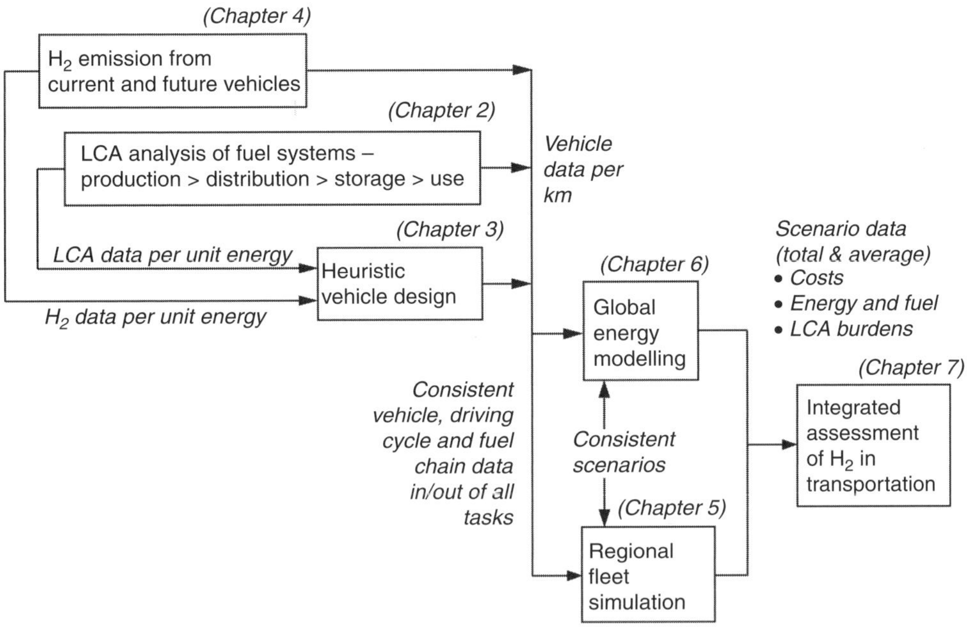

Figure 1.1 Structure of the research programme.

- Market dynamics of introduction and penetration of hydrogen vehicles into the fleet (Chapter 5).
- Comprehensive energy–economic analysis of the role of hydrogen in the global energy system (Chapter 6).
- Integrated assessment and analysis of hydrogen's potential use and impacts in the transportation sector (Chapter 7).

1.4 Organisation of book

The book is organised by chapters corresponding to each of the individual research tasks shown above. A brief description of each task and chapter is given below.

1.4.1 Life cycle assessment

Energy technologies generally involve a related chain of steps including resource extraction, transportation, refining or conversion, distribution, use, and disposal. A life cycle assessment (LCA) is necessary in order to integrate the potential environmental impacts of each step in the chain. For transportation systems, this is usually called a well-to-tank analysis if the chain ends with the fuel and a well-to-wheel analysis if the chain includes final use in the vehicle. Use of LCA guarantees that all the relevant material, resource, and energy flows throughout the complete chain are considered, which is a strict prerequisite for comparative environmental assessment of energy systems. LCA is particularly important for hydrogen because as an energy carrier most of its burdens are associated with the production step rather than final end use.

Chapter 2 presents an overview of the environmental performance of various current and future routes to hydrogen production, including fossil, renewable, and nuclear energy as the primary sources. Production using fossil energy is based on chemical conversion from coal and natural gas, as well as production of electricity for electrolysis. The direct production of hydrogen from renewable energy focuses on innovative solar processes, and production via electrolysis is based on a range of renewable generation technologies. The assessment considers centralised production with transport and distribution to the point of sale, as well as smaller, distributed, and on-site (fuelling station) production methods.

The chapter compares the various production pathways for producing 1 kg of H_2 with the combustion of energy-equivalent quantities of conventional fossil vehicle fuels in a EURO3 standard passenger car engine.

The LCA performed was based on the ecoinvent database (ecoinvent, 2009), which is the world's most comprehensive and transparent LCA database. The impact assessment was conducted using a range of indicators: cumulative primary energy demand (fossil and nuclear) and the cumulative emissions of GHGs, suspended particulate matter (PM), volatile organic compounds, nitrogen oxides, and sulphur dioxide.

1.4.2 Heuristic vehicle modelling

Well-to-wheel analyses have in the past typically been performed based on a rather limited number of generic or typical vehicle types. In Chapter 3 the approach that was used instead was to automate the generation of vehicle designs based on a set of rules to ensure consistency between component weights, power, etc. The design variables (or options) that were combined to form different designs included vehicle class and weight, type of drivetrain and fuel, degree of electrification (from zero for conventional designs to one for electrical vehicles, with hybrids in between). In addition to these exogenous variables, a range of dependent variables endogenous to the vehicle design were also determined. By automating the design process and integrating it with a drivetrain simulation of each vehicle, it was possible to perform repetitive modelling of a very large number of designs composing a 'virtual fleet'.

A range of characteristics (multi-attribute analysis) was calculated for each vehicle design, including not just drivetrain attributes (e.g. energy use and emissions), but also characteristics including vehicle performance, utility, range, safety, materials use, and cost. These attributes were then used to perform trade-off analyses and multi-attribute analysis for different preference profiles.

Areas of analytic emphasis included study of the synergistic and competitive relationship between hybrid and fuel cell vehicles (FCVs), impacts of variations in the degree of electrification, lightweighting, and correlating vehicle weight to energy use.

1.4.3 Hydrogen emissions

Molecular hydrogen (H_2) present in the atmosphere is not a direct GHG, but it is removed from the atmosphere through a chemical process that competes with the process that removes methane, a potent GHG. H_2 is therefore effectively an indirect GHG. Present internal combustion engine vehicles (ICEV) are the dominant source of anthropogenic hydrogen to the atmosphere. Large-scale use of hydrogen as a vehicle fuel, in parallel with increasing H_2 production for non-transport-based end uses, could lead to an increase in atmospheric hydrogen.

Chapter 4 first provides an overview of atmospheric hydrogen, including the major sinks and sources, with particular emphasis on the anthropogenic contribution from technological processes (i.e. motor vehicles and H_2 production, distribution, storage, and non-transportation-based end uses). It then relates the influence of vehicle technology and driving patterns to hydrogen emitted in vehicle exhaust, and qualitatively and quantitatively compares emissions from current and future vehicle technologies. These emissions are scaled using various scenarios for time periods throughout the twenty-first century to quantify global H_2 emissions from road-based transportation. H_2 emissions from non-transportation sources are also scaled up based on various leakage and loss rates, using these same scenarios. The scaled results from transportation are then combined with results from the non-transportation-based sources to find overall global, anthropogenic H_2 emissions from technological processes, based on various vehicle fleet composition scenarios and H_2 production strategies for industrial and direct energy-based end uses.

1.4.4 Fleet modelling

People purchase cars based on a wide range of factors, including cost, mileage, fuel availability (infrastructure), and the behaviour of other people around them. The aggregate choices of people that are reflected in vehicle fleet composition dynamics are therefore non-linear and exhibit a range of feedbacks. This makes the use of systems dynamics particularly interesting for the purposes of modelling market penetration of different fuels and technologies, and the impact that different policy options can have upon the rates and ultimate levels of market penetration into the vehicle fleet.

Chapter 5 focuses on understanding the market dynamics of technology penetration into the vehicle fleet, and, in particular, on the interplay between technology introduction and the availability of fuel (or energy carrier) infrastructure. These dynamics reveal a range of behaviours, such as hybrid vehicles prolonging dependence on fossil fuels, technology lock-in that can delay bridging to ultimately cleaner vehicles, and rebound effects (increased driving) due to improved fuel economy. This modelling includes current technologies (gas and diesel vehicles), as well as competing future technologies (natural gas and hydrogen vehicles). It also shows that, in addition to zero-emissions vehicles, reductions in fleet size and vehicle travel are important in reducing total fleet emissions. The effects of different policy options on fleet penetration have also been examined, as well as the levels at which they can tip technology adoption into alternate system states.

1.4.5 Energy–economic modelling

The energy–economic analysis in Chapter 6 expands the consideration of vehicle energy carriers and technologies by setting them in the wider context of the whole energy sector. It uses a least cost optimisation model (MARKAL) to determine constrained capital investment in all stages of the competing energy chains. These chains include primary resource extraction (e.g. oil), conversion to energy carriers (e.g. hydrogen, electricity, or synfuels), transportation (e.g. hydrogen pipelines versus pressurised tank trucks versus local production) to the final demand technology that converts energy into consumer services (e.g. vehicles that provide transport). By making investments in all steps of the many competing energy chains to minimise net present costs, the global, multi-region model can incorporate energy system interactions, technical change, and economic and demographic driving forces. Such a model assumes perfect foresight (i.e. certain forecasts), so it serves as a complementary comparison to the system dynamics modelling of the previous chapter that models without foresight.

The model can include policy scenarios by imposing emissions caps, or modifying costs by adding taxes or subsidies. The results of this chapter show hydrogen can be an important option for achieving stringent GHG targets over the long term, but strict emissions caps or other strong incentives are necessary to achieve vehicle and infrastructure deployment. Hydrogen transport is identified as a potential bottleneck that could reduce hydrogen's attractiveness, and the mid-term investment that is required to keep the long-term option for hydrogen open is found to be moderate compared to other power sector investments.

1.4.6 *Integrated assessment of hydrogen in transportation*

Chapter 7 integrates the results of the various research tasks above, and synthesises the implications into overall conclusions regarding the role of hydrogen-fuelled vehicles in the future transportation system and the potential of various policy options to affect their implementation.

1.4.7 *Appendices*

Several of the chapters have included appendices expanding certain technical analysis or providing detailed input assumptions and data. Of particular interest, Appendix A lists key assumptions for each chapter and highlights instances where these assumptions may have differed between chapters. This allows differences in results to be explained, and provides useful insight into the factors driving many of the results.

1.5 Influencing future transportation systems

The transportation sector is subject to many different types of regulations that differ between different markets. These include safety standards, emissions standards, and fuel taxes. The regulations of most interest to this study are fuel- and emissions-related, based on the primary concerns expressed above. There are a significant number of ways to control these areas, including emissions caps, trading, taxes, feebates, and scrappage incentives. This book explores some of these regulatory options as scenario parameters in Chapters 5 and 6 noted above. These options are posed as what-if analyses, and are not intended as predictions of the future. Evaluation of such regulatory options and of the book's chapters and overall conclusions are left to the reader as an individual stakeholder.

1.6 Background of research

This book is the result of an academic research partnership between two Swiss Federal research institutes, the Paul Scherrer Institute (PSI) and the Swiss Federal Laboratories for Materials Science and Technology (Empa), and the Sloan Automotive Laboratory at the Massachusetts Institute of Technology. This collaboration provided a rewarding structure for sharing research and results from European and American perspectives through regular meetings and active project collaboration. The differences in European and American vehicle fleets, policy options, and many other data provided opportunities to explore differences and commonalities in research results.

The results of the MIT research conducted as part of this joint research programme have been reported in the book, *On the Road in 2035* (Bandivadekar *et al.*, 2008a).

The individual research projects within the Swiss half of this collaboration were funded through the Swiss Competence Centre Energy and Mobility (CCEM). The overall sponsorship for the research was from a group of vehicle manufacturers, fuel suppliers and industry research groups. These sponsors supplied not only financial support, but also regular and critical advice on the direction and progress of the research.

1.7 Intended audience

This book is intended to serve as a technical reference for policy development and research. It should be of use to researchers, technical advisors, and policy makers who are interested in the economic, energy, and environmental issues related to transportation issues, including:

- Government research and regulatory agencies
- Non-governmental organisations and interest/advocacy groups
- Public policy commissions and panels
- Academic research groups in the areas of energy and transportation policy
- Private sector firms such as original equipment manufacturers (OEMs), including their strategy, R&D, and marketing sections.

The aim of this book is to focus on using hydrogen to fuel LDVs, and to compare them to competing fuels and technologies, rather than to be an encyclopaedic compendium of either the broader hydrogen or transportation sectors. The goal has been a comprehensive analysis from LCA of energy production and emissions, to vehicle design, to fleet penetration, and energy–economic interactions. This focus should not prevent the conclusions reached from being applied in a more general way, with due consideration of the clearly stated assumptions.

2

Life cycle assessment of hydrogen production

ANDREW SIMONS AND CHRISTIAN BAUER

2.1 Introduction

During combustion or electrochemical conversion, the reaction of hydrogen with atmospheric oxygen produces energy and pure water. Thus, in itself, it is an environmentally benign process. However, when judging the environmental merits of any fuel or energy carrier, it is important to consider the other, often complex processes related to additional stages of its life cycle, particularly its production, transport, and storage. Only by considering the cumulative burdens on the environment and the total consumption of primary energy resources can a fair comparison be made between the available and proposed options.

LCA is a central element in the assessment of technologies and production routes, enabling a comparative evaluation of the environmental performances of different options. LCA methodology, sometimes referred to as 'cradle-to-grave' analysis, considers cumulative burdens stemming from the entire lifetime (construction, operation, and dismantling) of the directly and indirectly related industrial activities involved in the production of goods or the provision of services. By considering the consequences of an option according to these principles, a fair and unbiased evaluation of the merits and shortcomings can be achieved. In turn, an LCA forms a crucial aspect of a broader understanding from the perspective of sustainability, and can inform a decision-making process of the true level of either energy dependency or security that an option is likely to represent. For options which are 'renewable' or 'non-fossil' with regard to the direct conversion processes, a complete account of the background processes will determine whether a reduction of the reliance on conventional fuels will actually occur. Particularly for emerging energy systems, such as the use of hydrogen in transport, conducting such assessments at an early stage of development is important in order to

Transition to Hydrogen: Pathways Toward Clean Transportation, ed. E. Wilhelm and A. Wokaun. Published by Cambridge University Press.

identify the most promising options as well as determine environmentally significant aspects of an individual production route.

2.1.1 Objectives

The research outlined in this chapter is a well-to-tank LCA of gaseous hydrogen (H_2) as an energy carrier, where H_2 is intended for use in the fuel cell of a road vehicle. The aim is to present and compare the environmental performances of a range of H_2 production routes considered representative of current (2005) and future (2035) technologies. In this sense, the assessment aims to reflect the broad range of possible routes and processes which are currently available, as well as those foreseen to be commercially viable within the coming decades.

As a basis for calculation, all LCA results refer to 1 kg of purified H_2 at a pressure of 45 MPa available at the vehicle fuelling station pump.

In order to gauge the results against conventional vehicle fuels, the LCA results for hydrogen production are also compared to those for producing and combusting, on the basis of equal energy contents, the same quantity of gasoline, diesel, and natural gas.[1] The emissions from the combustion of these fuels are calculated with consideration of an ICE of EURO 3 standard (Spielmann *et al.*, 2007). The impact assessment is conducted on two methodological levels: one is the use of aggregated indicators which characterise relevant inventory data, whereas the other considers cumulative quantities of specific emissions. With regard to aggregated indicators, in this chapter we use GHG emissions determined on the basis of each gas's global warming potential (Solomon *et al.*, 2007), and cumulative non-renewable energy demand (CED) (Frischknecht *et al.*, 2007b), which quantifies the primary energy contained in fossil and uranium resources. For the cumulative quantities of specific emissions, we consider the air pollutants defined by the World Health Organization (WHO, 2006) as being of particular relevance to transportation. These are taken directly from the inventory results, being suspended particulate matter ($PM_{2.5}$ and PM_{10}), non-methane volatile organic compounds (NMVOC), nitrogen oxides (NO_x), and sulphur dioxide (SO_2).

It is not the aim of this chapter to produce a definitive solution as to how H_2 should be provided, or to suggest an 'H_2 mix' of the techniques available. Rather, the chapter deals with one of the first steps in such an assessment, which is to

[1] 1 kg of H_2 is able to release 120 MJ energy if we consider the lower heating value (LHV). We therefore compare this with the equivalent production and combustion of 120 MJ gasoline, diesel, and natural gas. The LHV is used in order to be consistent with the methodology of the ecoinvent database (Frischknecht *et al.*, 2007a) and also because the reaction products are typically vapour.

inform the stakeholders involved of the merits and drawbacks of each production and transport pathway.

2.1.2 Scope

The application of LCA methodology ensures the inclusion of direct and indirect environmental burdens associated with all industrial processes in the production and use of fuels. It also includes the entire production chains of all the goods and services consumed in the life cycles of the fuels by using inventory data from ecoinvent (ecoinvent, 2007), the leading commercial LCA database.

In order to enable a feasible assessment and comparison of environmental burdens, the following work includes only those production pathways dedicated to obtaining purified H_2 for use as an energy carrier; it does not consider H_2 produced as a by-product from industrial processes (for further explanation on this see Section 2.3.2).

The range of H_2 production pathways included in the assessment considers both fossil and renewable (biomass) resources as hydrogen-containing feedstocks. While, for some pathways, H_2O in the form of water or steam is the main source of H_2, it acts as an important additional source of H_2 to all the production pathways assessed. Separation of the H_2 from these feedstocks uses fossil, nuclear, and/or renewable resources as primary energy inputs.

In order to understand how technological improvements and developments may change the environmental performance indicators, this chapter looks at the two time frames of 2005 and 2035. The year 2005 forms the base for many important background datasets and is considered representative of current technologies. Projecting up to 2035 allows for the development of energy chains and the integration of technologies currently at the pilot phase, without increasing uncertainties beyond reasonable levels.

Both natural gas and coal are included as direct fossil feedstocks, with the application of CCS taken into account for 2035. Hydrogen production via electrolysis is considered with the European electricity mix as well as selected individual power generation technologies, exploiting either fossil, nuclear, or renewable energy resources. This is in order to demonstrate the sensitivity of this route to particular electricity generating sources, although it was not within the scope of this assessment to differentiate between peak and off-peak generation capacities and therefore the availability of electricity, or to consider the intermittency of some renewables. The direct production of H_2 from biomass gasification and solar-powered, thermochemical pathways are included as important innovative technologies for use in 2035, and serve to increase the comparative range of options for H_2 from renewables. For the time horizon considered, nuclear technologies are limited to generation two and three (Gen II and Gen III) reactors for electricity generation.

Only electrolysis is considered to be located at the vehicle fuelling station, meaning that all other pathways require the transport of hydrogen to the fuelling station. In order to consider the additional potential impacts due to this factor, we include a range of transport distances, using either road or pipeline delivery.

The processes modelled are for European boundary conditions.

2.2 The life cycle assessment methodology

This section provides an overview of general LCA methodology and defines the specific inventory results and indicators used in the impact assessment.

2.2.1 Basic principles

LCA quantifies the potential environmental impacts of a specific product or service over its whole life cycle, beginning with the extraction of resources and covering the intermediary processing stages, the use phase, and final disposal. It includes fossil and mineral resource consumptions, land uses, and emissions to air, water, and soil.

The International Organization for Standardization (ISO) has specified international standards on LCA (ISO, 2006). Four main steps are distinguished: goal and scope definition, inventory assessment, impact assessment, and interpretation (see Figure 2.1). Due to the comprehensive approach and the interdependent assessment steps, conducting an LCA is usually an iterative process.

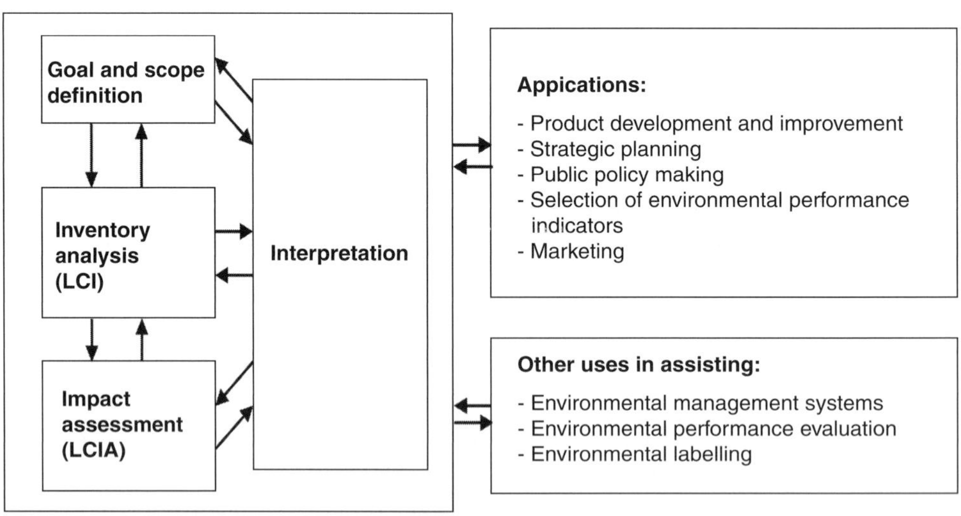

Figure 2.1 Main steps and applications of a life cycle assessment.

2.2.2 Goal and scope definition

The goal and scope define the fundamental characteristics and constraints of the LCA being conducted. Although the definitions of these aspects can also be iterated upon throughout the duration of a study, they provide a framework and guidelines for the collection of the inventory data. The inventory data are elementary (mass) flows of resources and emissions, the use and occupation of land, or the provision of services in the form of unit processes. Some of the main aspects for consideration when setting out the objectives and scope of the study include (ISO, 2006):

- Functional unit: the quantifiable unit of a product or service, and according to which the inventory data are then related.
- System boundaries: the definition of the processes included, ideally considering all elementary flows into and out of the processes under focus. The level of detail considered in the study depends on the capabilities of the tools employed as well as the resources available. At the data level, system boundaries are defined according to temporal, geographical, and technological constraints and should reflect the goals of the study.
- Cut-off criteria: the criteria for selecting the inputs and outputs according to the system boundary definition are an iterative process throughout an LCA, determined as various aspects are tested for their sensitivity to the final results.
- Data quality requirements: the description of requirements concerning temporal, geographical, and technological aspects in order to enable the goal and scope to be met and to properly interpret the results of the analysis.
- Definition of the type of critical review: this depends on the framework of the study and the audience addressed.

2.2.3 Life cycle inventory analysis

Life cycle inventory (LCI) analysis quantifies all elementary flows associated with single processes, i.e. mass (materials and resources) and energy flows, land use, emissions to air, water, and soil, and products of the processes as outputs. By linking the single processes, the elementary flows per desired functional unit (e.g. in this chapter, per kg of H_2) are quantified.

The ecoinvent LCA database contains more than 4000 individual processes for the production of goods and the provision of services, with a focus on European production chains. For important globally traded goods (e.g. energy carriers, such as oil, gas, coal, and uranium), regions outside of Europe are also considered. The inventories include more than 1000 individual elementary flows, i.e. emissions to air, water, and soil, consumption of renewable and non-renewable energy and mineral resources, and various types of land use. These LCI data mainly refer to conditions existing between 2000–2005; for projection to the year 2035, the European electricity mix and Russian natural gas production were modified according to expected developments in the power generation and fuel supply sectors. Cumulative LCI data results for a range of specific air pollutants of most relevance to road transport vehicles are presented in this chapter.

2.2.4 Life cycle impact assessment

LCIA is the third step within an LCA and focuses on the aggregation of specific or total environmental burdens to compare competing technologies. LCIA can also be used to analyse the contributions from system parts or for use in product optimisation. Figure 2.2 shows the main aspects or steps of the LCIA: it contains several mandatory elements and certain additional options for a higher level of aggregation and a more comprehensive assessment. Implementation of all optional elements allows aggregation of all environmental burdens to a single indicator.

The concept of category indicators for environmental impacts is the basis for LCIA. Each category has its own environmental mechanism, such as infrared radiative

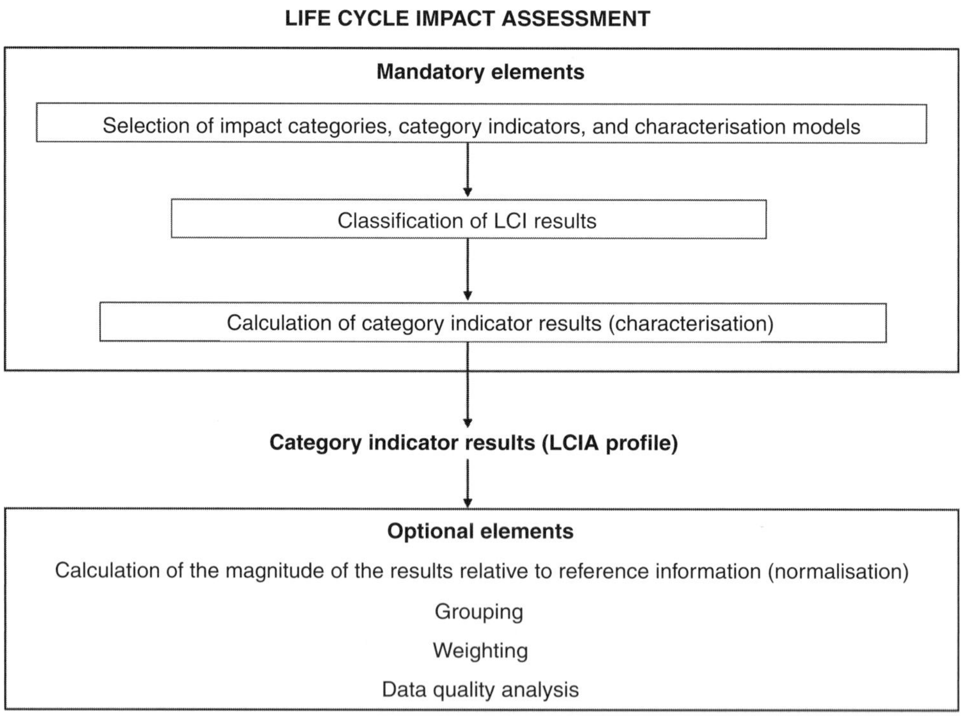

Figure 2.2 Elements of the life cycle impact assessment (ISO, 2006).

forcing for climate change or proton release for acidifying emissions. All mass flows taken into account in the LCI are classified and multiplied with specific characterisation factors concerning the environmental impact for a particular impact category.

Implementation of the optional elements depends on the intended assessment. Normalisation allows comparison of different category indicators by dividing the values with a selected reference value, such as given emissions or resource uses for a certain area, e.g. the whole of Europe. Grouping of impact categories allows a specific comparison depending on the interests defined in the goal and scope description. Certain characteristics can be grouped, such as emissions or resource consumption; or impact categories can be ranked depending on value-based priorities. Weighting the indicators allows for a differentiated rating of impacts, such as for an aggregation of the categories to a total environmental indicator. As the choice of weighting factors depends on personal value judgement and is not solely based on scientific criteria, weighting involves a strong element of subjectivity. Therefore, different weighting schemes as well as sensitivity analysis may be helpful for consolidation of results and conclusions.

LCIA methods are implemented in the ecoinvent database (Frischknecht *et al.*, 2007b) and in the SimaPro LCA software (Geodkoop *et al.*, 2008), both of which were used to conduct the LCA presented here. To analyse and interpret cumulative lifecycle inventories, the chapter uses two LCIA methods to quantify specific environmental burdens; these are contributions to climate change and the consumption of non-renewable energy resources. The two methods are described below.

GHG emissions

Due to their impact on the world's climate, GHG emissions are a central element of today's energy policy, and their complete quantification is therefore crucial to any comparative technology assessment. The assessment applies the IPCC infrared radiative forcing values of different GHG species relative to CO_2 in terms of CO_2-equivalents (CO_2-eq.) (Solomon *et al.*, 2007). Three different time perspectives (20, 100, and 500 years) are available for an evaluation of the potential impacts of the emissions. In accordance with common practice, the time frame of 100 years is used in this chapter.

Cumulative Energy Demand (CED), non-renewable

This indicator quantifies the total consumption of non-renewable energy resources considering their primary energy content in terms of MJ-equivalents (MJ-eq.) and therefore addresses the long-term availability of energy resources. According to the implementation method in the ecoinvent database, energy resources are split into five different categories: fossil, nuclear, hydro, biomass, and other renewables, i.e. solar, wind, and geothermal energy. The characterisation factors used in this assessment method use the higher heating values (HHV) of fossil fuel resources in order to represent the maximum theoretical energy that they contain. This was done for comparison, and does not affect the use of LHV in the calculations of process energy flows. The definition of the adequate energy value for 1 kg of natural uranium is not straightforward and requires some subjective decisions, since the energy conversion of uranium strongly depends on the technology and fuel management of the system used. The energy value is calculated based on the nuclear fuel chain as modelled in ecoinvent (Dones, 2007).

CED values should not be aggregated but rather compared separately for each of the individual categories, as resources have varying reserves and technical potentials, and can be used in different ways. Therefore, the same energetic amount used, whether renewable or non-renewable, can have a distinctly higher or lower impact on long-term global energy supply. Even though CED results are interesting and relatively comprehensible, they should not be used independently of other LCIA methods as they do not provide any direct indication of environmental impacts or burdens.

In this chapter, we consider the non-renewable primary energy resources of fossil and uranium only. This therefore gives an indication of the sustainability of the options considered with regard to the long-term availability of energy resources.

2.2.5 Database and software

The LCA performed here uses background data from ecoinvent which is currently the most comprehensive and transparent LCA database (ecoinvent, 2007) for industrial processes that are not directly part of the analysed energy chains, e.g. steel production and processing, heat and electricity supply, etc. This allows a fair and balanced comparison of the different fuel chains, taking into account all relevant environmental burdens, i.e. emissions to air, soil and water, land use, and consumption of energy and material resources.

The construction of inventories and processes specific to the hydrogen production scenarios determined for this chapter was done using the SimaPro software (Geodkoop *et al.*, 2008). With this software, a number of indicators and impact assessment methodologies are available, which allowed us to use an iterative procedure between results and inventory data.

2.3 Hydrogen production

2.3.1 Background

Hydrogen is currently produced around the world in a wide variety of industrial processes as either a primary or secondary product and, in some instances, even

Table 2.1. *The characteristics of hydrogen use in Europe (Maisonnier et al., 2007).*

Market Sector	Production (bn Nm^3/yr)	As % of total
Captive	57	65%
Merchant	8	9%
By-product	23	26%

simply as a waste by-product to be vented or flared to the atmosphere. The majority is produced as an intermediary input to on-site industrial processes (e.g. in the production of fertilisers and refrigerants, as well as in the upgrading of crude oil and petroleum products) and is referred to as 'captive' hydrogen. Where viable, surplus captive hydrogen may be made available as 'merchant hydrogen' and thus supplied to external markets. Additionally, hydrogen occurring as a secondary or by-product may either be used as a source of on-site energy to industrial processes, be made available as merchant hydrogen, or exist as a waste due to the inexistence of infrastructures for it to become merchant (Maisonnier *et al.*, 2007). Table 2.1 shows the approximate quantities of current hydrogen production according to the above categories.

Although hydrogen can be produced from all primary energy resources, not all avenues are equally advanced with regard to commercial operation. In the petroleum industry, the hydrogen necessary to upgrade fuels has been obtained from the hydrocarbon by-products of refining. In other industrial sectors, coal is a cheap and abundant fuel which can be easily transported, stored, and handled without significant risk. Alternatively, the electrolysis of water is a simple process enabling hydrogen to be produced on all scales opening up a range of possibilities to other primary energy resources, such as nuclear and renewables, to generate the electricity necessary for this process.

Today, however, approximately 96% of H_2 production is achieved using fossil fuels as the main feedstock in direct conversion processes where, overall, the production of H_2 accounts for around 1.5% of global primary energy use (Hirschberg, 2004). This is largely via steam reforming, autothermal reforming, or partial oxidation of the fuel, with the latter being mainly used for lower quality feeds (Kothari *et al.*, 2008; Holladay *et al.*, 2009). Using oil as a direct feedstock is mostly a practice of the oil industry for hydroprocessing applications in refineries, whereas hydrogen production from coal is often a stage in the production of ammonia (for refrigerants and fertilisers). Natural gas is also used in both of these industries and thus accounts for the largest share of direct feedstocks, being most commonly processed by steam methane reforming (SMR) (see Figure 2.3; Stiegel

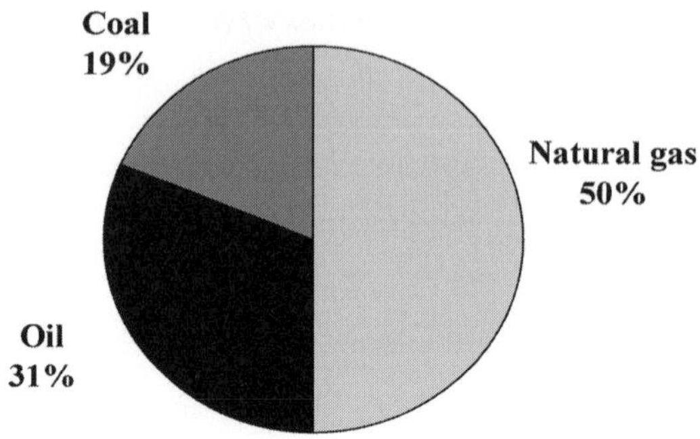

Figure 2.3 Shares of direct feedstocks to hydrogen production (Stiegel and Ramezan, 2006; Kothari *et al.*, 2008).

and Ramezan, 2006). Hydrogen use in ammonia production accounts for more than 60% of global production, with petroleum refineries consuming around one-quarter of this amount (Spath and Mann, 2001). Indirectly, fossil fuels are often also the main primary energy resources for the generation of electricity used in the electrolysis of water.

The amount of surplus industrial hydrogen in Europe could already be sufficient for between 1 and 6 million passenger vehicles per year (Maisonnier *et al.*, 2007). These quantities would be available from a combination of spare capacity in the industrial sectors and H_2 as a by-product. Although the H_2 occurring as a waste by-product could theoretically be used without consequence (other than the installation of infrastructure to acquire it), marketing useable by-product H_2 as a high-quality energy carrier for the transport sector would, within the industrial processes previously consuming it, most likely lead to a substitution for a less cleanly combusting fuel, i.e. natural gas. On the whole, therefore, it is not likely that these sources will be a long-term solution if hydrogen-fuelled vehicles achieve a significant market share. Where they could prove of importance is during a transition period until sufficient production capacities for transport demand are in place. Any degree of longer term reliance on by-product or surplus H_2 may also be ill-advised if, for instance, improvements to industrial processes such as chlorine manufacture lead to the reduction of merchant H_2 from this route. Furthermore, if the intention of using H_2 is to move away from oil-based fuels, then it would be contradictory to consider global oil refineries as providers.

Initially, the growing demand for H_2 outside of industrial facilities and its gradual appearance in vehicles may also be aided by small plants, such as on-site electrolysers and natural gas reformers serving single fuelling stations. The desire to break away entirely from fossil fuels and achieve an acceptable level of energy autonomy and security could also promote the use of renewable energy resources for electricity generation and as a fuel for small-scale gasifiers. On-site

H_2 production also means that electricity, natural gas, or perhaps biomass must be transported rather than the hydrogen itself. Whether H_2 production in fact develops along the lines of centralised facilities with associated distribution networks, or along the lines of decentralised plants at the point of demand is yet to be seen. For this reason, the following assessment considers a range of technologies representing different commercial scales of current and future production.

2.3.2 Production routes

Although a fraction of a future FCV fleet could be supplied with H_2 made available as a secondary product, apportioning of the various environmental burdens between the primary product and the H_2 becomes a complex and controversial procedure. Therefore, in aiming to enable an assessment and comparison of the potential environmental burdens caused by the production of H_2 for specific use in vehicles, the following overview includes only dedicated production routes to obtain H_2 as an energy carrier.

As H_2 production and fuel cell technologies develop and become more competitive, arguably, so too will the technologies and infrastructures for H_2 transport, therefore increasing fuel availability. As the deployment of one depends to a large part on the deployment of the other, and all depend on technical advancement and consumer demand, this chapter assumes a largely nodal spread of facilities which are mainly characterised by on-site H_2 production in order to negate H_2 transport. Larger centralised facilities are assumed to be likely to supply major, single point areas of demand such as cities. Here, transport via pipeline is mainly considered a technique for 2035. A complex network of pipelines linking multiple producers and areas of load is not expected to be in place within the scenario boundary of this environmental assessment.

The most important characteristics of the selected technologies for production and transportation will be described from the perspective of LCA and according to the objectives outlined above.

2.3.4 Technology overview

Using the feedstocks and general methods outlined above, there are many possible techniques to produce hydrogen, and so this chapter focuses on technologies which are broadly representative of the different pathways. All processes modelled are done so on the basis of European conditions, which are of greatest relevance to the fuel and electricity mix assumptions. Additional reflection of European conditions was also taken into account when considering transport modes and distances.

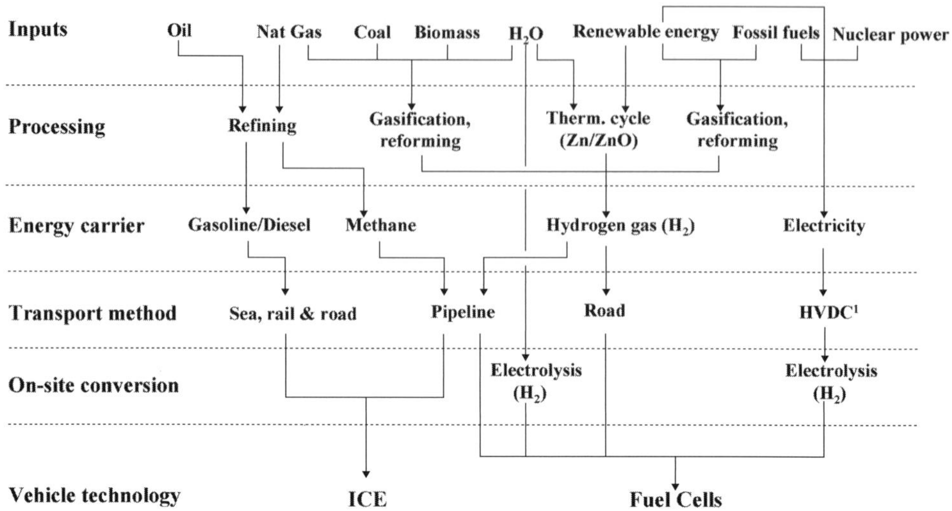

Figure 2.4 Overview of assessed fuel production and supply pathways. [1]high-voltage direct current.

The final stage of the modelling procedure (and therefore a system boundary of the LCA) for all production processes under consideration is H_2 available at the fuelling station pump at a pressure of 45 MPa. This means that the same basic fuelling station infrastructure is used for all processes, with on-site electrolysis being an exception.

Figure 2.4 shows an overview of the H_2 production and supply pathways addressed in this chapter.

Both natural gas and coal are included in direct conversion and reforming processes, as well as being fuels for the generation of electricity for H_2 production at the vehicle fuelling station via electrolysis. Here, fossil fuels are considered both individually and as contributors to electricity production mixes. In addition, the use of renewables considers individual sources of electricity for electrolysis, but also direct feedstocks such as biomass, and as reactionary heat sources using concentrated solar radiation.

In this chapter, the harnessing of nuclear energy for the production of H_2 is limited to the generation of electricity for electrolysis. For this purpose, Gen II technologies are used in 2005, whereas development to 2035 sees the construction of only Gen III technologies. Other techniques employing more direct pathways are not expected to be at a stage of development that would allow commercial deployment by 2035. The use of nuclear power plants to provide heat for the thermochemical separation of H_2 is a consideration for Gen IV nuclear reactors which are not expected to be operating before 2040. It is also possible that nuclear reactors

Table 2.2. *Energy resources and methods for hydrogen production for current (2005) and future (2035) time frames.*

Source	Year	Method
Fossil fuels as direct feedstock	2005	Natural gas steam methane reforming (SMR)
		Coal gasification and reforming (CGR)
	2035	Natural gas SMR
		Natural gas SMR with carbon capture and storage (CCS)
		CGR
		CGR with CCS
Biomass as direct feedstock	2005	Wood gasification and reforming (WGR)
	2035	WGR
Electrolysis	2005	Average European mix
		Hydro, river, average for Europe
		Nuclear, average for Europe
		Coal, average for Europe
		Natural gas, average for Europe
		Photovoltaics (PV), average mix in Spain
		Wind, onshore, average for Europe
		Solar thermal, average mix in Spain
	2035	Average European mix (no CCS)
		Hydro, river, Switzerland
		Nuclear, European pressurised reactor (EPR)
		Coal, integrated gasification combined cycle (IGCC)
		Natural gas combined cycle (CC)
		PV, average mix in Spain
		Wind, offshore, North Sea
		Solar thermal, average mix in Spain
Concentrating solar	2035	Thermochemical dissociation of ZnO (STD) and hydrolysis
		Carbothermic reduction of ZnO (SCR) and hydrolysis

will be constructed specifically to provide high-grade heat, thus enabling the thermal splitting of water into its primary elements. By 2035, the demand for H_2 might support the development of such facilities but it is not expected that a commercially operating example will exist within Europe (Hirschberg, 2004; Gomez *et al.*, 2007; Vitart *et al.*, 2008; Marques, 2010).

Table 2.2 outlines the specific hydrogen production methods analysed in this chapter for the years 2005 and 2035. More detailed descriptions of each are furnished in the following sections.

2.3.4 Natural gas steam methane reforming

Although SMR can also be small scale for on-site H_2 production, the pathway presented here is representative of larger, centralised technologies.

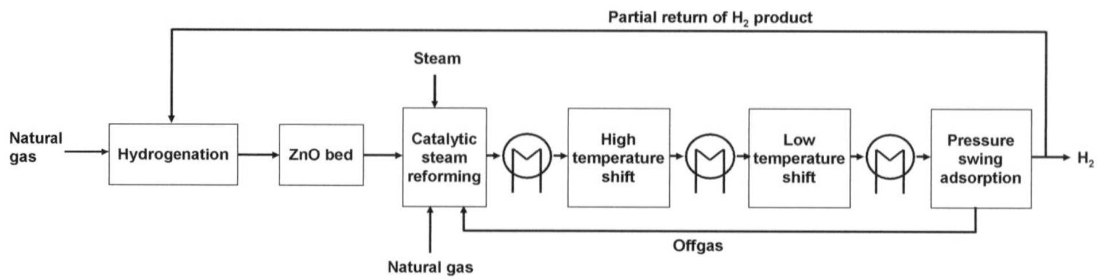

Figure 2.5 Block flow diagram of hydrogen production from natural gas (Spath and Mann, 2001).

Consisting primarily of methane (CH_4), natural gas can be reacted with steam (H_2O) to form a carbon monoxide- (CO) and hydrogen-rich synthesis gas (syngas). This hot syngas is then fed through high and low temperature shift reactions to further convert steam and the CO to H_2 and carbon dioxide (CO_2). In a final stage, the H_2 is separated from the CO_2 using chemical adsorption and the H_2 stream is purified in a pressure swing adsorption (PSA) unit. The tail gas (offgas) from the PSA unit is used to preheat the natural gas, as shown in Figure 2.5 (Spath and Mann, 2001; Molburg and Doctor, 2003; Maack, 2008).

The conversion processes can be summarised by the following reactions:

$$CH_4 + H_2O \rightarrow CO + 3H_2 \tag{2.1}$$

$$CO + H_2O \rightarrow CO_2 + H_2 \tag{2.2}$$

Without a secondary utilisation of the excess steam the SMR of natural gas can achieve HHV efficiencies[2] of above 70%, rising to more than 80% with a secondary utilisation of waste streams (Spath and Mann, 2001; Simbeck and Chang, 2002; Holladay *et al.*, 2009). SMR is currently the main method employed for the production of H_2 and differs only marginally to that of autothermic reforming (Stiegel and Ramezan, 2006). It is therefore a key route for representation in this chapter.

Modelling of H_2 from SMR in 2005 is done using an overall conversion efficiency of 79.2% (or 67% based on the LHV) and therefore without a secondary use of the steam (Simbeck and Chang, 2002). The output pressure at the PSA is 30 kPa.

Natural gas, which also powers the catalytic steam reforming unit, is delivered to the SMR plant as a supply mix consisting of long-distance pipeline transport from diverse international gas fields, as shown in Figure 2.6.

Development to 2035 considers improvements in the energy supply chains, characterised most significantly by an improvement in Russian infrastructure to reduce leaks in natural gas pipelines (Bauer *et al.*, 2008a). The shares of natural gas supply from each country remain the same. Electricity consumption, which

[2] Defined as the ratio of the energy in the hydrogen energy carrier to the energy in fuels and feedstocks used.

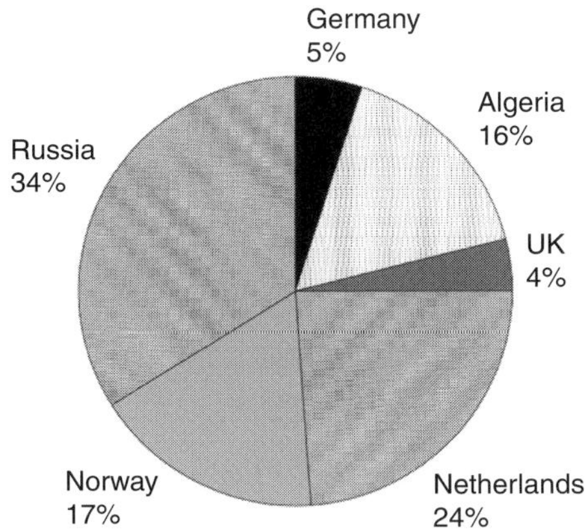

Figure 2.6 The European natural gas supply mix (Faist-Emmenegger *et al.*, 2007).

showed noticeable effects on the results, was replaced with the European mix for 2035 (see Section 2.3.7). The same SMR efficiency is used because this already represents a mature technology and the additional modelling of allocation factors due to a secondary use of the steam was outside the scope of this comparison. Due to the very high CO_2 emissions from natural gas SMR, the main advancement to 2035 was seen as being the incorporation of CCS, applied to the CO_2 exiting the PSA unit. This stream accounts for only 55% of the total direct CO_2 produced from the SMR process, with the remaining 45% being contained in the flue gas of the steam generator (Molburg and Doctor, 2003). Of the 55%, the CO_2 capture rate is 95%. This is then compressed and transported in a 400 km pipeline for storage in a depleted gas field 2.5 km below the surface (Bauer *et al.*, 2008b).

Data for the processes and infrastructures required in the SMR of natural gas, as well as the gasification and reforming of coal are based on the work of Simbeck and Chang (2002) and Felder (2007), with adjustments made to suit the functional unit and required condition of the hydrogen.

2.3.5 Coal gasification and syngas reforming

The gasification of coal using oxygen and steam produces a hydrogen-rich syngas which can be processed using shift reactions and PSA processes similar to those of natural gas SMR. Gasification of carbonaceous materials involves a complex set of chemical reactions due to the formation of both pyrolysis and combustion products within the gasification chamber. Pyrolysis is the anaerobic conversion of the feed material into volatile compounds and char using heat. The volatile products and some of the char combust in the presence of a limited

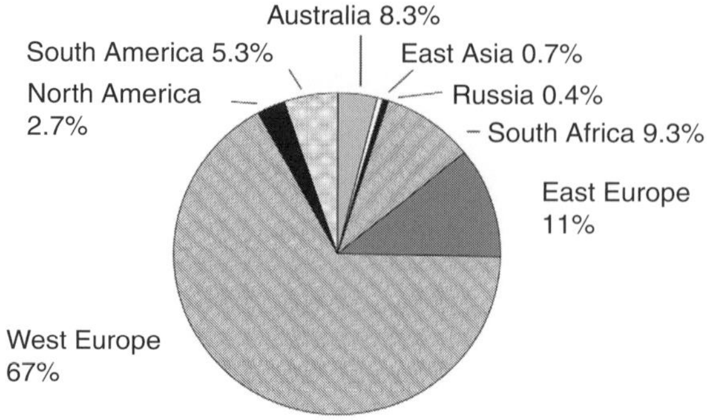

Figure 2.7 The German hard coal supply mix (Dones *et al.*, 2007).

oxygen supply. This provides heat for the gasification of the remaining char, with steam being injected as both a reactant and further source of hydrogen. The high operating temperature means that the non-combustible constituents of the feedstock form a slag rather than fusing as ash. Gasification uses very similar mechanisms to those of partial oxidation of heavy oils (Stiegel and Ramezan, 2006; Kothari *et al.*, 2008).

The overall conversion can be summarised by the following reactions:

$$C + H_2O \rightarrow CO + H_2 \tag{2.3}$$

$$CO + H_2O \rightarrow CO_2 + H_2 \tag{2.4}$$

The natural abundance of coal, the efficient and reliable qualities of gasification as a technology and its compatibility with carbon capture mechanisms highlight this pathway to hydrogen production as a very important one for consideration in this comparison. A German hard coal supply mix is used, which consists of contributions from the global regions indicated in Figure 2.7.

The selected gasifying procedure prioritises the production of H_2 rather than a co-production of H_2 and electricity. The coal enters an entrained flow gasifier as a dry and pulverised feedstock. It is gasified with steam and purified oxygen which results in a syngas consisting primarily of CO and H_2. With an operating temperature of more than 1000°C in the gasifier, the syngas must be cooled prior to being scrubbed of trace elements before undergoing further reforming treatment similar to that for natural gas SMR (Stiegel and Ramezan, 2006). The LHV efficiency of the complete process is 72.7% and results in purified H_2 with an output pressure at the PSA unit of 30 kPa (Simbeck and Chang, 2002). Figure 2.8 shows the main stages of this process.

Again, it was not within the scope of this work to include co-products or to account for the fraction of the syngas being used to generate electricity. For coal gasification and reforming (CGR) technology in 2035, the changes include the use

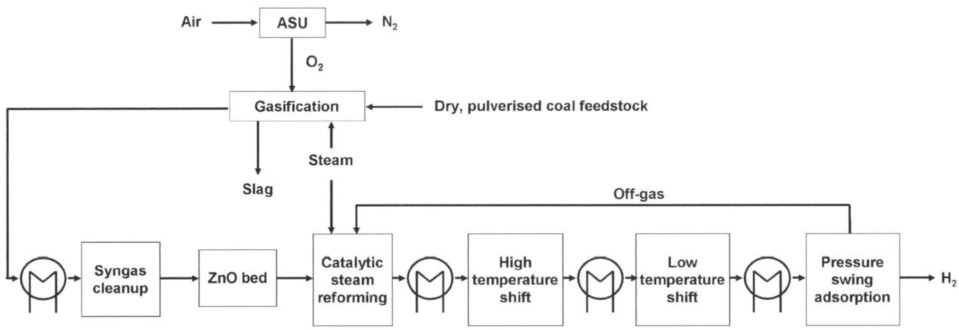

Figure 2.8 Block flow diagram of hydrogen production from coal gasification.

of 2035 electricity mixes (see Section 2.3.7) for all noticeably influential uses, as well as the inclusion of CCS. Here a significant advantage of CGR over SMR is that all CO_2 produced in the process is available for capture. A capture rate equal to the natural gas scenario was used but, instead of storage in a depleted gas field, the CO_2 from this process is transported 400 km by pipeline and injected into an aquifer 800 m below the surface (Bauer *et al.*, 2008b).

2.3.6 Biomass gasification and syngas reforming

Biomass gasification on the scale considered in this chapter is currently being tested in demonstration plants and with sufficient experience of the processes to allow a representative scenario for 2005. The gasification of biomass with product gas reforming has a high potential to become commercially viable due to synergies of the technology with those described above for coal gasification and syngas steam reforming. Individual technologies differ widely, however, depending on the biomass fuel used and the intended function and product streams. For the production of H_2, the syngas exiting the biomass gasifier must be suitable for use in steam reforming processes which require a high temperature (800–900°C) and a pressure of 20–30 kPa (Felder, 2007; Koroneos *et al.*, 2008). This process is shown in Figure 2.9, and follows the same reactions shown in Equations (2.3) and (2.4).

For the gasification process, we use data from the demonstration plant near Güssing in Austria (Bolhàr-Nordenkampf *et al.*, 2002; Rauch *et al.*, 2004). The biomass feedstock modelled is mixed wood chips, sourced as a waste material from industry (Bauer, 2007). This is transported to the plant by a combination of rail and road freight services using standard transport distances as assumed in the ecoinvent database (Frischknecht *et al.*, 2007b). Waste heat from gasification is used to dry the biomass to the required humidity level of 15%. The Güssing plant is an indirectly heated, fluidised bed gasifier (see Figure 2.10) operating at atmospheric

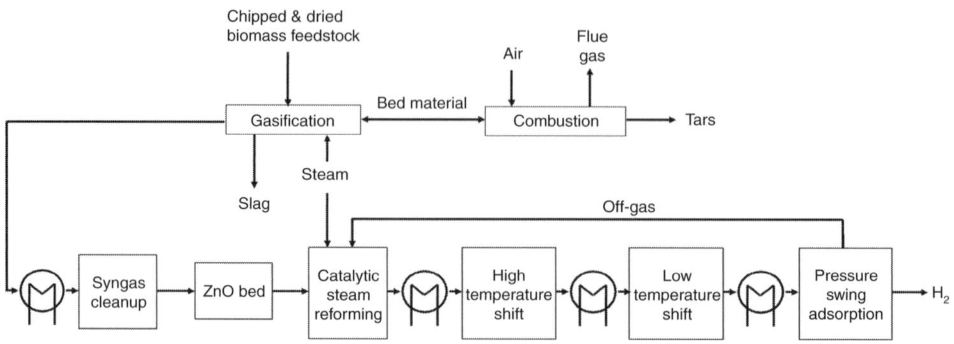

Figure 2.9 Block flow diagram of hydrogen production from biomass gasification.

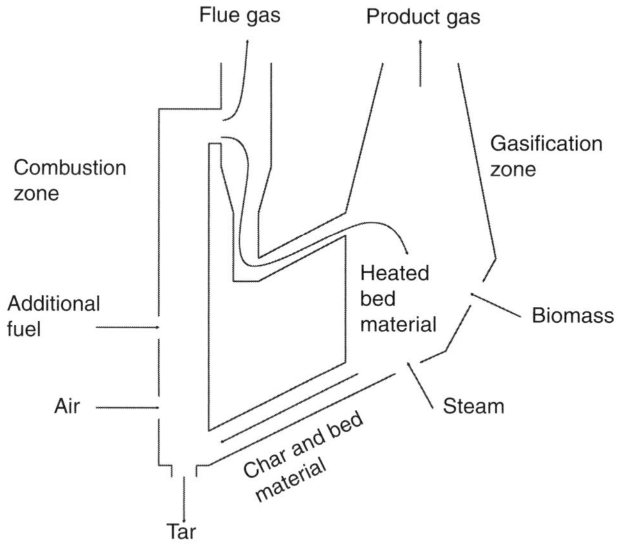

Figure 2.10 Schematic diagram of the indirectly heated fluidised bed reactor at Güssing (Duret *et al.*, 2005).

pressure (1 kPa). The indirect heating of the gasification zone is achieved by the combustion of char and bed material in a separate chamber, which means that air rather than purified oxygen can be used to feed this combustion, thereby reducing the nitrogen and tar contamination of the product gas. Gasification is conducted with steam (generated using waste heat from the gasification), resulting in a product gas consisting of 39% by volume H_2, followed by CO (27%), CO_2 (20%), and CH_4 (14%). For the 2005 scenario, the gasification process itself is characterised by a cold gas efficiency (CGE) of 73.6% (Felder, 2007; Bauer *et al.*, 2008a) and a temperature of roughly 850°C. With a pressurisation stage, this output is then suitable for the reforming and purification of the product gas.

Following gasification, the product gas is scrubbed of tars and other particles using methyl ester from rapeseed. The cleaned gas then undergoes reforming and purification processes similar to that in the SMR of natural gas to produce a purified H_2 stream at 30 kPa pressure. The efficiency of this stage is 76.3%, which gives an overall LHV efficiency of 56%.

The scenario in 2035 uses an increase in gasification efficiency to 85% (Simbeck and Chang, 2002) to reflect the scope for improvement of a currently immature technology, but uses the same steam reforming efficiency. Although the gasification of biomass and the steam reforming of the product gas leads to larger quantities of CO_2 emissions per unit H_2 than for either of the fossil reforming scenarios, both the smaller scale of the biomass plant and the biogenic source of most of the carbon mean that the application of CCS is not considered in the 2035 scenario.

2.3.7 Electrolysis

Electrolysis currently accounts for the majority of production not achieved through the direct reforming of fossil fuels, and represents approximately 4% of total H_2 produced worldwide (Kothari *et al.*, 2008). Electrolysis is the splitting of water molecules into H_2 and O_2 by passing direct electric current between two electrodes separated by an ion exchange membrane. Oxygen is produced at the anode and hydrogen at the cathode, both in a very pure form (Ivy, 2004). The modularity of electrolysis equipment makes it possible to locate production at the point of consumer demand, meaning that the inputs to production are the more readily available and easily transportable resources of electricity and water.

Electrolysis represents the conversion of one energy carrier (electricity) into another (H_2) and therefore carries with it the efficiency penalties of the preceding conversion stages. The effectiveness of the system is thus not only dependent on the conversion efficiency of the electrolyser but also on the conversion of primary energy to electricity, as well as the conversion of H_2 to motive force, causing motion along the road. According to the vehicle manufacturer Honda (Honda, 2010), a fuel cell passenger car has tank capacity for approximately 4 kg of H_2, which allows for a range of around 380 km, equivalent to 0.0105 kg H_2/km. The direct use of energy in the vehicle is thus around 0.35 kWh H_2/km if using the LHV of H_2 (33.3 kWh/kg). If the electricity used to produce this amount of H_2 by electrolysis is included, then this becomes 0.6 kWh_{el}/km. For the production of the electricity, a thermal power plant of 40% electrical efficiency can be assumed, meaning that the primary energy demand, or indirect energy use for 1 km, becomes 1.5 kWh_{PE}/km. This is without considering further losses due to electricity distribution. This case

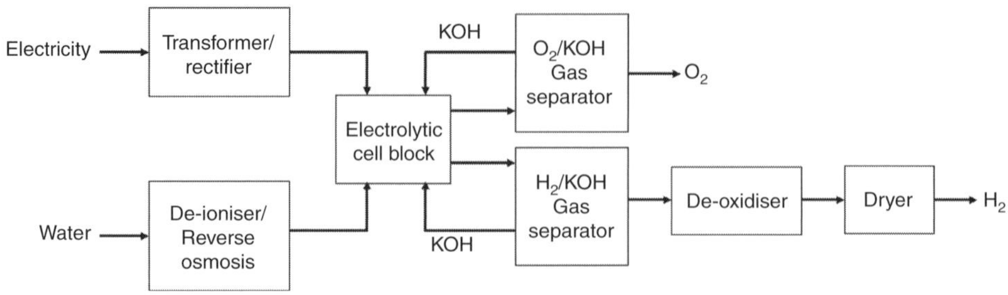

Figure 2.11 Schematic diagram of on-site electrolysis (Roads2HyCom, 2009).

therefore shows that the primary energy needed to achieve 1 km of vehicle travel is more than four times the end energy extracted from the H_2.

For the scenario in 2005, we model a low temperature, bipolar, alkaline electrolyser. Here, potassium hydroxide (KOH) is added to the water to about 30 wt % in order to improve its conductivity as an electrolyte. Although this is the least efficient form of electrolyser, it is the most developed and currently has the lowest capital costs (Holladay *et al.*, 2009). The main processes involved in this method of H_2 production are shown in Figure 2.11.

In the 2005 scenario, electrolysis requires almost 47 kWh electricity per kg H_2, but, with auxiliary services and compression to 45 MPa, this increases to 57 kWh. This represents a conversion efficiency of 58%, which is relatively good for current systems (Ivy, 2004).

We follow Ivy (2004) in assuming that a small fuelling station has a demand of around 100 kg/day and a large station around 1000 kg/day. At a maximum production capacity of 5.4 kg/h, the electrolyser would need to work almost continuously to supply the demand of the former if working on its own. Therefore, in order to fuel vehicles in a reasonably short time, a fuelling station must either have sufficient storage capacity or a number of electrolyser cells working in parallel. For the 2035 scenario, it is assumed that demand will be sufficient to require both of these measures.

While there is potential for efficiency improvements in the future, by moving to high temperature electrolysis in solid oxide electrolysis cells (SOEC), the technology is still encumbered by a number of technical hurdles (Holladay *et al.*, 2009). For development to the 2035 scenario, we therefore remain with the alkaline electrolyser but include a target of 50 kWh/kg (H_2) which translates to a system efficiency of almost 67% (Ivy, 2004).

Electrolysis also consumes significant amounts of water. In 2005, the amount is 10 kg/kg H_2, which, in 2035, falls to 8.8 kg/kg H_2 when reducing water demand in proportion to the electricity. Currently, the sources of water, the quantities

Table 2.3. *European electricity production mixes in 2005 and 2035.*

	Nuclear (%)	Fossil (%)	Hydro (%)	Other renewables (%)	Pumped storage (%)	Other (i.e. waste incineration) (%)	Total production (TWh)[a] (%)
2005[b]	31.6	51.1	11.4	3.1	1.3	1.4	2 681
2035[a]	21.6	50.6	17.6	8.8	0.8	0.7	3 866

[a] Frischknecht *et al.* (2008)
[b] Frischknecht *et al.* (2007c)

used and the resources expended through providing it are not consistently and consequentially modelled, introducing small uncertainties in the results.

2.3.8 Electricity mixes

It has been shown that the production of H_2 using electrolysis requires significant electricity consumption. Therefore, the results of an LCA of this method will be very much guided by the generating source of the electricity, whether from fossil fuels, nuclear fission, or renewable energy; and by the specific technologies used. For the two time frames of 2005 and 2035, and consistent with the geographical region of the H_2 production, we used the average European electricity mix. In 2005, this is based on the 22 member countries of the UCTE[3] (Frischknecht *et al.*, 2007b) and, in 2035, on the mix determined by Frischknecht *et al.* (2008), see Table 2.3. Here we employ the analysis of Bauer *et al.* (2008a) and Roth *et al.* (2009) to determine suitable technological development. Where appropriate, and for consistency, these two electricity mixes are used throughout the inventories in the process chains of other H_2 production scenarios, and where the electricity use is significant to the results. For 2005 and 2035, we refer to both as simply the European mix.

In 2035, it is expected that both the capture and storage of CO_2 will have been proven on pilot plants using fossil fuel but that a wide-scale application across those European countries considered will not yet be a reality. Therefore, CCS has not been applied to generating technologies forming the mix in 2035. Fossil fuel technologies present in the mainstream in 2035 include integrated gasification of hard coal with combined cycle electricity generation (IGCC) and the assumption that all natural gas will be combusted either in combined cycle (CC) or in combined heat and power (CHP) plants. Synthetic natural gas (SNG) from the

[3] Union for the Co-ordination of Transmission of Electricity; since 2009, replaced by the European Network of Transmission System Operators for Electricity (ENTSO-E).

gasification and methanation of wood used in CHP plants, onshore, and offshore wind turbines are three renewable energy technologies which together account for more than 8% of the 2035 mix. Production from other renewables, such as solar photovoltaic, solar thermal, and ocean wave remain minor contributors on the European level.

Individual electricity sources (hydro, nuclear, coal, natural gas, wind, photovoltaics (PVs), and solar thermal) are modelled as a means of determining the sensitivity of the results to the electricity used for electrolysis. Table 2.4 outlines the main characteristics of each type, as well as providing the data sources used.

The assessment does not differentiate between peak and off-peak generation capacities and therefore the availability of electricity, and does not consider the intermittency of some renewables. As a storable energy carrier, it is possible that producing H_2 via electrolysis could be done as and when surplus capacity exists within the electricity network. This would allow it to be produced during off-peak periods as well as according to the availability of renewables such as wind and solar, thereby offering possibilities for load-levelling and load-shifting. If, from the results of this chapter, electrolysis pathways are considered to have a good environmental performance compared with other H_2 production routes, then these aspects may well prove to be of interest to future studies.

2.3.9 Concentrated solar, thermochemical

This is the production of H_2 via thermochemical cycles, using concentrated solar radiation as the source of high temperature process heat. The focus for the purposes of this chapter was the two-step zinc/zinc oxide (Zn/ZnO) cycle in which the zinc functions as an energy carrier, having an energy density of 5.37 MJ/kg. It is either formed by the dissociation of ZnO or by the carbothermic reduction of ZnO. In both cases H_2 is produced via the hydrolysis of the Zn (Felder, 2007). This technology is still under development at the PSI, but has already seen small-scale pilot plants, most notably in Israel; plans exist for larger demonstration plants and progress toward realisation on a commercial-scale (Wieckert *et al.*, 2007; Epstein *et al.*, 2008). It is therefore expected that remaining development constraints could be overcome in order for commercial scale plants to exist by 2035. A possible advantage of these processes is that the hydrolysis stage must not be conducted at the concentrating solar plant. Instead, the zinc itself can be transported to the site of demand or regional distribution point and the hydrolysis conducted there. In order to constrain the number of scenarios under assessment, we remain with the concept of energy being carried in the form of either electricity or H_2. For more details on the two solar thermochemical scenarios please refer to Felder (2007).

Table 2.4. *Individual electricity sources used in the impact assessment.*

Technology	Description	Data reference
2005		
Nuclear	All requirements based on two Swiss nuclear plants (pressurised water reactor (PWR) and boiling water reactor). Emissions are based on UCTE (BE, CH, DE, ES, FR, NL) nuclear power plants.	Dones (2007)
Coal	Average of hard coal power plants in UCTE in year 2000.	Dones *et al.* (2007)
Natural gas	Average of natural gas power plants in AT, BE, DE, ES, FR, IT, LU, NL, GR, PT. Fuel input from high pressure European network.	Faist-Emmenegger *et al.* (2007)
Hydro	Representative mix of four Swiss and one Austrian run-of-river power plants. Lifetime is assumed to be 80 years.	Bauer *et al.* (2007)
Wind	800 kW onshore wind turbine operating under average European conditions (20% capacity factor). The lifetime of moving and fixed parts is assumed to be 20 and 40 years, respectively.	Burger and Bauer (2007)
Photovoltaic (PV)	Production mix of photovoltaic electricity in Spain. Annual output for roof and façade installation: 1282 and 813 kWh/kWp respectively.	Jungbluth and Tuchschmid (2007)
Solar thermal	Average for technologies installed in Spain.	Viehbahn *et al.* (2008)
2035		
Nuclear	European pressurised reactor (EPR) with average burn-up 42.5 MWd/kg heavy metal, 90% uranium and 10% MOX fuel elements. Plant material requirements extrapolated from the Swiss PWR using capacity factors.	Bauer *et al.* (2008a)
Coal	Integrated gasification combined cycle (IGCC) plant representing average European conditions in 2035. Operating lifetime 150 000 hours at mid-load.	Bauer *et al.* (2008a)
Natural gas	Combined cycle (CC) plant operating at base load. Infrastructure based on new power plant in Germany. Fuel input from high pressure European network.	Bauer *et al.* (2008a)
Hydro	Based on data from one Swiss run-of-river power plant, 50 MW. Lifetime is assumed to be 80 years.	Bauer *et al.* (2007)
Wind	20 MW offshore wind turbine. Average wind speed 10 m/s at 60 m (capacity factor 0.457). The lifetime of moving and fixed parts is assumed to be 20 years.	Bauer *et al.* (2008a)
PV	3 kWp polycrystalline-silicon laminates as a slanted-roof installation. Note: installation in Switzerland.	Bauer *et al.* (2008a)
Solar thermal	Average of solar thermal technologies and efficiencies from European concentrated solar power plants.	Viehbahn *et al.* (2008)

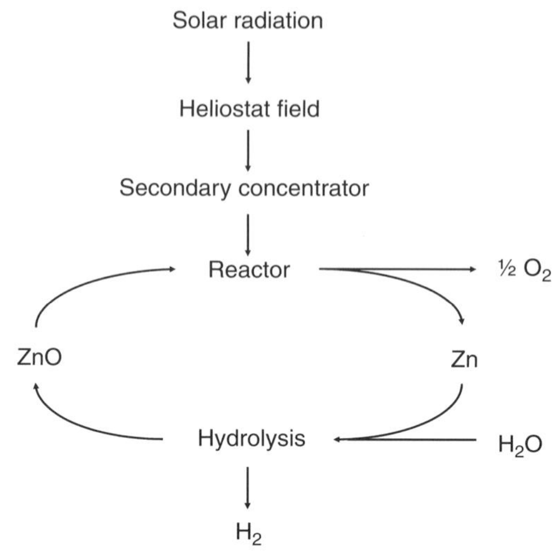

Figure 2.12 H_2 from the solar thermal dissociation of ZnO/Zn (Felder, 2007).

Solar thermal dissociation of ZnO

Using a concentrating array of heliostat mirrors, solar radiation is focused onto a secondary concentrator and, via a quartz window, enables temperatures exceeding 2000K to be achieved in the reaction chamber. At these temperatures, the ZnO is separated into its compound elements, which must be immediately quenched in order to avoid recombination. To obtain H_2, the Zn is reacted with steam in a hydrolyser (see Figure 2.12). We assume an H_2 pressure of 1 kPa after hydrolysis.

The overall process is represented by the following reactions:

$$ZnO \rightarrow Zn + \tfrac{1}{2}\,O_2 \tag{2.5}$$

$$Zn + H_2O \rightarrow ZnO + H_2 \tag{2.6}$$

An important characteristic of the solar thermal dissociation (STD) process as modelled here, and one which has significant impact on the results, is the large consumption of argon by the current system. The production of argon is energy-intensive, and inhibits the economic feasibility of the STD process. It is mostly used to quench the Zn and O_2 leaving the reactor in order to avoid recombination and to protect the reactor window against extreme temperatures (Felder, 2007). Currently, it can be neither recycled within the system nor the process achieved without it. Therefore, rather than speculating on a future situation where the argon is recycled (requiring an as yet unspecified process), the current level of demand

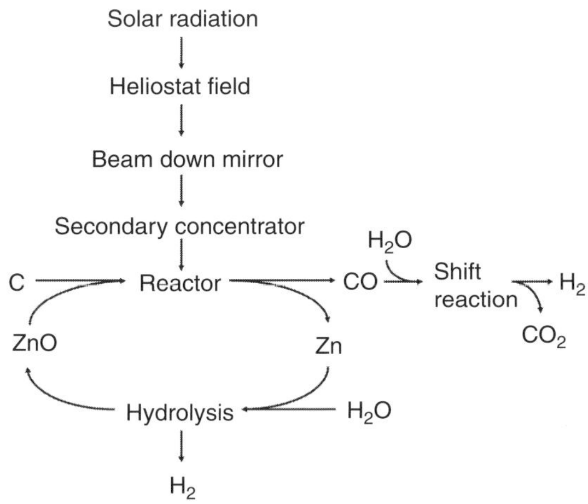

Figure 2.13 H_2 from the solar carbo-thermal reduction of ZnO/Zn (Felder, 2007).

is included in this assessment. However, in order to understand the sensitivity of the results to this uncertainty factor, a discussion on a scenario in which only 10% of the argon is used in quenching is included. This figure is intended to represent a significant reduction through recycling, but one for which argon is still necessary.

Solar carbo-thermic reduction of ZnO

In this process the solar carbo-thermic reduction of ZnO (SCR), a carbonaceous material such as coal (or charcoal for a renewable source) is added into the reaction chamber to provide carbon as a reducing agent. The resulting effect is that the separation of the ZnO produces Zn and CO rather than Zn and O, as shown in Figure 2.13. The reaction temperature is lower at around 1500K and the process does not require argon to quench the resulting gases.

Although the hydrolysis of the Zn is then the same as for the STD procedure, an additional step is that the CO is put through shift reactions to yield H_2 and CO_2. This means that a higher quantity of H_2 per kg of Zn is generated.

2.3.10 Hydrogen distribution

The transport of H_2 to the fuelling station (where appropriate) is accounted for both under contemporary and future scenarios suitable for central Europe. For 2005, the focus is on road transport, for which a range of distances are modelled. In 2035, a range of both road and pipeline distances are considered.

Road transport

For the road transport scenarios, H_2 is delivered to the fuelling station in pressurised bottles. Yang and Ogden (2007) state a figure of 300 kg H_2 as representative for a single truck transport capacity, but this does not reflect the additional weight of the bottles. Data from Airliquide 2010) describe H_2 at 150 kPa in 50-litre steel bottles, each weighing 65 kg (65.7 kg including H_2). Tubes are transported in bundles of 12, with 42 bundles possible per truck. This results in a total weight of 335 kg H_2 and a total freight of 33 t. However, approximately 20% of the H_2 remains in the tanks as a cushion gas after dispensing, thus, only 270 kg H_2 is actually delivered (Bossel *et al.*, 2003; Mueller-Langer *et al.*, 2007). The transport of 1 kg H_2 is therefore accompanied by around 123 kg of necessary infrastructure, not including the truck itself.

Pipeline transport

Today, hydrogen pipelines exist to provide a means of transport between industrial facilities. There are currently roughly 1500 km in Western Europe and approximately 900 km in the USA (Mueller-Langer *et al.*, 2007). One method of enabling a much wider distribution is to mix H_2 with natural gas and transport it through the much larger network of natural gas pipelines. However, re-separation of the two products imposes a further energy demand, as well as significant additional costs (Bedel and Junker, 2006). In the present study, H_2 is transported in pure form in a natural gas pipeline and we assume an equal gas flow rate as for natural gas, rather than an equal energy flow. The latter would require a far higher flow speed for H_2 and hence a higher energy demand for pumping it through the pipeline (Bossel *et al.*, 2003). For 2005, a distance of 200 km is used in order to complement the larger types of production technologies assessed and link them with large, central demand sites, such as cities. This also serves to provide a useful comparison to the road transport scenarios. Pipeline leaks of H_2 to the environment are not taken into consideration.

The results for the transport scenarios are presented separately from production scenarios in order to avoid arbitrary combinations and to allow the effects of different transport scenarios to be viewed against those of potentially suitable production pathways.

2.3.11 Conventional fossil fuels for internal combustion engines

To enable comparison, the environmental impacts from the production of 1 kg H_2 are compared with the production and combustion of equivalent quantities of fossil ICE fuels, quantified on the basis of equal energy content (shown along

Table 2.5. *Physical characteristics of the assessed gaseous (g) and liquid (l) fuels.*

	Hydrogen (g)[a]	Natural gas (g)[b]	Gasoline (l)[c]	Diesel (l)[c]
Density (kg/Nm^3)	0.0899	0.763	–	–
Density (kg/l)	–	–	0.75	0.84
Energy content (LHV) (MJ/kg)	120.2	47.84	42.5	42.8
Energy content (LHV) (MJ/Nm^3)	10.8	36.5	–	–
Energy content (LHV) (MJ/litre)	–	–	31.875	35.952
Quantity in MJ/kg (LHV), equivalent to 1 kg H_2 (kg)	–	2.499	2.824	2.804

[a] HARC (2006)
[b] Faist-Emmenegger *et al.* (2007)
[c] Jungbluth (2007)

the bottom row of Table 2.5). The characteristics of the fossil fuel chains in this assessment are based on published literature on the LCA of oil (Jungbluth, 2007) and natural gas (Faist-Emmenegger *et al.*, 2007). Average European import mixes are used for each fuel, reflecting the actual shares of worldwide production regions to the European fuel supply. For the measurement of emissions, the combustion of the conventional fuels is done in a EURO 3 ICE (Spielmann *et al.*, 2007).

2.4 Life cycle assessment results

This section contains the results of the LCA and (aggregated) specific environmental burdens in order to compare the environmental performances of the different technologies. To compare H_2 with conventional vehicle fuels as described in Section 2.2.4, the two impact assessment methods of total GHG emissions and CED were used.

As a means of comparing the potential quantities of air pollutants emitted from the different scenarios, and in accordance with the World Health Organization (WHO, 2006), the individual pollutants of $PM_{2.5}$ and PM_{10}, NMVOCs), NO_x, and SO_2 emissions are also given in terms of their cumulative amounts from the complete life cycle. Although WHO specifically lists ozone (O_3), otherwise known as photochemical smog, as an important air pollutant at low altitudes, we do not directly quantify it here due to its very minor existence as a direct emission from industrial processes and its non-existence as a direct vehicle emission. O_3 is produced as a result of vehicle operation via the secondary effect of sunlight acting on other pollutants, specifically NO_x and VOCs. Although WHO defines nitrogen dioxide (NO_2) rather than NO_x individually as one of the four critical air pollutants, in this chapter we do not attempt the complex modelling of secondary pollutants (i.e. the formation of particulates and low altitude ozone after initial emission of the substances). Instead, we show the quantified inventory results in order to reflect the potential to cause these secondary pollutants and therefore use NO_x, consisting of NO and NO_2, and NMVOC in place of O_3. We use NMVOC instead of VOC because methane (CH_4) does not contribute to ozone formation. CH_4 is accounted for according to its significance as a GHG.

2.4.1 Cumulative non-renewable energy demand

As described in the previous section, the pathways to separate H_2 from other elements are not 100% efficient and it requires more energy to separate H_2 than can be regained from it. The production of fossil ICE fuels, on the other hand, involves their extraction, refining, and upgrading, and it is therefore possible that less energy will be invested in producing the fuels than is released upon combustion.

Primary energy resources can be characterised as either renewable or non-renewable, a critical factor in the assessment of a scenario being its level of dependence on non-renewable resources. These are defined as being either fossil or uranium in nature. The CED is therefore a quantification of the total non-renewable energy resources consumed over the life cycle of each pathway, and follows the methodology described in Frischknecht *et al.* (2007b).

Figure 2.14, shows that the CED for ICE fuels is equivalent to between 1.2 and 1.4 MJ, meaning that, in reality, the processing of these fuels is around 25% more energy intensive than the energy contained within the end-product fuel itself. More than 98% of the primary energy used in production of the fuels is fossil-based.

For H_2 production, the efficiencies of thermal power plants in the European electricity mix (highlighted by the individual coal, gas, and nuclear scenarios) mean that it is very inefficient to produce H_2 via electrolysis from fossil and uranium fuels. The wood gasification and reforming pathway (WGR) shows relatively high CED due to fossil fuel demands for transport of the biomass as well as oil fuel used to generate steam for product gas reforming. The use of renewables as a source of electricity for electrolysis is the best possibility for reducing non-renewable primary energy use significantly below levels for the ICE fuels.

The application of CCS leads to additional energy being required for the capture, transport, and injection of CO_2 and has the effect of counteracting any of the improvements seen in process and supply chain efficiencies of the SMR and CGR pathways by 2035.

The solar thermochemical scenarios require significant amounts of electricity in the operation of the heliostat mirrors and the hydrolysis stages, as well as for the shift reaction stages of the SCR. This electricity is from the European mix, a

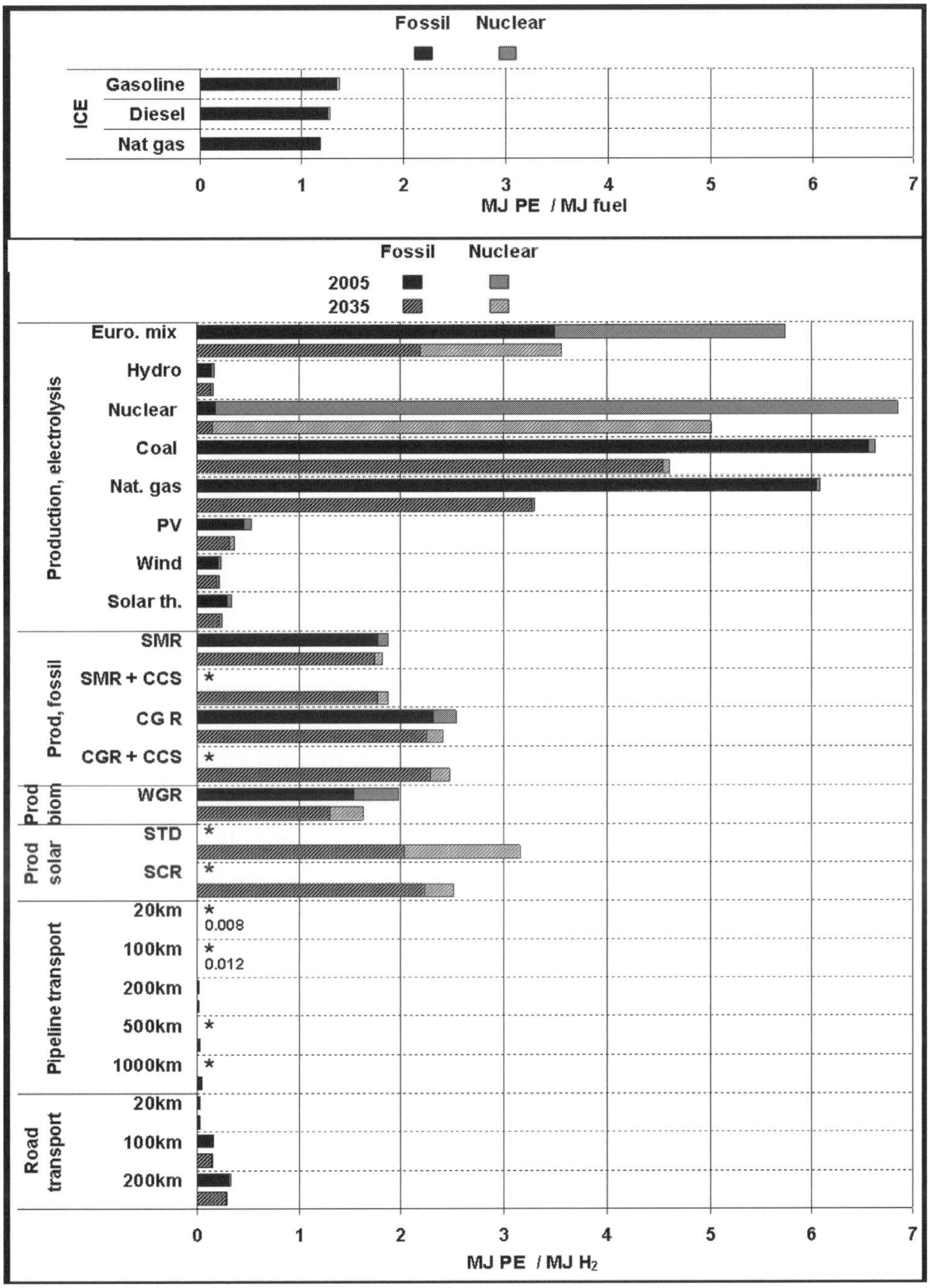

Figure 2.14 Life cycle cumulative primary energy demand, non-renewable (CED). For each H_2 production and transport scenario, 2005 is the higher column and 2035 the lower column.

Note: * indicates that data are not available for 2005.

factor which is reflected in the results. However, it is the production of argon for the STD pathway and the use of coal as a source of carbon in the STD pathway which contribute most significantly to the results. For the latter, the use of charcoal instead of coal would defer some of the demand to renewable biomass sources. In consideration of the uncertainty surrounding STD and argon, a scenario for which 90% of the argon is recycled would reduce the total fossil and nuclear CED to 0.75 MJ/MJ, 24% of that shown in Figure 2.14.

For the pathways to which it is relevant, the transport of H_2 does not increase the results by a significant margin according to the options considered.

2.4.2 Greenhouse gases

For conventional fuels, emissions occur during the extraction, refining, and transportation of the fuel as well as during combustion in an ICE. Figure 2.15 shows the total GHG emissions for production and combustion of ICE fuels. The quantities of the fuels are equivalent to 1 kg of H_2 in terms of their energy content.

The cumulative GHG emissions for the H_2 production and transport scenarios in 2005 and 2035, as well as the conventional fossil fuels as reference and benchmark values, are shown in Figure 2.16.

The results show that, for the production of H_2 and ICE fuels, the most significant GHG is CO_2. For 2005, only those technologies exploiting renewable or nuclear energies as the primary feedstock or source of operational energy achieve lower GHG emissions than the use of conventional ICE fuels. The high proportion of fossil fuels in the European electricity mix leads to emissions more or less three times higher than those of diesel or gasoline for the electrolysis pathways. This is reflected in the levels that would be emitted if single source fossil fuels were used to generate the electricity. Although it is a slightly less efficient process than CGR, the combustion of natural gas in the SMR process leads to lower GHG emissions, due to the lower CO_2 emissions released from combusting natural gas as opposed to coal. These processes reflect technologies by which most H_2 is

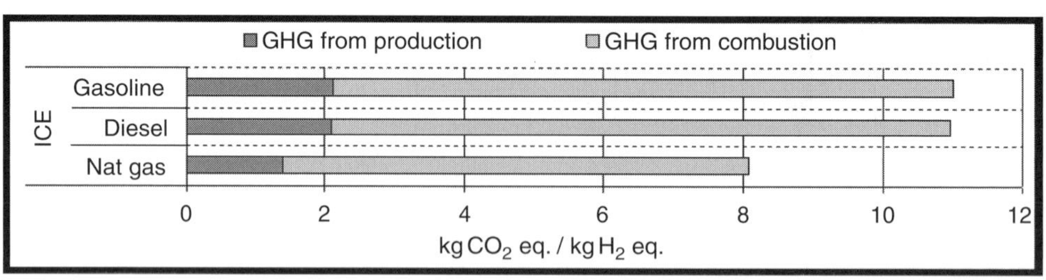

Figure 2.15 Total greenhouse gas (GHG) emissions from production and combustion of 1 kg H_2 equivalent quantities of internal combustion engine (ICE) fuels.

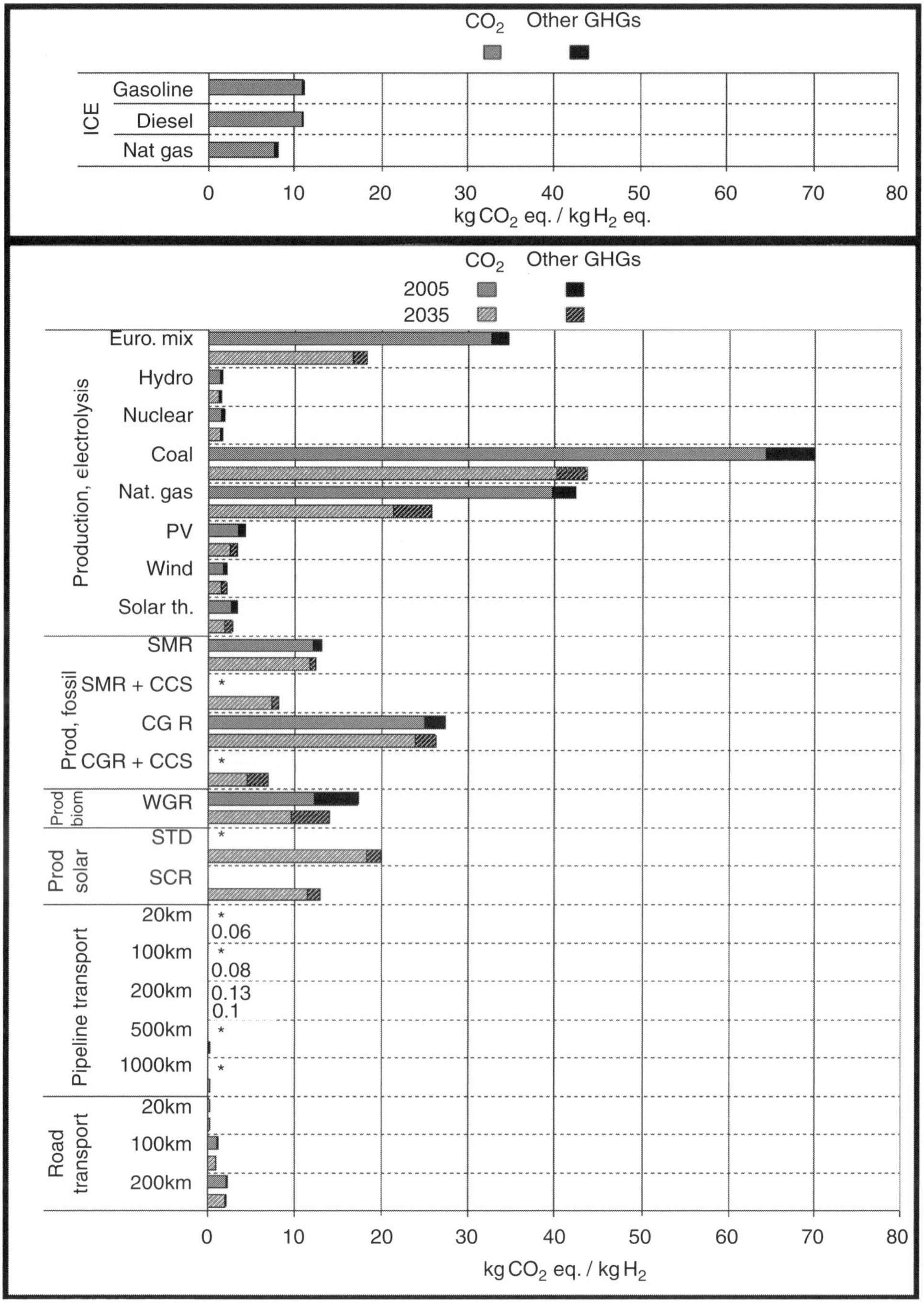

Figure 2.16 Life cycle greenhouse gas emissions from fossil sources only, i.e. without biogenic CO_2. For each H_2 production and transport scenario, 2005 is the higher column and 2035 the lower column.
Note: * indicates that data are not available for 2005.

currently produced, but the results show that using natural gas to produce H_2 generates twice the level of GHG emissions than if natural gas was used directly in an ICE, whereas the CGR process yields more than twice the amount of GHG emissions than combusting either gasoline or diesel in an ICE. Due to the majority of GHG emissions from the biomass gasification pathway being biogenic in their origin (and thus not shown in Figure 2.16), the impacts of this chain are marginally below those of natural gas in an ICE, which has the lowest emissions of the ICE fuels.

The renewable and nuclear electrolysis scenarios already show good performance in 2005, with only small improvements being determined in relation to the other scenarios in 2035. The European electricity mix shows a significant improvement for this time frame but still remains above the levels of the ICE fuels.

Improvements in the energy supply chains of the fossil-based reforming pathways (SMR and CGR) have only minimal influence on the results for 2035. The competitiveness of the pathways is only improved through the application of CCS, most noticeably for the CGR pathway. As explained in Sections 2.3.4 and 2.3.5, only slightly more than 50% of the CO_2 from the SMR process is captured, whereas for CGR the capture rate is 95%.

The two solar thermochemical pathways show relatively high GHG emissions. For the STD pathway, this is mostly due to the requirement for argon gas. Recycling 90% of the argon used in quenching the Zn and O_2 would have a large impact on the results and reduce total GHG emissions to around 7.8 kg CO_2 eq./kg H_2 which equals 40% of the amount in Figure 2.16. If this were achieved, it would place STD on a level equal to natural gas in an ICE. For the SCR pathway, the shift reaction of CO with steam creates CO_2 as an off-gas, thereby ranking this technology slightly lower than the ICE fuels.

Figure 2.16 also shows that only through long transportation by road will the GHG emissions of an H_2 production pathway be appreciably increased by the inclusion of the delivery stage.

2.4.3 Particulates

PM is classified by its size, a factor which affects its ability to remain airborne and to be inhaled and absorbed by the human body. WHO (2006) distinguishes PM of most significant risk to human health as being all that is less than 10 μm (PM_{10}). This is further delineated as fine PM ($PM_{2.5}$) versus coarse PM (between 2.5 μm and 10 μm). Figures 2.17 and 2.18 illustrate the cumulative emissions of each of these PM classifications from production and combustion of the ICE fuels.

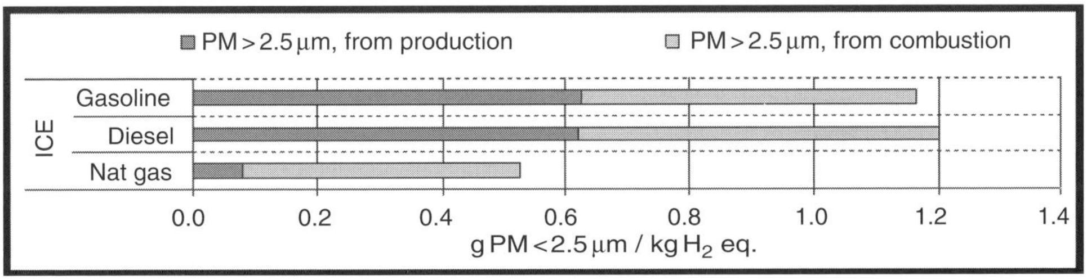

Figure 2.17 Emissions of PM less than 2.5-μm diameter from the production and combustion of 1 kg H_2 equivalent quantities of the internal combustion engine (ICE) fuels. ICE standard is EURO 3.

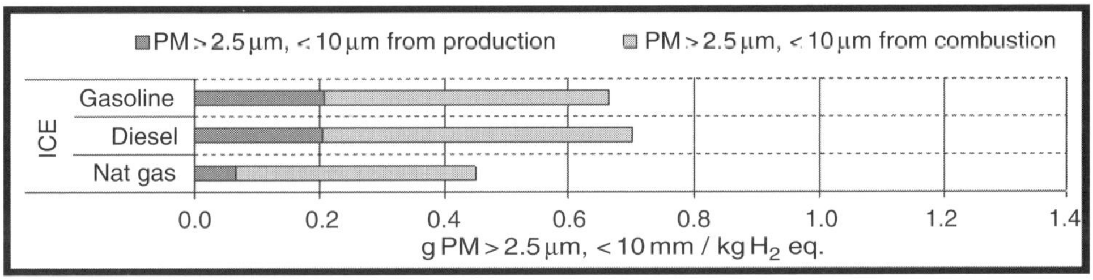

Figure 2.18 Emissions of PM between 2.5 μm and 10 μm in diameter from the production and combustion of 1 kg H_2 equivalent quantities of the internal combustion engine (ICE) fuels. ICE standard is EURO 3.

PM emissions for most of the different H_2 pathways are shown to be very similar (Figure 2.19) and about two to three times the levels for the production and combustion of conventional ICE fuels. The electrolysis processes using a high proportion of coal or lignite in the electricity mix are notable exceptions. Improvements for all production scenarios by 2035 are visible, with the slight exception of electrolysis using hydropower-generated electricity. Here, the specific power plant in 2035 is associated with higher PM than the average of plants in 2005 (see Table 2.4). Production methods using the direct conversion of fossil fuels are slightly lower in their PM emissions than the electrolysis pathways, due in part to the low particulate emissions from the fossil fuel technologies and in part to common demands of the electrolyser plants. If the STD scenario recycled 90% of the argon in its quenching process, this would result in a more or less one-third reduction in overall PM_{10} emissions. This pathway would then have the lowest PM emissions of any H_2 pathway, but remain marginally higher than for any of the ICE fuels.

Transport of H_2 must be hundreds of kilometres by road for the emissions from this stage to have any significant effect on the production scenario.

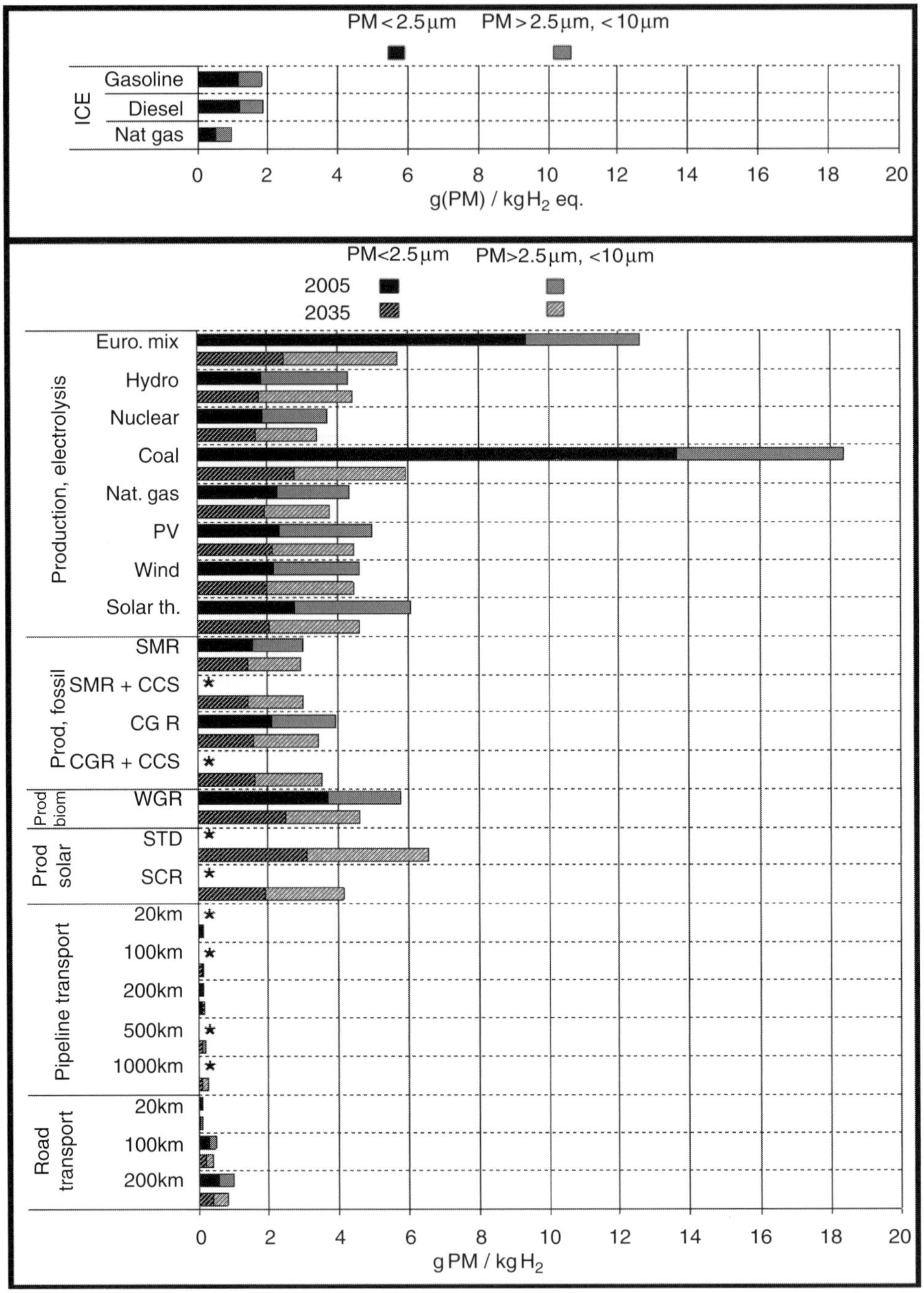

Figure 2.19 Life cycle emissions of airborne particulate matter. For each H_2 production and transport scenario, 2005 is the higher column and 2035 the lower column.

Note: * indicates that data are not available for 2005.

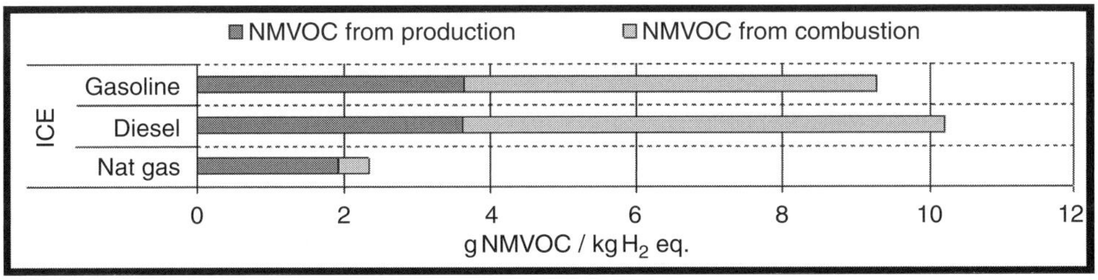

Figure 2.20 Emissions of non-methane volatile organic compounds (NMVOCs) from the production and combustion of 1 kg H_2 equivalent quantities of the internal combustion engine (ICE) fuels. ICE standard is EURO 3.

2.4.4 Non-methane volatile organic compounds

NMVOCs are particularly present in the emissions from the production and combustion of oil-based fossil fuels, with those from diesel combustion being marginally higher than for gasoline, as shown in Figure 2.20. For natural gas, the majority of emissions occur during raw gas sweetening.

The NMVOC emissions from the gasoline and diesel ICE fuels are significantly higher than all others, except the electrolysis scenario using natural gas-generated electricity, as shown in Figure 2.21, which uses an average of currently installed power plant technology. By referring to Figure 2.14, showing the CED, it is clear that the quantity of natural gas required for this method is far larger than the energy equivalent quantity of natural gas combusted in the ICE. The emissions from raw gas sweetening are therefore reflected in the results for H_2 production via this method as well as by the SMR method. A marked improvement is shown from the 2005 to the 2035 scenario, which is largely due to anticipated improvements in the Russian gas infrastructure and a reduction in gas leaks.

WGR is also characterised by relatively high NMVOC emissions and is mainly attributable to two factors: one is the road transport of the biomass to the gasifier, and the other is the production of steam at the reforming stage, which uses an oil-fuelled furnace. A natural option might be to use biomass to fuel the furnace, but this in turn would increase the NMVOC emissions from the biomass transport.

In terms of NMVOCs, several of the H_2 production pathways are seen to perform better than the ICE fuels, in particular, some of the electrolysis pathways. The improvement in the results of the electrolysis pathway using natural gas is primarily caused by a reduction in the natural gas leakage during long distance transport. It can also be seen that H_2 produced via electrolysis and solar photovoltaic (PV)-generated electricity also produces relatively high levels of NMVOC emissions. A large part of these can be traced back to the manufacturing process of the PV cells.

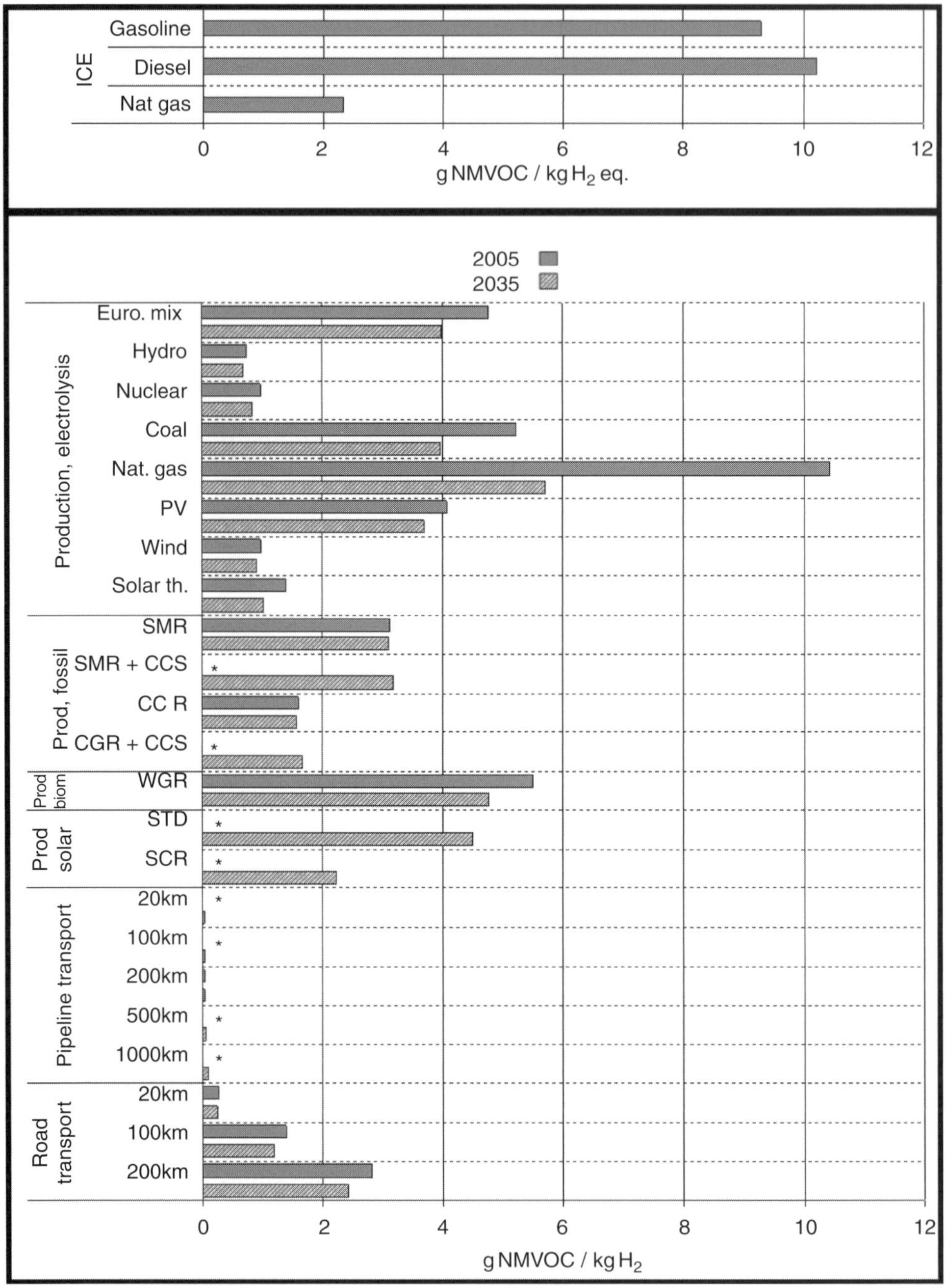

Figure 2.21 Life cycle emissions of non-methane volatile organic compounds (NMVOCs). For each H_2 production and transport scenario, 2005 is the higher column and 2035 the lower column.

Note: * indicates that data are not available for 2005.

The STD pathway again shows relatively high levels of emissions. The recycling of argon would also have a large impact on the results, reducing emissions to about 1.8 g/kg H_2 and, thus, less than for the SCR pathway.

Considering the results for the ICE fuels, it is understandable that transport of H_2 by road can add significantly to those production pathways for which this method of transport might be relevant.

2.4.5 Nitrogen oxide emissions

In an ICE, the higher the temperature of combustion, the more NO_x are formed within the combustion chamber, where both the nitrogen and oxygen are taken from the air entering the chamber. As can be seen in Figure 2.22, diesel engines have a higher temperature of combustion than gasoline-fuelled Otto engines and therefore produce a higher quantity of NO_x emissions.

This difference is significant in the comparison with the H_2 production pathways in Figure 2.23, where the performance of the individual pathways depends very much on the ICE fuel used as the reference. So, whereas none of the H_2 production pathways perform better than the natural gas used in an ICE, many of the pathways perform better than when using diesel. The H_2 pathways show relatively high emissions improvements based on 2035 assumptions which, with the exception of coal, bring their NO_x emissions down to lower than or equal to that of the ICE using diesel.

Low NO_x emissions are achieved by those H_2 pathways not using fossil fuel combustion as a significant aspect of the production method, as well as those where the nitrogen-entering combustion processes are kept to a minimum so that the product gas is not polluted with NO_x. The WGR pathway shows higher emissions, due in part to the combustion of chars in order to provide heat to the pyrolysis, and in part to those arising from the biomass transport. Improvements in the gasification process and ICE standards go some way to reducing NO_x emissions from this

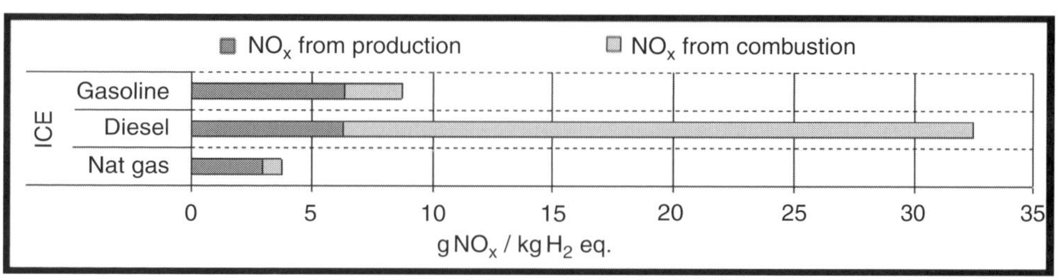

Figure 2.22 Emissions of nitrogen oxides (NO_x) from the production and combustion of 1 kg H_2 equivalent quantities of the internal combustion engine (ICE) fuels. ICE standard is EURO 3.

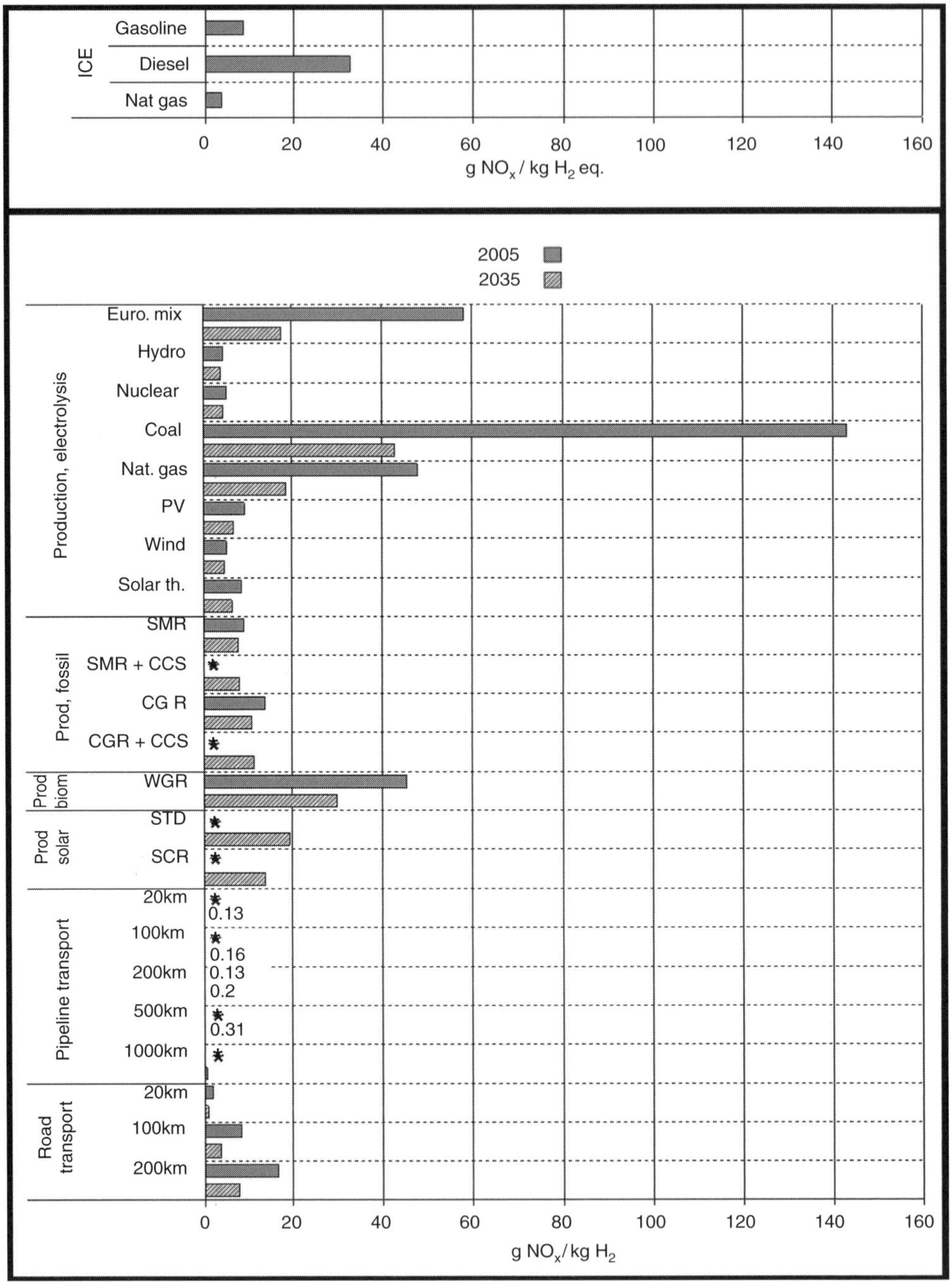

Figure 2.23 Life cycle emissions of nitrogen oxides (NO_x). For each H_2 production and transport scenario, 2005 is the higher column and 2035 the lower column. *Note*: * indicates that data are not available for 2005.

pathway by 2035, although they remain higher than renewables-based pathways using electrolysis.

Certainly in comparison with WGR, the solar thermochemical pathways show relatively low NO_x emissions, again with the SCR performing better than the STD. However, the sensitivity analysis on the argon demand of the STD demonstrated that, by recycling 90% in the quenching process, the total NO_x emissions would be reduced by 60% to almost half that of the SCR pathway, comparable with H_2 from electrolysis using PV or solar thermal-generated sources of electricity.

Road transport uses diesel-fuelled trucks and can therefore be considered an influential factor for those production pathways for which this could be an option.

2.4.6 Sulphur dioxide emissions

SO_2 is emitted during the extraction and processing of fossil fuels, and in significantly larger quantities for oil-based fuels, as shown in Figure 2.24. Sulphur must be removed in order to improve the quality of the fuels and the process energy required is often supplied by lower quality fuels with higher sulphur content, thereby exacerbating the levels emitted.

The gasoline and diesel ICE fuels are shown to have similar life cycle SO_2 emissions to many of the H_2 production pathways and scenarios. Figure 2.25 also indicates a threshold for the H_2 production pathways of around 15 $gSO_2/kg\ H_2$, which is caused by emissions occurring in the production processes of the various metals necessary in the operation of the hydrogen fuelling stations, as well as in the production of copper cables for the electricity distribution network.

SO_2 is produced in large quantities by the combustion of coal to generate electricity, but the emergence of gasification technologies offers a much cleaner way to convert the coal. With the increased efficiency of a combined steam cycle, the production of electricity from this source performs far better by 2035. The influence of coal power plants is also reflected in the European mix scenario. The emissions from the WGR 2005 pathway show a higher value than most, due in large part to

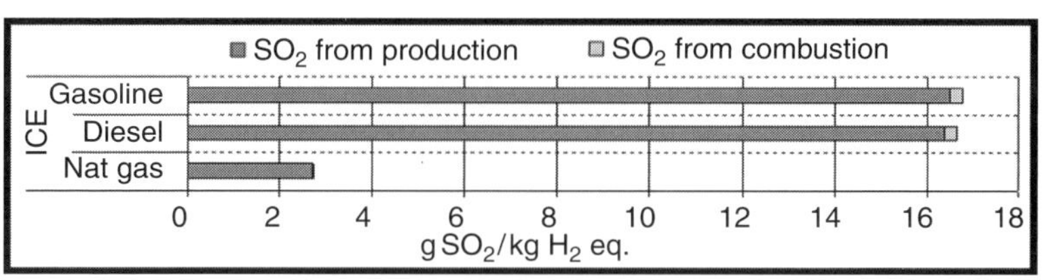

Figure 2.24 Emissions of sulphur dioxide (SO_2) from the production and combustion of 1 kg H_2 equivalent quantities of the internal combustion engine (ICE) fuels. ICE standard is EURO 3.

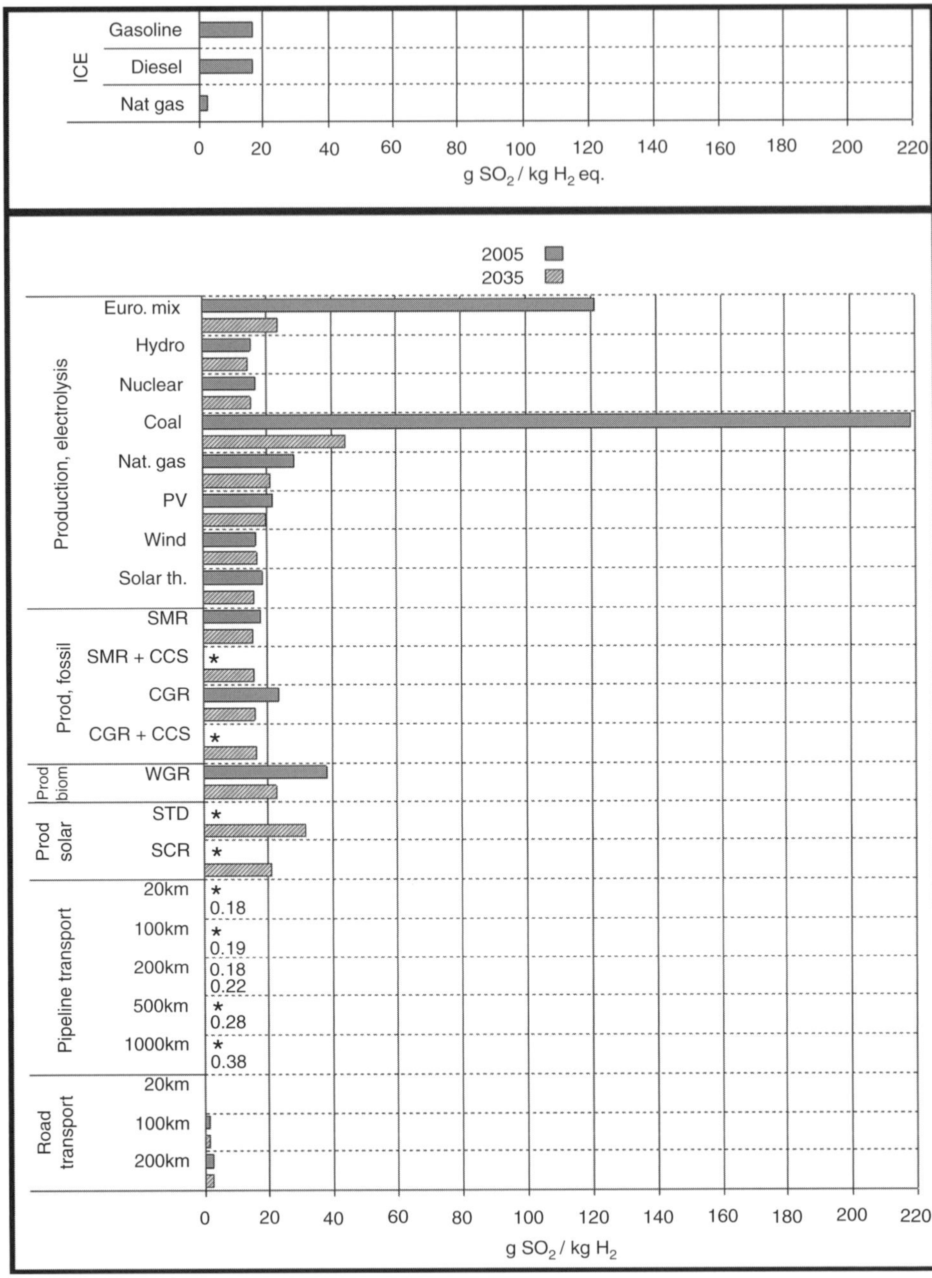

Figure 2.25 Life cycle emissions of sulphur dioxide (SO_2). For each H_2 production and transport scenario, 2005 is the higher column and 2035 the lower column. *Note*: * indicates that data are not available for 2005.

the necessity of compressing the product gas following gasification, and where this uses electricity from the European electricity mix. Other SO_2 emissions are produced by the transport of the biomass (requiring diesel production) and from the fuel used to generate steam for the product gas reforming stage.

While the SCR pathway shows relatively similar total emissions of SO_2 to those of the better performing pathways, the STD is again affected by the high consumption of argon. Were it able to achieve production of H_2 and recycle 90% of the argon, then the overall emissions of SO_2 would reduce to levels virtually equal to those of the SCR pathway, and to a very similar levels as many other pathways.

Transport of H_2 via pipelines would not influence the results of any pathway by an appreciable amount and, although SO_2 emissions occur in the production of diesel fuel, transport by road of up to 200 km would not influence the results by any significant amount.

2.4.7 Validation

In Figures 2.26 and 2.27, some of the critical results of this chapter are validated against those of previous studies. Not all of the studies model the same H_2 production pathways and, when cross-checking GHG-related results, they also do not present the same form of results (i.e. some only calculate CO_2 emissions; others use total GHG emissions). Figure 2.26 shows the results of studies which concentrated only on CO_2, whereas Figure 2.27 reflects all GHG emissions.

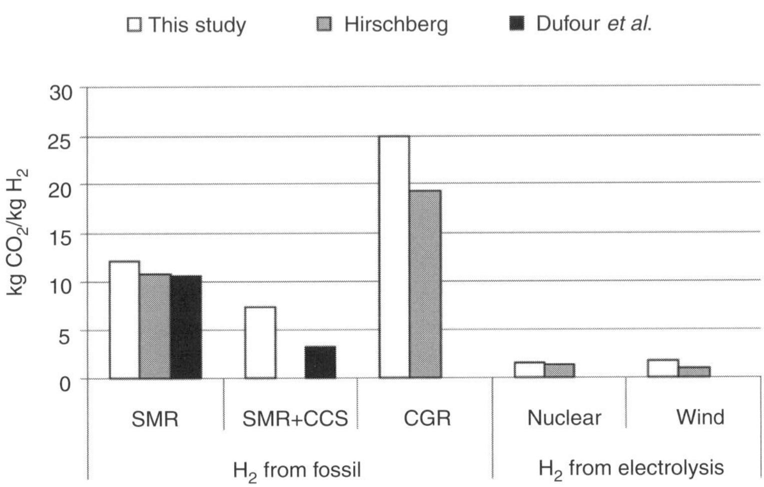

Figure 2.26 CO_2 emissions from H_2 production pathways compared with results of Hirschberg (2004) and Dufour *et al.* (2009).

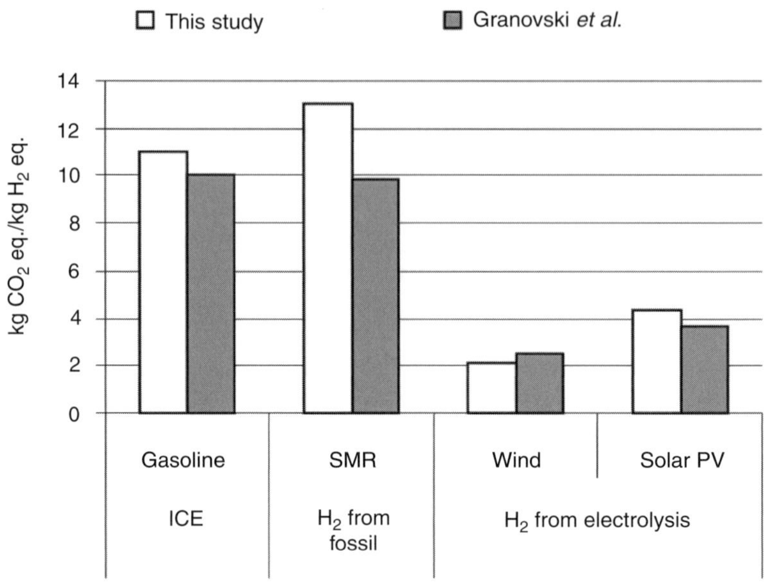

Figure 2.27 GHG emissions from H_2 production pathways compared with results from Granovski *et al.* (2006).

Generally, the results of this chapter are higher than those of other studies, but certainly within the same order of magnitude. The other studies do not allow a deep analysis of the reasons behind the differences in the results. However, in general, higher values would be expected in this analysis, due to differences in the system boundaries and the accumulation of emissions from the more comprehensively covered background processes included in the pathways considered in this chapter, a result of using the ecoinvent database (Frischknecht *et al.*, 2007a).

2.5 Conclusions

This assessment has considered a broad range of present and future pathways for the production and transport of H_2 for use in the fuel cells of road vehicles. The pathways were assessed on the basis of their complete life cycles, made possible by applying the most comprehensive LCI database (ecoinvent, 2007). This ensured methodological consistency in the use of background data and the system boundaries, so that pathways could be meaningfully compared. Using 2005 as the base year for the current state of technologies, an understandable and realistic progression over a period of 30 years was applied in order to assess possible conditions in 2035.

The potential impacts of each pathway were determined with regard to their total GHG emissions and CED. Additionally, cumulative quantities of specific air pollutants were determined which reflect transport-related emissions of significant influence to human health. These were PM_{10}, NMVOCs, NO_x, and SO_2.

By considering the complete energy chains of production pathways, the results by no means determine that H_2 for use in fuel cells is automatically more environmentally friendly or leads to potentially lower impacts on human health than using conventional fuels in ICE. Even though the only direct emission from the conversion of H_2 is water vapour, the various pathways to produce H_2 show widely varying results.

The SMR of natural gas is the most widely used method in industry today, and is shown to be more energy intensive than using ICE fuels, although almost equal in terms of GHG emissions. Because most of the natural gas used is reformed rather than combusted, it performs well with regard to emissions of NMVOCs, NO_x, and SO_2, although it results in higher PM_{10} emissions over the whole life cycle. For 2035, without CCS, this particular pathway shows only relatively small improvements. As outlined in the technology descriptions, applying carbon capture to SMR is not extremely effective, and therefore the potential for reductions is limited. On the other hand, although the CGR route produces higher GHG emissions, it lends itself more effectively to CCS. Thus, applying CCS to this technology would transform it from one of the worst-performing scenarios to one of the best, resulting in only slightly higher GHG emissions than some of the renewables-based electrolysis pathways. For both SMR and CGR, the application of CCS would reduce GHG emissions to below those of the ICE fuels, but maintain CED at a higher level than the ICE fuels.

Electrolysis, although an energy-intensive method of production, is a process with relatively few stages and supply chains, and for which the possible sources of GHG emissions are limited to the electricity supply chain. Therefore, it is this process which allows the production of 1 kg H_2 at fuel pump pressure to have lower impacts than producing and combusting the same quantity, in terms of energy content, of conventional fossil fuel for an ICE. The condition here is that 'clean' power is used as an input for the electrolysis. The added advantage of electrolysis is that the H_2 can be produced in smaller, decentralised, on-site units, thereby possibly negating the need for transport from the production site to the fuelling station. The indicators used in the assessment reflect badly on the production of H_2 via electrolysis using the European electricity mix due to the heavy reliance on fossil fuels, particularly coal. By 2035 it is expected that efficiency improvements in both the electrolysis procedure and the electricity generation technologies will significantly reduce many of the emissions and the non-renewable CED. The improvements in coal-fired generation technologies will have a leading influence.

WGR leads to lower fossil-derived GHG emissions than the gasification of coal, but both processes have higher emissions than SMR. Due to the scale of the plant and the nature of the GHG emissions, CCS is not considered for the WGR pathway. The WGR pathway also shows comparatively high emissions of NMVOCs, NO_x, and SO_2 but across all indicators it shows a good potential improvement by 2035. As a decentralised method of H_2 production, however, the benefits of this over electrolysis depend on the source of the electricity powering the electrolysis process. WGR is further influenced by the distance necessary to transport the H_2 to the user as well as the transport mode used.

As they have been modelled, the solar thermochemical pathways are seen to be less preferable than either the SMR, CGR, WGR, or the nuclear and renewable electrolysis scenarios. However, due to the fact that these technologies have yet to be proven on a commercial scale, some adjustments to the LCI would be expected, thereby influencing the results. Most specifically, in order to be commercially viable, the consumption of argon to quench the zinc and oxygen leaving the reactor in the STD process must be drastically reduced due to the energy intensiveness of its production. This could be via the development of an efficient recycling process or by negating the need for it all together. If such a reduction were achieved, the STD pathway may become preferable to the SCR pathway due to the latter's inclusion of carbon into the reaction chamber and the ensuing need for a shift-reaction process. Using biomass instead of coal for the source of carbon, so that the CO_2 released is balanced with the CO_2 absorbed, would then impose indirect impacts from its harvesting, processing, and transport, which may not have the visible reduction of impacts anticipated. This can be judged to some degree by referring to the WGR pathway. In consideration of these points, however, on the whole, the solar thermochemical pathways have the potential to be very competitive.

Regarding the transport of H_2, the results showed that this aspect should not have a significant influence on the environmental performance of an individual production route to which it is appropriate. The analysis considered road transport distances of up to 200 km, assuming that road transport will primarily be used to serve smaller, decentralised production plants and the earlier phases of the move to a hydrogen transportation infrastructure. The modelling of pipeline transport shows consistently low results even for distances of up to 1000 km, assuming that the pumping will be powered by electricity from the European mix. It is therefore a very efficient method to transport the gas and, similar to natural gas pipelines, the transport could be powered by the product itself. Even though it is based on a low-capacity, natural gas pipeline, the results do reflect a substantial and continuous flow of H_2, meaning that both the production and consumption must be of significant capacity, and of relevance, perhaps, only to the fossil reforming pathways.

Overall, and according to the majority of the indicators used, if H_2 is produced by electrolysis and the electricity for this process can be obtained from either renewable or nuclear sources, then the use of H_2 as an energy carrier in vehicles would be a more environmentally benign option than the combustion of conventional fossil fuels in an ICE. An exception would be with regard to the emissions of PM_{10}, for which all H_2 pathways perform worse than the ICE fuels. Using the European electricity mix to power the electrolysis process would largely lead to an increase in environmental impacts compared with using an ICE, due to the high proportion of fossil fuelled generating capacity in Europe. Although this chapter has not considered the effects of intermittent electricity production from some renewable sources (i.e. solar and wind) or peak and off-peak electricity generation, these are undoubtedly relevant issues when considering the use of H_2, which is a storable energy carrier. As such, the use of H_2 therefore offers the ability to cushion the intermittent nature of renewables by production with off-peak surplus power of base-load capacities (i.e. nuclear). Therefore, countries with existing or potentially sufficient renewable or nuclear electricity production capacities clearly stand in a more favourable position to incorporate hydrogen as a transport fuel over those already relying heavily on fossil fuels.

3

Technical characterisation and multi-criteria analysis of light-duty vehicles

ERIK WILHELM

3.1 Introduction

Over half of all oil is used in transportation, and global demand is steadily increasing (EIA, 2008a). The exclusive dependence of our mobility on fossil fuels raises serious concerns over energy security, cost, environment, and human health. Creating a more sustainable transportation system has been the subject of much attention from stakeholders in the governmental, consumer, and industrial arenas (Kasseris, 2006; Stern, 2006). Technological developments are necessary to move toward sustainability in personal transportation while meeting dramatically increasing demand for mobility from developing countries, particularly India and China (IEA, 2008b: 393, Table 16.4). Carmakers understand that consumers will not readily tolerate any reduction in the quality of service provided (Stremler, 2008) while at the same time government is expected to regulate industry to protect human health and mitigate climate change. There are no easy answers to the transportation energy question, and there is a clear need for an accurate description of the inherently difficult trade-offs associated with vehicle technology options.

In the near term, improvements in vehicle efficiency can be achieved by incremental modifications in combustion engines, accessory load reduction, low-rolling-resistance tyres, and downsizing/lightweighting (Weiss *et al.*, 2000; MacLean and Lave, 2003). In the long term, fuel cells and batteries with high-energy densities promise to drastically reduce transportation emissions at constant performance and safety. The trend toward powertrain electrification introduces not only a new set of technical challenges but also many new supply chain and social acceptance challenges. Biofuel is another alternative that is being actively researched and developed, but often faces hurdles when considering land-use change (Hertel, 2009) among other things. There tends to be bias in the various whitepapers and

Transition to Hydrogen: Pathways Toward Clean Transportation, ed. E. Wilhelm and A. Wokaun. Published by Cambridge University Press.

technical reports released by the lobby groups of the respective transportation technology developers, each attempting to present the facts in a way that promotes their own technology over their competitors.

3.1.1 Problem definition

A multitude of new technologies are being researched and developed with the goal of reducing the environmental impact of personal transportation and dependence on fossil fuels while increasing performance and maintaining or reducing cost. Decision makers in industry and government, as well as consumers, lack a comprehensive, unbiased comparison of these options based on life cycle data combined with rigorous powertrain simulations.

3.1.2 Objectives and scope

The objective of this chapter is to present a systematic evaluation of LDV technology using heuristic combinations of vehicle powertrain options. The impact of selected new and future powertrain, material, and fuel technologies across a broad range of vehicles is analysed using the approach shown in Figure 3.1. There is

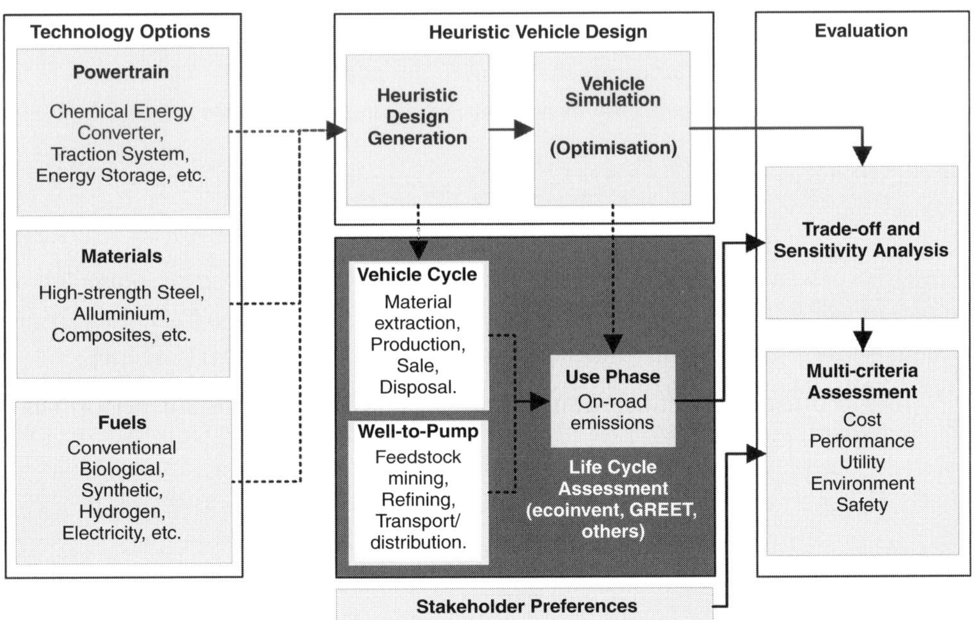

Figure 3.1 Procedure for the heuristic design and evaluation of light-duty vehicle technology.

special emphasis in the design set on hydrogen and fuel cell technology options. Multi-criteria decision analysis (MCDA) is used to evaluate the effect of stakeholder preference on technology choice, and to identify technologies which are robust over a wide range of criteria weightings.

The explicit goal of this work is to assist stakeholders in understanding the complex interaction between vehicle technologies and criteria of interest, and to support the decisions required to achieve a sustainable energy future. No single strategy will ever be optimum because different decision makers will disagree on how to balance conflicting objectives. An inherent advantage of the heuristic design approach is that natural laws rather than prejudicial technical assumptions allow the most viable solution(s) to be found. Some specific research questions addressed are:

1. Which technologies are being considered for improving current and future personal transportation modes?
2. What effect do different powertrain technologies have on key stakeholder criteria? For example, what are the trade-offs inherent in advanced powertrain concepts, such as ICE-electric and fuel cell hybrids?
3. How does introducing lightweight materials and technologies affect target criteria?
4. Which fuel options allow stakeholder preferences to be balanced, and what are the tipping points?
5. What can MCDA tell us about the robustness of specific technology options?
6. How sensitive are these results to model input assumptions?
7. What role can hydrogen play as a transportation energy vector?

3.1.3 Heuristic vehicle design

Heuristic rules are commonly applied when systems cannot be easily modelled or when models result in computationally expensive functions. The behaviour of the system is then separated into rules which are used to characterise it (Gigerenzer and Todd, 1999). Heuristic methods are often used when attempting to model human decision making; a pioneering example is the use of heuristics in competitive computer chess programs to optimise points per move within time constraints (Mueller, 2006). For this work, heuristics are applied to model the vehicle design process, with the primary function being to eliminate infeasible option combinations and to appropriately size and combine components. For example, diesel engines are typically smaller displacement than Otto engines, can only use diesel fuel, and are less likely to be mild hybrids. The heuristic design algorithm (HDA) uses design choices made in the past century of evolutionary vehicle design (Weaver *et al.*, 2005). The HDA then uses the available technology options to combinatorially compose a set of self-consistent vehicle designs which are representative of the current vehicle pool, but are not sales weighted (i.e. no type of vehicle is represented more than any other). This bottom-up, technology-centred approach differs fundamentally from the way manufacturers have historically designed and are currently designing vehicles, where a top-down approach is used with market demand driving technology choices.

The heuristic design tools applied in this work also differ from the conventional approach taken to vehicle technology assessment in several important ways. Researchers typically limit themselves to varying technology on one basic platform to simplify data gathering and model development. This almost always results in an extrapolation of results, shown in Figure 3.2a, and can lead to unrealistic application of technology and unnecessarily limit the analysis. The inherent advantage of the heuristic method, shown in Figure 3.2b, is that the large number of vehicle sizes, types, and configurations allows interpolation of performance within the design set, and hence a greater understanding of the higher order interaction between technology options and a broader scope of analysis. For this reason, however, care must be taken when attempting to isolate the individual influence of different technologies on stakeholder criteria. For example, it is often the case in the analysis presented in this chapter that no performance specifications are held constant, i.e. the acceleration, consumption, cost, etc. all change, based on the technologies chosen by the design heuristics.

3.2 Stakeholder criteria

The criteria used in the evaluation of heuristically designed vehicles were chosen based on their importance to stakeholders in the transportation energy debate. Stakeholder criteria were assigned to the five categories summarised in Table 3.1. Some criteria are characterised using powertrain simulation, others are modelled using LCA techniques, while yet others are surveyed from the literature.

3.2.1 Cost

The cost criteria are divided into the costs which are of interest for various stakeholder groups. The purchase cost of the vehicle typically makes up a significant fraction of the total cost of vehicle ownership over its lifetime, and is a predominant concern to buyers. Keeping purchase cost low is therefore very important to manufacturers who attempt to maintain a steady mark-up of 40% without sacrificing sales (Bandivadekar *et al.*, 2008a). In order to calculate production and purchase costs, a relatively detailed bill of materials for vehicles designed in the heuristic set

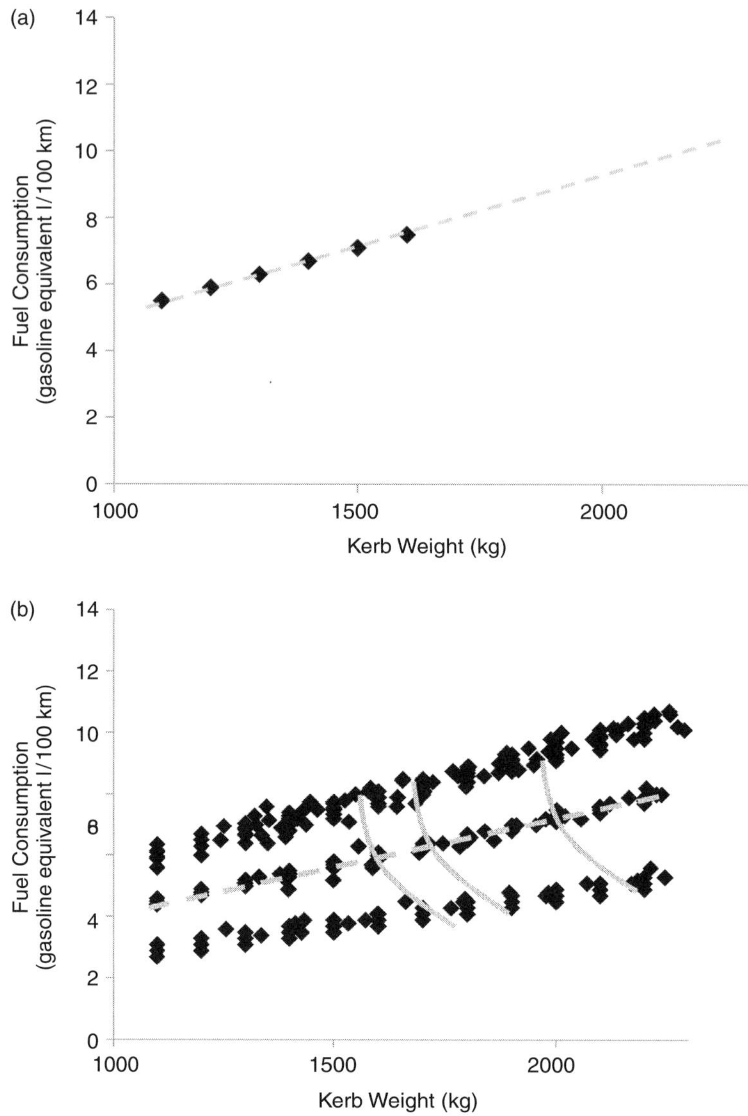

Figure 3.2 Comparison of case study technology analysis (a) with heuristic design set technology analysis (b). Light grey lines reflect how technologies are characterised in both cases. Heuristics design allows many technologies to be simultaneously compared for various trade-offs and allow technology family performance envelopes to be identified.

and manufacturer margins for the various components are assumed. Maintenance cost for both North American and European vehicles has been modelled using data from the German Automotive Club (ADAC, 2008), which may result in slight overestimation of North American maintenance costs in absolute terms, but does not influence relative technology assessment results. The total cost of ownership over

Table 3.1. *Main criteria of interest for stakeholders from industry, consumer, and regulatory areas.*

Category	Criteria	Units
Cost	Purchase	CHF
	Maintenance	CHF/year
	Resale value	CHF
	Total cost of ownership	CHF/life
Performance	Acceleration time (0–100)	s
	Acceleration time (80–110)	s
	Top speed	km/h
Utility	Driving range	km
	Passenger capacity	m^3
	Cargo capacity	m^3
	Towing capability	kg
	Turning radius	m
	Refuelling time	s
Environment	Criteria pollutants on a well-to-wheel basis	
	CO_2	g/km
	NO_x	g/km
	VOC	g/km
	CO	g/km
	PM_{10}	g/km
	$PM_{2.5}$	g/km
	CH_4	g/km
	N_2O	g/km
Safety	Fatality risk	fatality/km

a vehicle lifetime is calculated by summing the vehicle purchase price with annual fuel and maintenance costs as shown in the cost model diagram in Figure 3.3. Internalising costs associated with environmental damage is an important and often-discussed subject in the literature (Tol, 2005), but due to the significant uncertainty associated with such cost estimates they will not be treated in this chapter.

3.2.2 Performance

Vehicle performance is most often characterised by acceleration, both from standstill and at highway speeds. This characteristic is the most important for certain demographics (typically younger drivers), whereas safety takes priority for the majority of US vehicle consumers (McKinsey & Company, 2009). The performance criterion very often drives powertrain-sizing decisions, because peak engine power directly impacts acceleration time, although this peak power is seldom required over

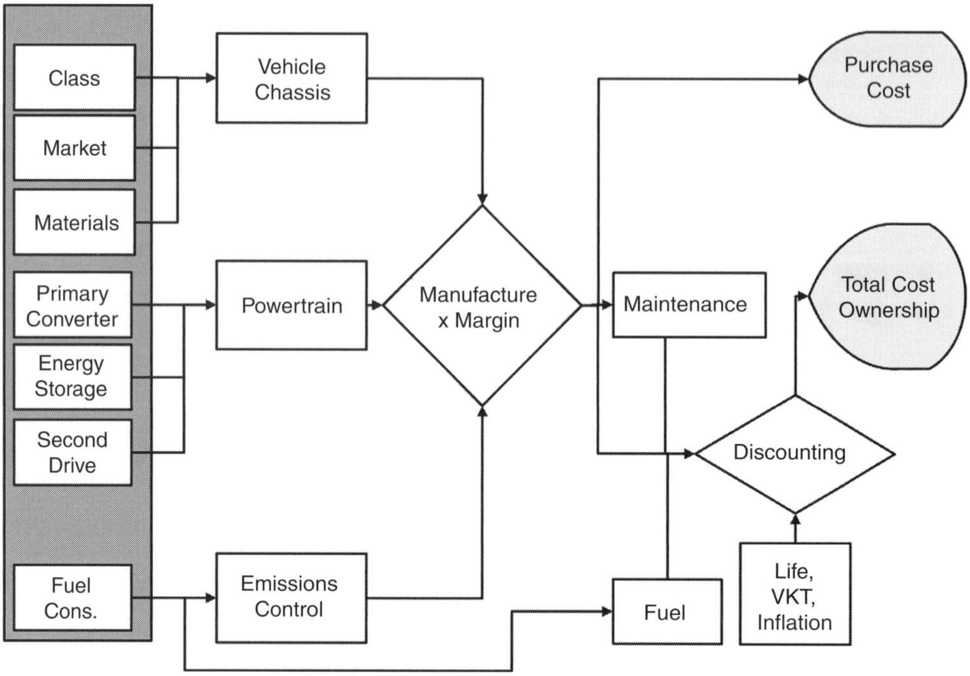

Figure 3.3 Components of the cost model used to estimate vehicle purchase price and ownership cost.

standard cycles or during normal daily driving. The trade-off between performance and environmental criteria is one of the focuses of this work.

3.2.3 Utility

In the discussion of alternative fuels, range is often the most important utility criterion; however, cargo and passenger space should not be neglected when assessing the impact of introducing large, secondary energy storage devices (e.g. batteries) or hydrogen storage technologies. It is often the case that utility criteria are designed in terms of a minimum acceptance level, and not optimised as aggressively as performance criteria. The exception to this rule is for cargo and haulage vehicles, where towing capability is of utmost importance. Inherent in the bottom-up, heuristic design approach is the fact that stakeholder criteria performance is a result of technology choice, and therefore there can be no explicit minimum design requirements met by the vehicles in the set. Inputs to the HDA are drawn from rules based on real-world engineering decisions, so it is to be expected that, with standard battery energy capacity, for example, the range achieved by modelled vehicles is similar to what may be seen on the road.

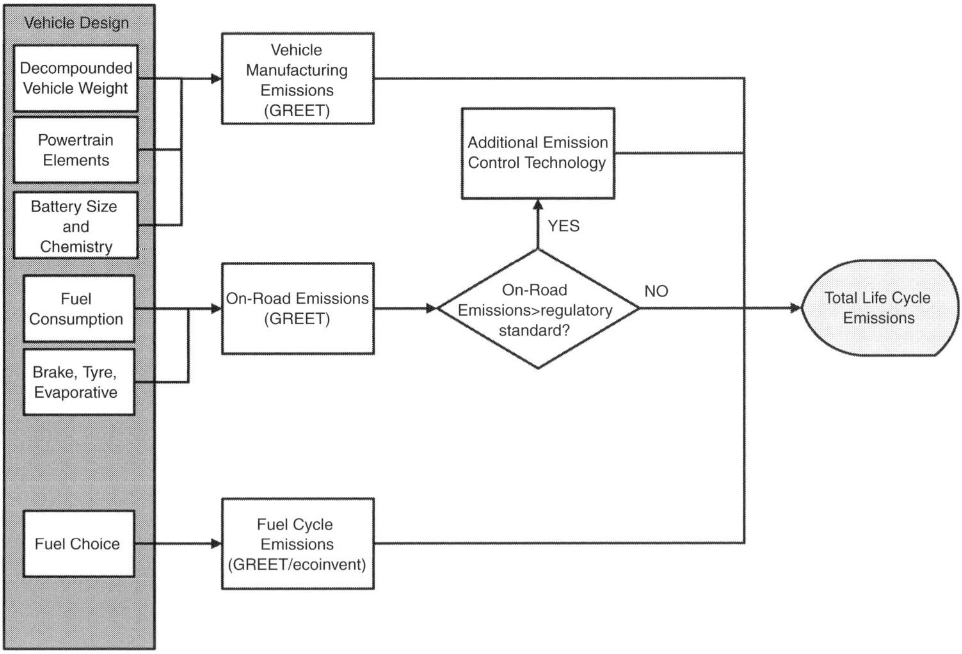

Figure 3.4 Components of the life cycle emissions model. Note the addition of exhaust after-treatment in situations where engine systems are predicted to exceed regulated emissions thresholds.

3.2.4 Environment

Policy makers and consumers are encouraging manufacturers to consider environmental criteria in their designs. The desire for personal mobility has often conflicted with sustainability goals due to the low thermal efficiency of ICEs at part-load, and the emissions generated during the combustion processes. Due to the geographic variation in emissions and material use from fuel chains and manufacturing, two distinct life cycle databases are used as sources of LCA data in this work: ecoinvent for European vehicle design sets and the Argonne GREET database for North American design sets (ANL, 2009; ecoinvent, 2009). The modelling procedure used for both databases is shown in Figure 3.4.

3.2.5 Safety

The impact of technology choice on the safety of passengers and pedestrians is a very interesting question for all stakeholders. It remains, however, a very challenging criterion to analyse due to the complexity of interaction between vehicle subsystems in a collision. Manufacturers invest significant resources in the modelling

Table 3.2. *Technology options which are available to the heuristic design algorithm. With 68 options, over 145 million combinations are possible.*

Option Bin	Category	Options				
Classes	Exog.	3	compact sedan	sedan	pickup truck	
Markets	Exog.	2	passenger	sport		
Engines	Exog.	3	Otto	diesel	fuel cell	electric motor
Emissions control	Dep. Endog.	2	selective catalytic reduction		particulate filter (open/closed)	
Hybridisation	Exog.	5	none	mild	plug-in series	parallel
Body structures	Exog.	4	steel	HS steel	aluminium	composite
Transportation fuel	Endog.	7	gasoline natural gas	diesel hydro- gen	biodiesel electricity	ethanol
Primary energy	Dep. Endog.	4	conven- tional	biomass	renewable/nuclear (non-biomass)	
Displacements (l)	Endog.	12	1.0 1.1 1.7 1.9 2.0 2.1 2.7 2.9 3.0 3.1 3.7 3.9			
Fuel cell power (kW)	Endog.	3	80 90 100			
Electric path power (kW)	Endog.	10	3 30 40 50 60 70 80 90 100 110			
Secondary energy storage chemistry	Endog.	2	NiMH	Li-ion		
Secondary energy charge/average energy (Ah/kWh)	Dep. Endog.	10	(4.5 / 1.7) (10 / 3.8) (20 / 7.5) (30 / 11.3) (40 / 15) (50/18.8) (60/22.5) (70 / 26.3) (80 / 30)			
Total		$1.45 \cdot 10^8$				

and simulation of crash scenarios to reduce development time for new generations of vehicles. The only practical way to examine safety criteria is through the use of statistical data collected and analysed in seminal works by Ross and Wenzel (Ross and Wenzel, 2001; Wenzel and Ross, 2008). This approach is limited to analysing differences in a subset of vehicle classes as weight-reducing technologies are introduced, and will not be emphasised in this chapter.

3.3 Technology options

The technology options available to the design set generator in this work are summarised in Table 3.2, and are classified as exogenous, endogenous, or dependent endogenous. Endogenous choices are choices driven by exogenous selections, and dependent endogenous choices may depend on one or more exogenous and/or endogenous selections. The options are further divided to answer specific research questions about advanced hybrid architectures, materials, and fuels.

The characteristics of many of the technology options are not static, and are expected to improve with development effort. Time variant quantities such as fuel and component price are important to model, particularly when making total cost of ownership calculations, and a selection of the most relevant assumptions used in this chapter can be found in Appendix B.

To use the HDA effectively, it is important to be able to predict the impact that a particular rule will have on the number of designs in the resulting set. Equation (3.1) determines the maximum number of designs that will be present in a set, given the number of choices per category. This simple relationship explains how the algorithm combinatorially places each exogenous technology with every other. Introducing a rule between two exogenous options results in the removal of the number of designs given by Equation (3.2), where the variable d_i represents the option categories not affected by the introduction of the rule. The variable h_{dep} is an unknown to account for the fact that, when designing the set, it is difficult to analytically predict the final number of designs due to recursive rule interaction.

$$n_{tot} = \prod_{i=1}^{N} s_i \tag{3.1}$$

$$h_{endog} = \prod_{i=1}^{N-2} d_i \mid d_i \notin h_{endog-rule} \tag{3.2}$$

$$n_{set} = n_{tot} - \sum h_{endog} - \sum h_{dep} \tag{3.3}$$

n_{tot}	total number of designs
s_i	number of designs in a category
h_{endog}	number of endogenous designs removed
d_i	number of sets not affected by the heuristic
$h_{endog\text{-}rule}$	number of sets affected by the heuristic
h_{dep}	number of dependent designs removed
n_{set}	number of designs remaining in set

A selection of heuristically designed sets is shown in Figure 3.5 to demonstrate how quickly the number of vehicle designs becomes unmanageably large without the application of heuristics. Included in Figure 3.5 is the number of designs (n_{set}) from four distinct heuristic sets to illustrate how the application of heuristic rules affects the volume of designs. The final number of designs described by Equation (3.3) is typically 0.2 to 4.0% of the maximum number of designs.

An example of a selection process made by the HDA is shown in Figure 3.6, describing the selection of hybridisation and electric path power. This shows the tiered approach to dividing the options, as well as the vehicle weight, which is a

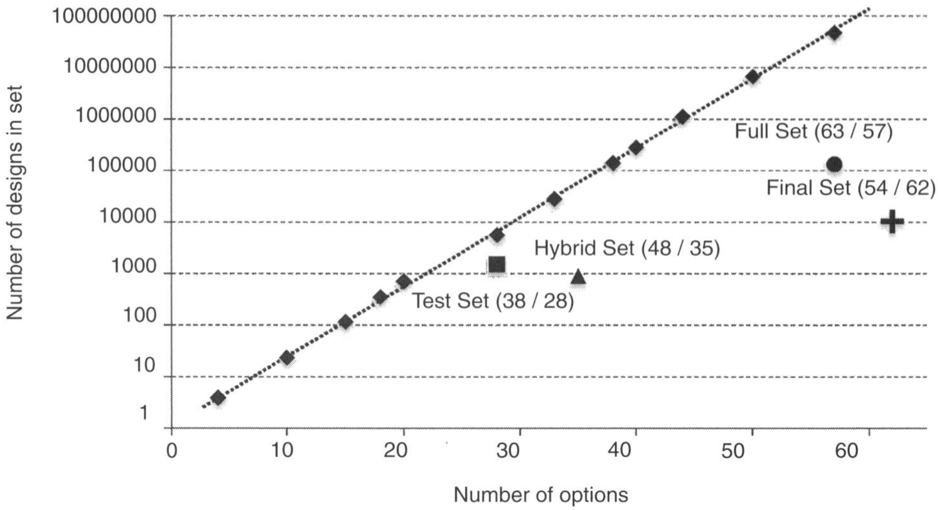

Figure 3.5 The factorial relationship between the number of options in a design set and the final number of designs (n_{set}). Set names are given with heuristic rules/ technology options.

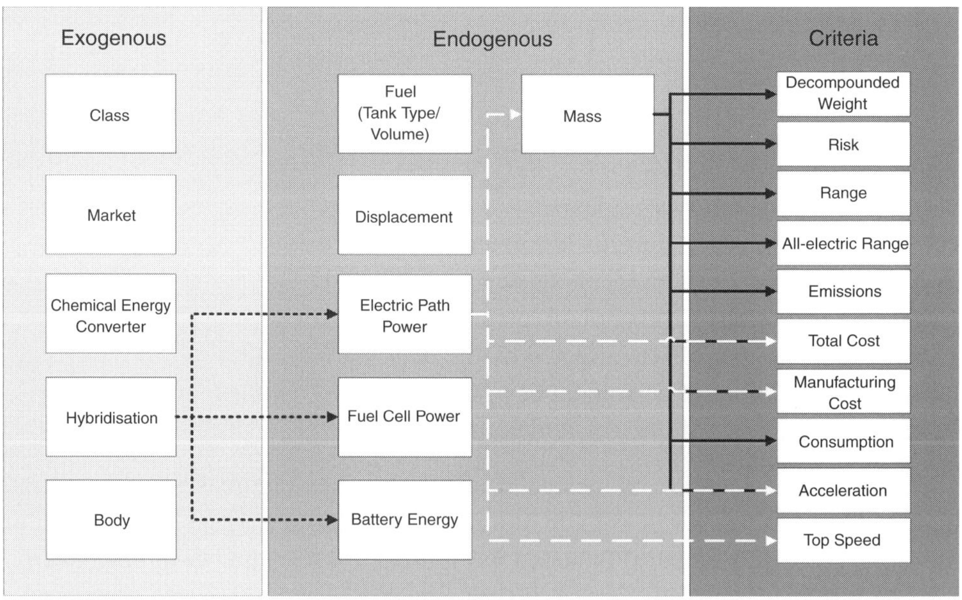

Figure 3.6 A subset of heuristic rules acting on exogenous and endogenous choices. Dashed lines represent the exogenous choice of hybrid architecture influencing various endogenous choices, which then influence specific criteria as represented by solid lines.

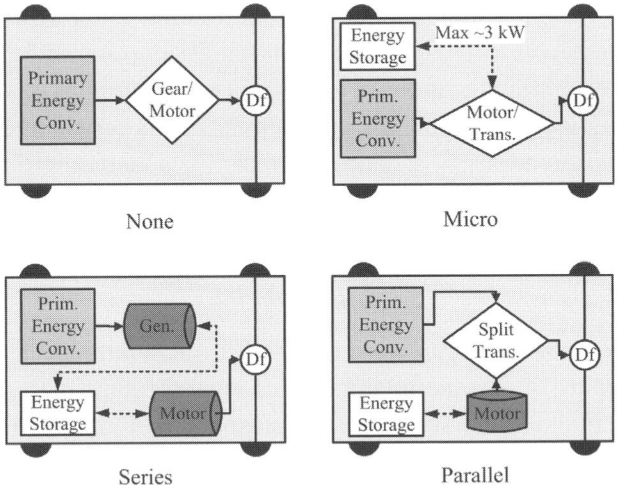

Figure 3.7 Powertrain architecture options for heuristic vehicle design. 'None' represents both electric vehicles as well as conventional internal combustion engine designs. Df. stands for differential.

result of technology combinations. The mass characteristic is an important design consideration, and mass decompounding effects are carried through the heuristically designed vehicle (Malen and Reddy, 2007).

3.3.1 Powertrain options

The exogenous hybridisation architecture options are shown in Figure 3.7. No hybridisation can either mean an ICE is the prime mover, or a battery/electric motor combination is used. Micro hybrids are restricted to limited acceleration boost and idle stop functionality. Heuristics are applied to ensure that series hybrids have batteries that can allow plug-in hybridisation.

Split-parallel hybrid architectures are not considered in this work because of the complexity involved in performing dynamic programming control optimisation on these architectures.

3.3.2 Lightweighting options

The lightweighting technology options that are used in this work are:

- High-strength steel applied in body structures (Shaw and Roth, 2002),
- Aluminium applied in body and powertrain components (Han, 1994),
- Composites applied to chassis and body components (Kang, 1998).

The cost and weight reduction inputs to the models come from three separate techno-economic studies linked by the common assumption of large volume production of lightweighting technology to account for economies of scale. The effect of lightweighting technology on life cycle emissions is considered using GREET lightweighting assumptions, which approximate the added emissions for reducing vehicle weight with more energy-intensive materials.

3.3.3 Fuel options

The transportation fuel options were selected to include a spectrum of the most common current fuels, and the most often discussed future fuels. Biodiesel from soya and corn ethanol were selected as representative biofuels, and conservative production emissions assumptions were used in the baseline cases. The well-to-pump emissions data for hydrogen from Chapter 2 were used as inputs to these models for the European scenarios. A validation of the life-cycle inputs showed good agreement between the data used in this chapter and that of notable, comprehensive LCAs performed on European (Joint Research Centre, 2007) and North American (General Motors, 2001) fleets.

3.4 Vehicle modelling and simulation

The modelling and simulation of vehicles has been well described in the literature (Rousseau *et al.*, 2007; Vinot *et al.*, 2008). There are two distinct methods of vehicle simulation used in this work; the first relies on the commercial vehicle system simulation software CRUISE, and the second follows the method described in detail by Guzzella and Sciarretta (2007), and is implemented in MATLAB.

3.4.1 Simulation methods

The two-model approach was taken in order to ensure that the vehicle simulations would achieve a high fidelity with constrained execution time. The MATLAB models are set up for 'backward-facing' simulation, where velocity at the road is assumed and energy consumption is calculated through the powertrain. AVL CRUISE software is used to build 'forward-facing' models where the driver 'sees' the velocity profile and interfaces with the powertrain as he or she would in a real vehicle, by actuating the gas and brake pedals and accordingly creating a fuel demand from the powertrain and a torque at the wheels. The backward-facing models lend themselves to control optimisation, whereas the forward-facing models are able to model the effects of powertrain limits and other disturbances more accurately. By creating a small number of benchmark models, the accurate CRUISE models were used to calibrate and validate the MATLAB models when on-road data were not available. All simulations, unless stated otherwise, are performed using the New European Driving Cycle (NEDC).

3.4.2 Comparison of advanced powertrains using dynamic programming

The main challenges in simulating advanced powertrains (defined here as parallel, series, and fuel cell hybrids) are to capture the appropriate power split between the chemical energy converter (engine or fuel cell) and the energy storage device (battery and/or super capacitor) and to handle all other control limitations which are present in hybrid powertrains (electrical limits, transients, clutch/gear state limits). Although many approaches to building accurate advanced powertrain models exist, most are not suitable for studying a wide variety of powertrains due to the time required to build unique control strategies for each architecture and component size.

In order to objectively compare vehicle technologies, it is critical to ensure that the variation in results caused by the control of the powertrain is minimised. Directly or indirectly, powertrain control impacts almost every criteria of interest. For example, automatic transmission shifting control affects acceleration, consumption, drivability, noise, etc. For this chapter, the control of the power split for hybrid vehicles is optimised using a deterministic dynamic programming approach (DDP), since the hybrid control strategy has the largest effect on the vehicles' energy consumption and therefore may influence the most parameters (Sundström *et al.*, 2008). The power split for a hybrid powertrain with battery charge x at time t is defined in Equation (3.4), where the ratio of the torque provided by the electric motor to the total torque required to drive the wheels is analogous to the power provided by the electrical path over the total power required to drive the vehicle.

$$U_{opt}(x,t) = \frac{T_{motor}}{T_{total}} \in \{-1...1\} \tag{3.4}$$

The DDP procedure is represented in Figure 3.8, where the state variable x (hybrid battery state of charge) is plotted on the y-axis, the drive cycle time t is plotted on the x-axis, and the control signal U is shown for various trajectories in grey. The grid defines a cost raster, with the final time T_f having the lowest cost (lowest fuel consumption) and where the powertrain state (power split) is calculated backwards to the start of the driving cycle, always choosing the minimum cost path. The cost function can be formulated to include multiple terms, including minimising the state of charge variation. However, the most effective way to implement the algorithm is to use a 'hard finish' approach to set the maximum variation that the state

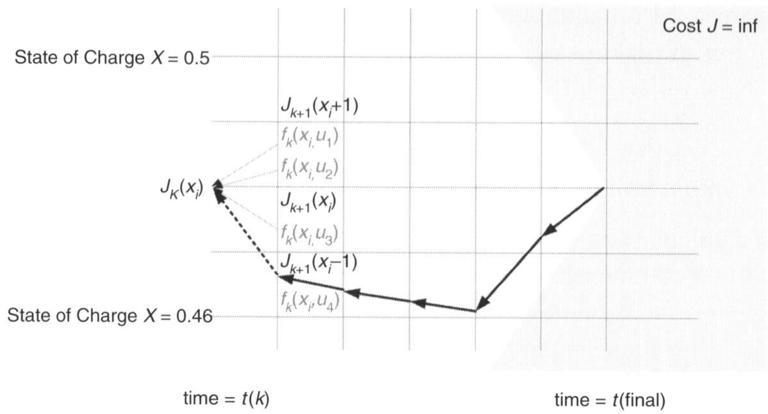

Figure 3.8 Dynamic deterministic programming algorithm schematic. The grid shows how the optimal control rule U is chosen by working backwards to find the path of minimum cost allowed by the mission.

of charge may have from its initial value, and to use fuel energy consumption as the only minimisation term. The advantage of ensuring that the final and initial states of charge are equal is that it allows different chemical energy converters to be objectively compared. This means that charge-depleting hybrid architectures, such as plug-in hybrids, do not operate as they were intended (Gao and Ehsani, 2010). This assumption has been justified by initial studies showing plug-in hybrid drivers tend to neglect to plug their vehicles in, and therefore often drive in charge sustaining mode (Morrow *et al.*, 2008).

The general dynamic deterministic programming protocol consists of minimising the cumulative cost as defined by the cost function $g(x,u,w)$ in the following manner (Bertsekas, 2005):

1. Assign final cost according to Equation (3.5):

$$J_N(x_N) = g_N(x_N) \tag{3.5}$$

2. Proceed backward by one step.
3. For each point, search for optimal input by calculating the minimum cost J according to Equation 3.6:

$$J_k(x_k) = \min_{u_k \in U_k(x_k)} (g_k(x_k, u_k, w_k) + J_{k+1}(f_k(x_k, u_k, w_k))) \tag{3.6}$$

4. Return to step 2 until $k = 0$.

An example result from a dynamic programming run is shown in Figure 3.9, where a parallel hybrid vehicle running the NEDC is simulated. Each pixel on the grey-scale colour map represents the optimal power split between the electric path

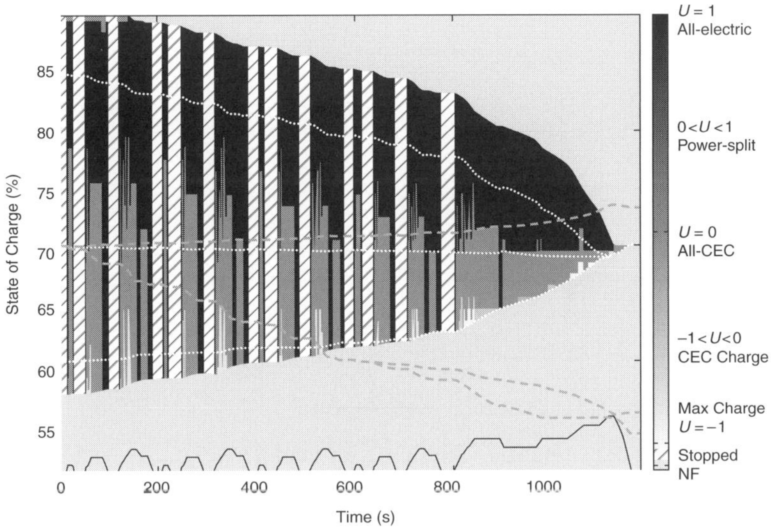

Figure 3.9 Optimal control applied to a parallel hybrid with various initial states of charge (dotted; 85%, 70%, 60%) compared with heuristic control strategies starting from 70% state of charge (dashed). The map represents the optimal control rule, with $U = 1$ (all-electric) in black, and $U = -1$ (full power chemical) in white (see also colour plate 1).

(black) and the chemical path (light). Note that the dotted lines following the interface between electric and thermal/electric operation represent the state of charge progression of the battery from various starting states of charge under optimal control. The dashed lines represent the state of charge progression for various non-optimal control schemes that would typically be found in real hybrid vehicles. Figure 3.10 highlights the importance of ensuring optimal control even more clearly. For two out of three rule-based control strategies, the fuel consumption results under arbitrary weight reduction measures vary in an undesirable way that can serve to severely distort the analysis.

For more details on the optimisation of hybrid vehicles, please refer to Sundström and Stefanopoulou (2007).

3.4.3 Well-to-wheel integration

Evaluating vehicle technology requires looking not only at the vehicle's use phase, which is treated using the simulation methods described in the last section, but also at the vehicle's manufacturing and fuel cycles. Figure 3.11 shows the LCA approach used to estimate the environmental impact of the heuristically designed fleet. GREET was used to model the vehicle manufacturing and disposal 'grey emissions' as well as the non-CO_2 vehicle use emissions. The ecoinvent database

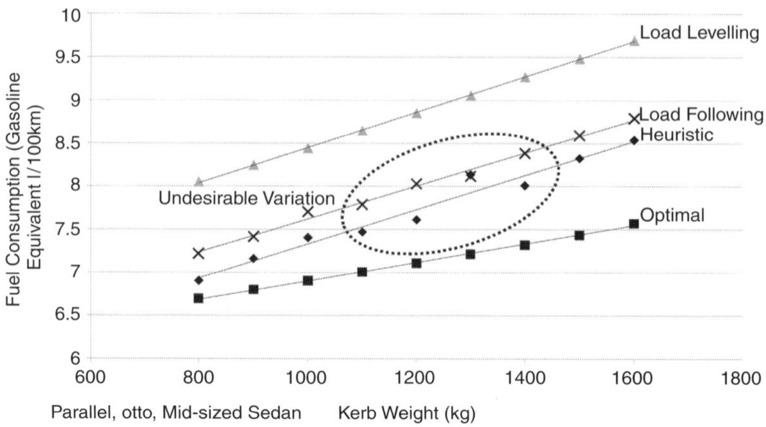

Figure 3.10 Fuel consumption variation with vehicle weight for various control strategies. Hybrid vehicles can have undesirable variation in simulation results if optimal control is not guaranteed (shown in the dotted oval).

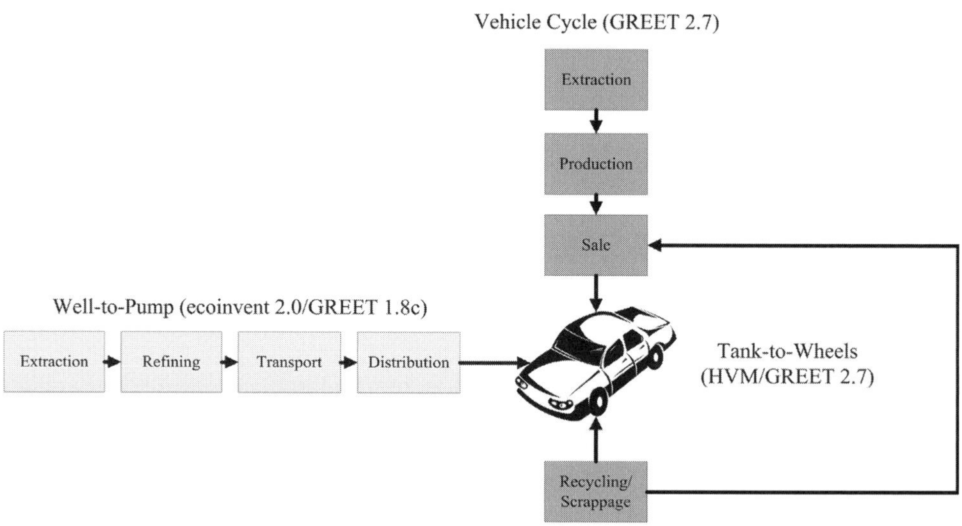

Figure 3.11 The life cycle assessment approach based on the GREET and ecoinvent models.

is used to provide European well-to-pump fuel energy chains to expand upon the data that are available in GREET.

Chapter 2 explored the methodology employed in performing LCA in detail. A subset of the life cycle assumptions that were used are stated explicitly in Appendix B, and reflect the fact that the analysis was primarily performed on the powertrain, fuel, and lightweighting technologies.

3.4.4 Validation results

Selected vehicle simulation results are presented in this section to demonstrate the successful model validation, and to highlight areas where uncertainty remains. As was outlined in Section 3.4.1, two distinct modelling methods were used to characterise the heuristically designed vehicle fleet. The CRUISE models have been validated against detailed, time-series test and driving data from fuel cell, hybrid, and conventional vehicles. After being calibrated using the CRUISE models, the MATLAB models were validated against data from vehicle manufacturers and government test reports to ensure a good representation of real-world behaviour is maintained.

A data-logging device was used to record trips taken by a series of test vehicles in order to validate conventional (non-hybrid) powertrain models. The speed trace was then used as an input to the model, designed to represent the vehicle driving the route. The validation results are shown in Figure 3.12, which illustrates how

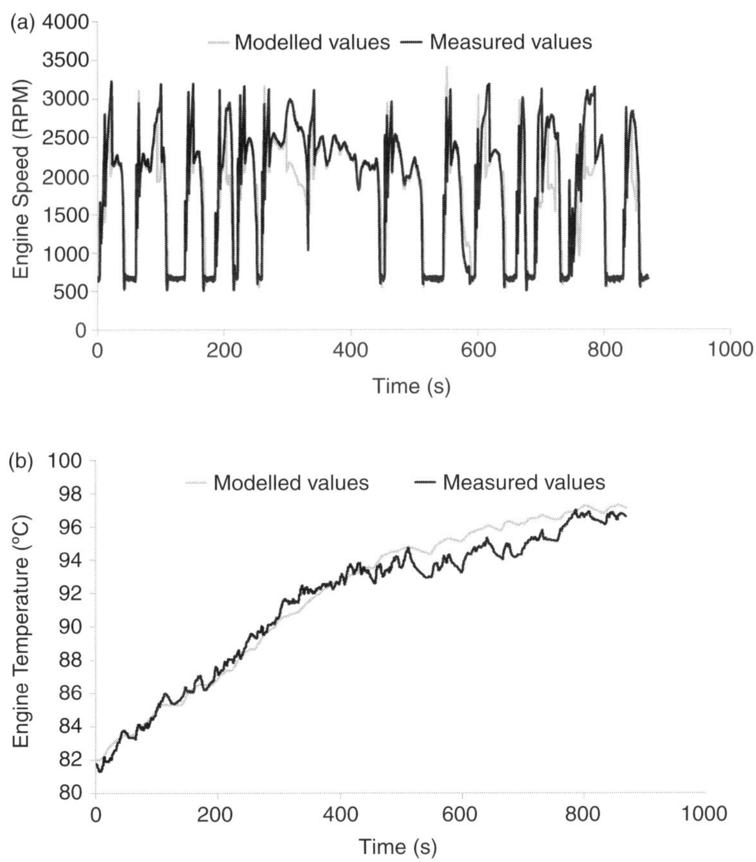

Figure 3.12 Validation of the CRUISE powertrain models with engine data. (a) Engine speed, (b) engine temperature.

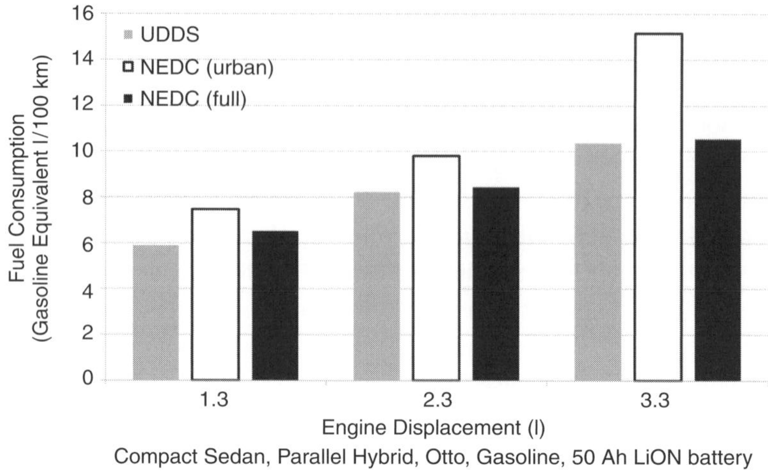

Figure 3.13 Comparison of fuel consumption with hybridisation ratio between various standard fuel consumption test cycles.

engine speed and temperature are accurately represented by the forward-facing CRUISE models.

Model sensitivity to input parameters

The main inputs to the vehicle simulations are standard driving cycles, which prescribe the velocity and gear of the vehicle over a given period of time. The cycles vary in terms of their 'aggressiveness' as defined by their acceleration and top speed characteristics. In order to establish the sensitivity of the models to driving cycles, a series of tests were performed over various standard cycles. The result for one parallel vehicle model, chosen as being representative of the wider set of advanced hybrids, is shown in Figure 3.13. The conclusion that can be drawn from this sensitivity analysis is that, while the driving cycle influences absolute energy consumption values, the relative values and trends do not depend strongly on the input cycle. The rank of vehicle design performance is unaffected by input cycle. The full NEDC was used to generate all the results presented in this chapter, a reasonable decision given how closely it matches the UDDS and urban NEDC cycle results (when engine sizes are comparable).

In order to ensure that the comparison of different vehicle technologies is as fair as possible, consistent life cycle data were taken from the ecoinvent and GREET databases, and care was taken to ensure the assumptions were harmonised between the models. When calculating vehicle emissions on a life-cycle basis, the well-to-wheel energy chain assumptions can exert a very large influence on results, as demonstrated by the large difference in CO_2 emitted by the same set of electric

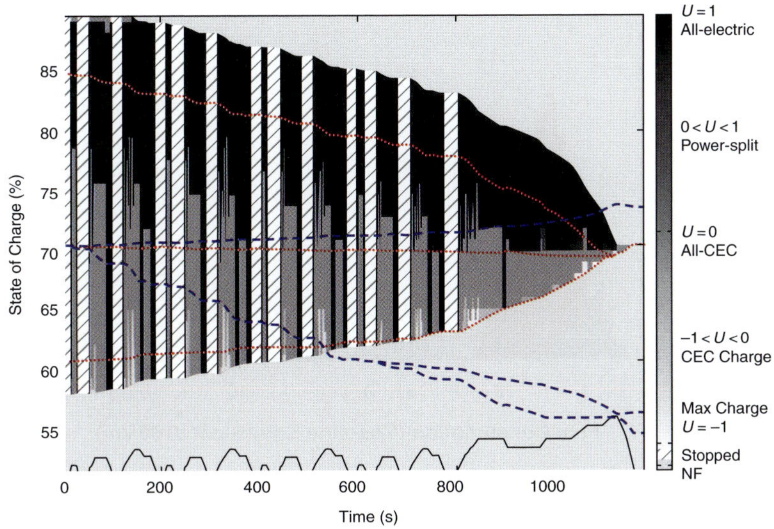

Plate 1 (Fig. 3.9) Optimal control applied to a parallel hybrid with various initial states of charge (dotted; 85%, 70%, 60%) compared with heuristic control strategies starting from 70% state of charge (dashed). The map represents the optimal control rule, with $U = 1$ (all-electric) in black, and $U = -1$ (full power chemical) in white.

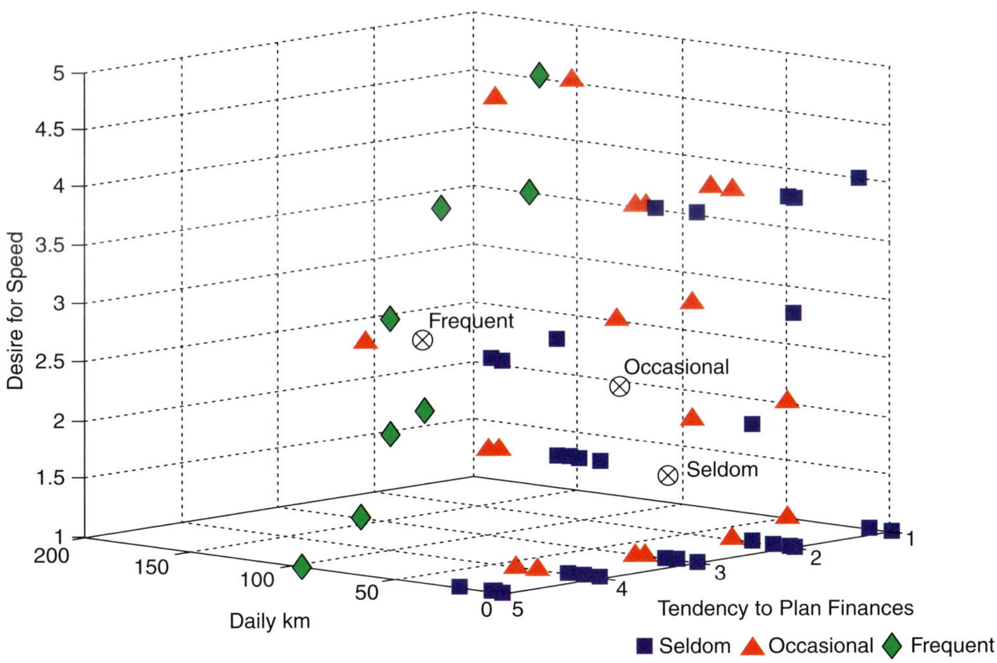

Plate 2 (Fig. 3.27) Survey responses clustered around stakeholder groups according to daily travel (km), tendency to plan finances (survey rank 1–5), and desire for acceleration performance (survey rank 1–5). The analysis shows a correlation between those who seldom drive tending not to value good acceleration performance, and frequent drivers, who are more likely to value detailed financial planning.

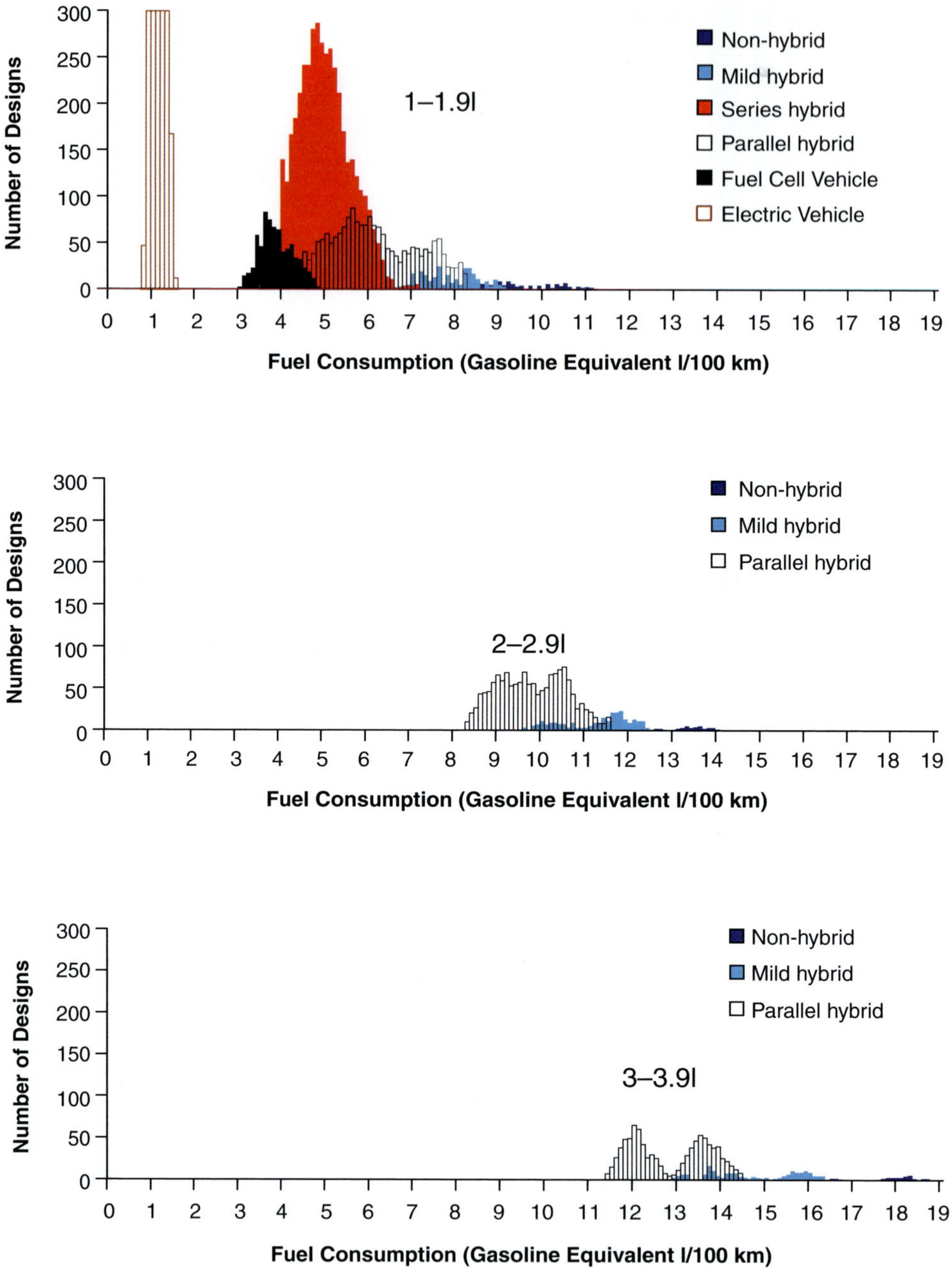

Plate 3 (Fig. 3.15) The distribution of fuel consumption of compact passenger vehicles in the design set. The consumption is primarily distributed according to hybrid architecture and displacement.

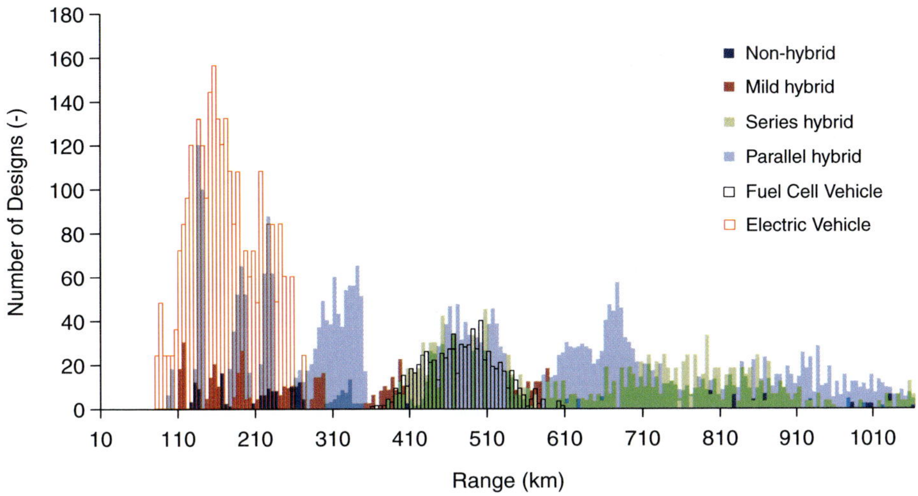

Plate 4 (Fig. 3.18) Distribution of driving range in the design set. Although all-electric vehicles emit very few life cycle emissions, their driving range is limited.

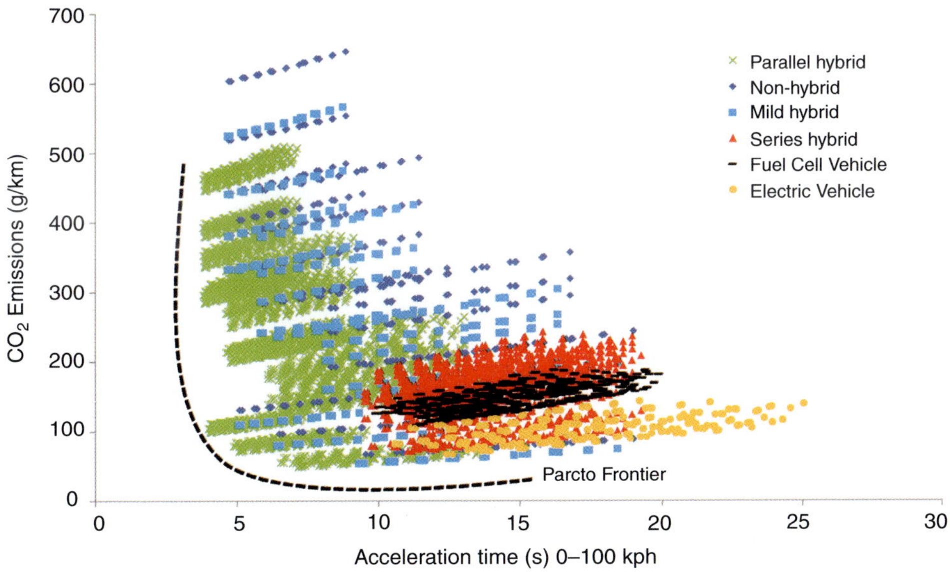

Plate 5 (Fig 3.19) Parallel hybrids dominate the performance versus life cycle CO_2 emissions trade-off. Note that the high life cycle CO_2 emissions for conventional, mild, and parallel hybrid powertrains come from the fact that the design set is composed according to heuristics, and not sales weighted, which would tend to mean smaller-displacement vehicles would bring the average down. The average CO_2 emissions for a 2010 vehicle in the European fleet are closer to 180 g/km.

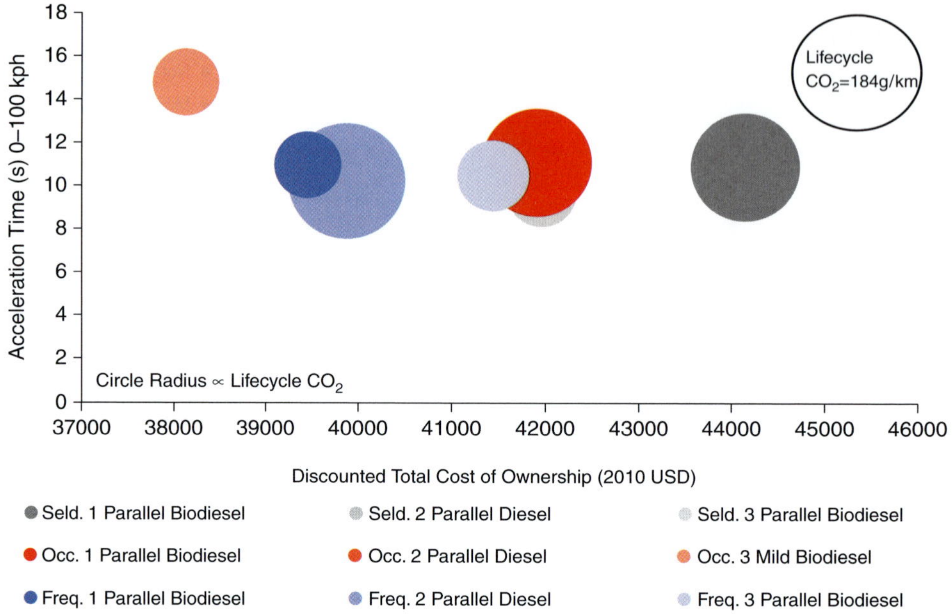

Plate 6 (Fig. 3.29) Multi-criteria decision analysis results for survey respondents using US 2010v technology characteristics. The CO_2 emissions reference of 184 g/km was selected because this represents the largest emissions for the present case, corresponding to the parallel gasoline vehicle.

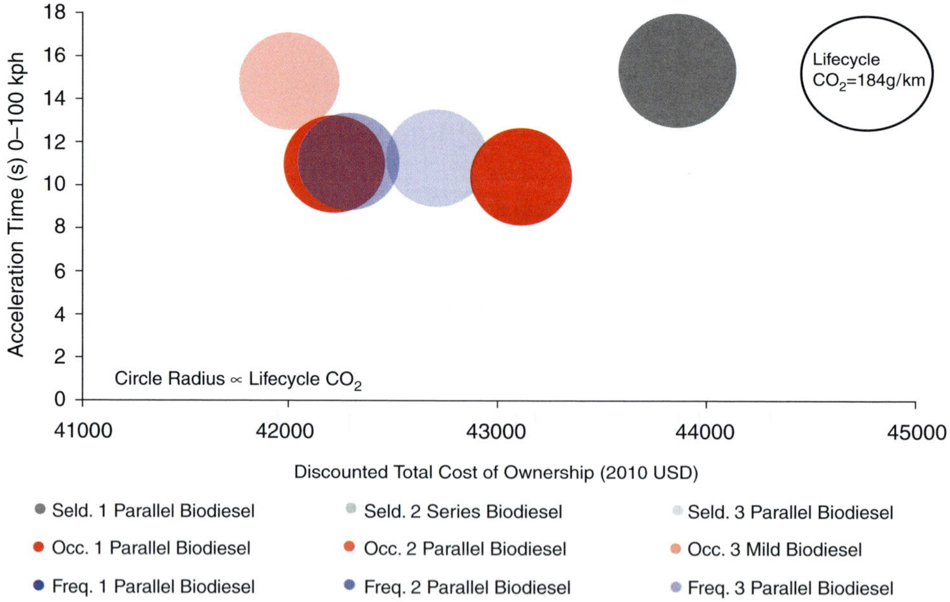

Plate 7 (Fig. 3.30) Multi-criteria decision analysis results for survey respondents using US 2035 renewable technology characteristics. The CO_2 emissions reference of 184 g/km was selected because this represents the largest emissions for the present case, corresponding to the parallel gasoline vehicle.

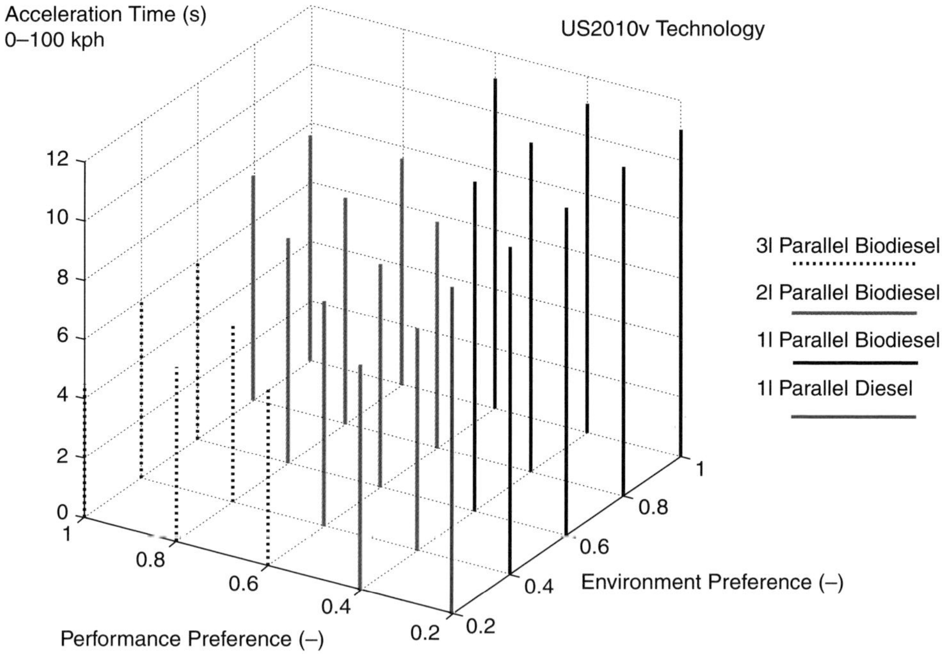

Plate 8 (Fig. 3.31) The sensitivity of design choice to environmental and performance preference, quantified by vehicle acceleration time. Acceleration is equally sensitive to performance and environmental preference, with biodiesel dominating the selected technology based on low life cycle CO_2 emissions.

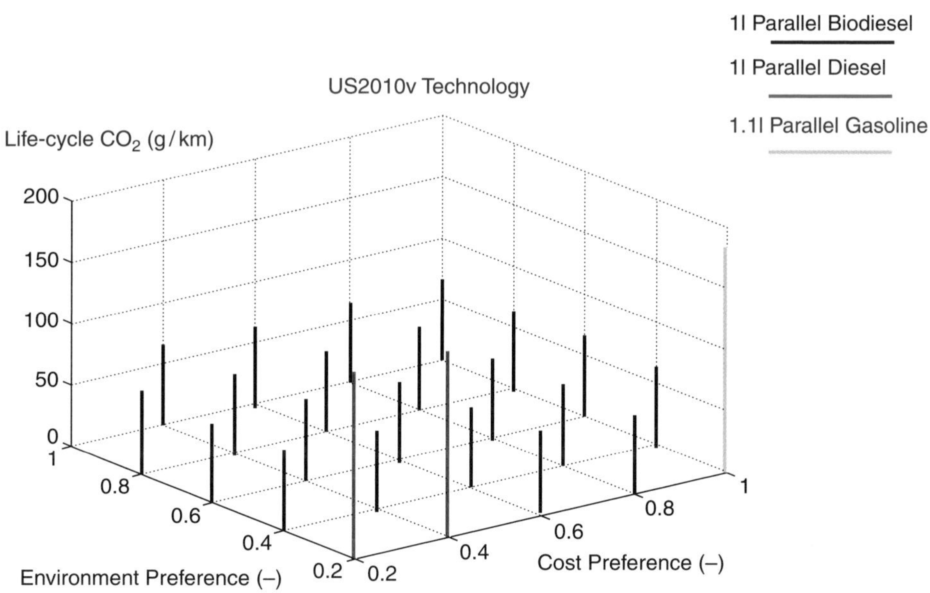

Plate 9 (Fig 3.32) Sensitivity of design choice to environment and cost preference, quantified by life cycle CO_2 emissions. Biodiesel options dominate for high environmental preference weightings; Otto powertrains dominate where cost preference is high.

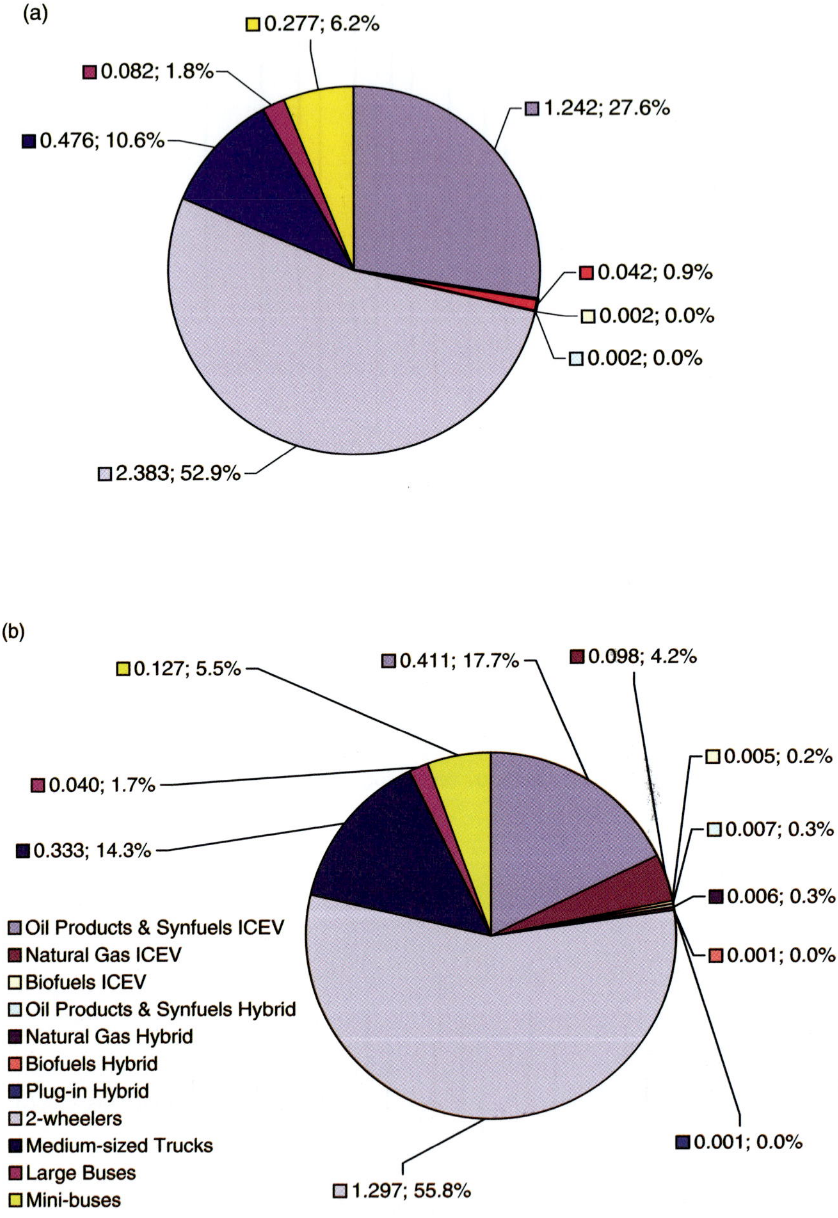

Plate 10 (Fig. 4.7) H_2 emissions from each vehicle technology in 2010 (a) and 2020 (b) separated by absolute emissions (in Tg) and percentage of total emissions.

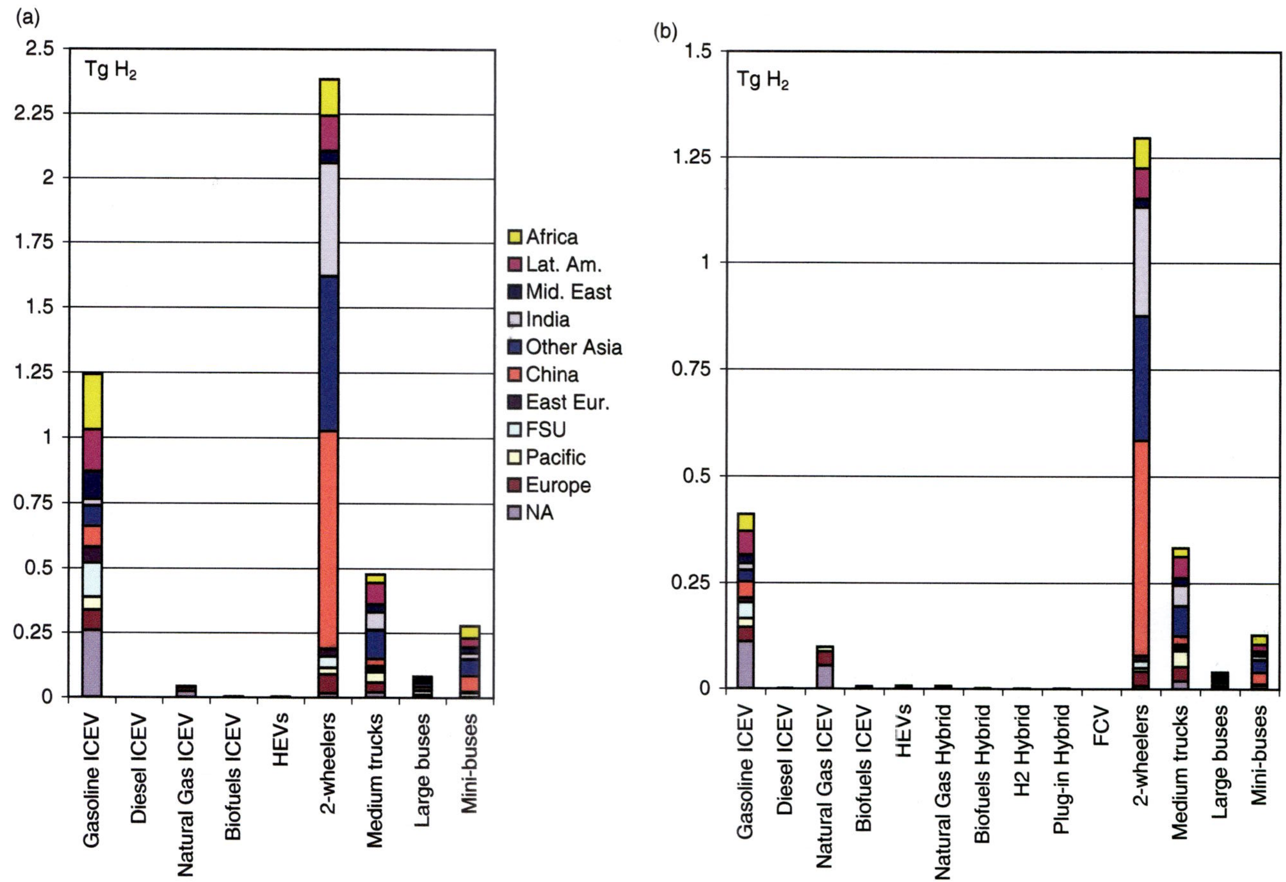

Plate 11 (Fig. 4.8) H_2 emissions from each vehicle technology in 2010 (a) and 2020 (b) separated by the absolute contribution (in Tg) from each world region (note the difference in scales).

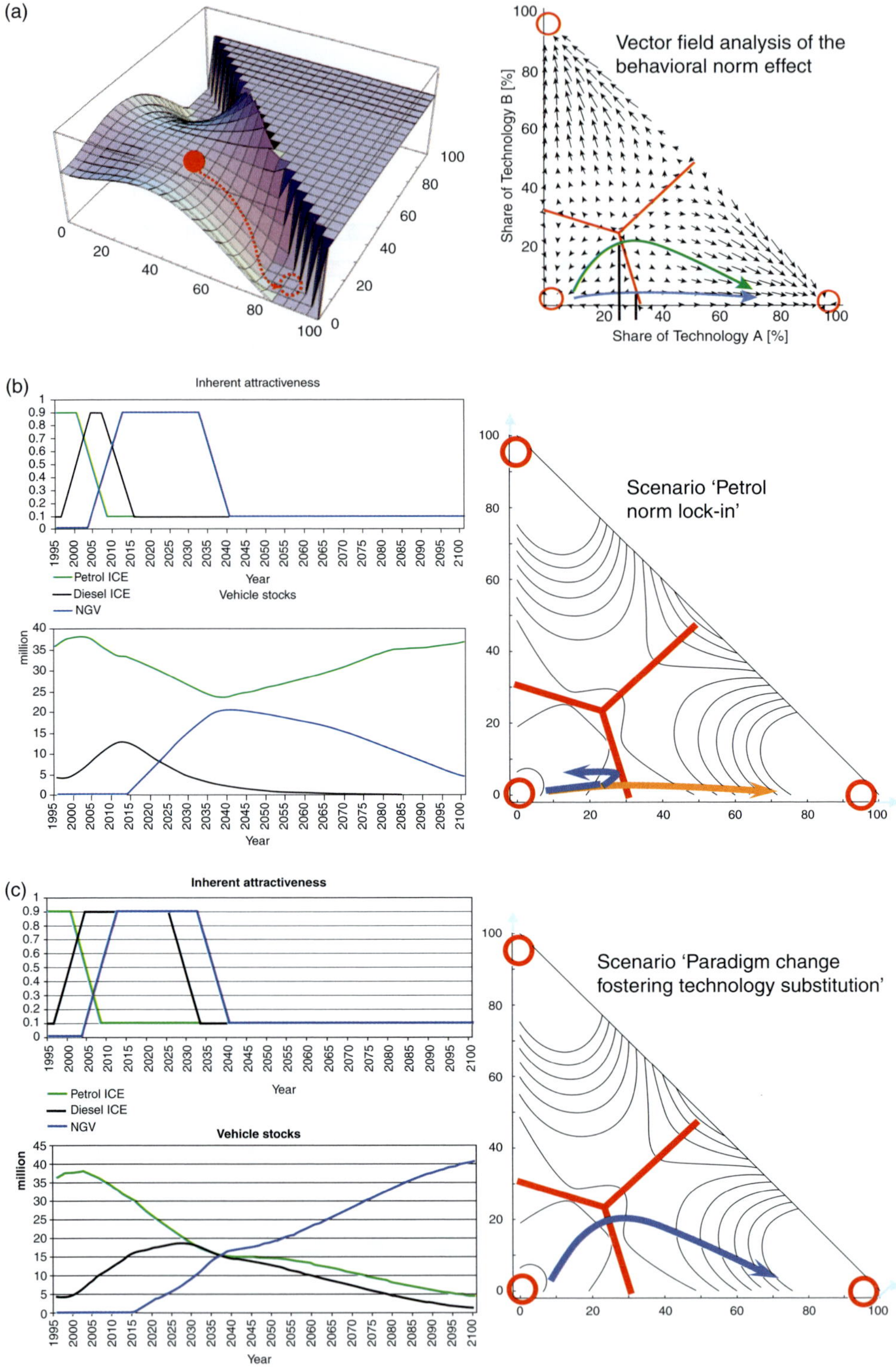

Plate 12 (Fig. 5.8) Explanation of the behavioural norm effect. The effect of a behavioural norm and the critical mass on the diffusion process of competing technologies is illustrated with the Germany case under the baseline scenario.

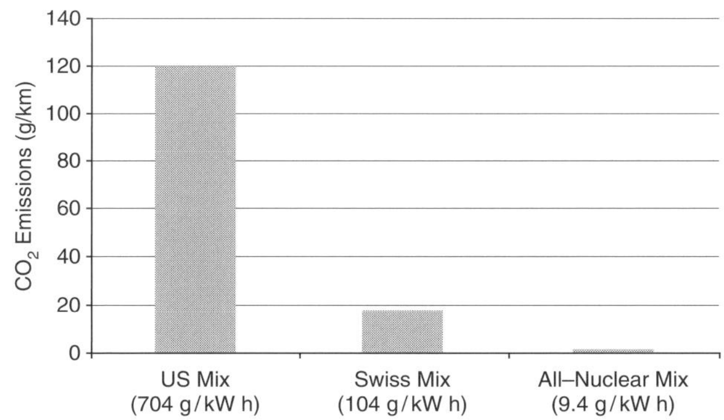

Figure 3.14 Influence of grid CO_2 intensity on the life cycle emissions for all-electric vehicles.

vehicles plotted in Figure 3.14. Nuclear and renewable fuel cycles have very positive effects on vehicle emissions, while charging from the US supply mix means the electric vehicle emits in the same magnitude as compact ICE vehicles.

Further sensitivity analysis will not be presented, although much has been explored in this area in order to identify key influence parameters. For example, in our analysis of fuel price, we determined that fossil fuels would have to be roughly eight times more expensive with fixed hydrogen and electricity prices (a highly unlikely scenario) in order to pay for the added powertrain cost of fuel cell and electric vehicles with current powertrain technology. For further analysis of the sensitivity of vehicles to various technology factors, see Liu (2007) and for sensitivity to usage factors see Kwon *et al.* (2008).

3.5 Results and discussion

The results of heuristic design modelling are used in two ways, starting with a trade-off analysis on key criteria, and finishing with a broad, multi-criteria assessment of technologies relating to the key research questions. On-road energy use is an important vehicle characteristic because of the number of stakeholder criteria that depend on it (operating cost, emissions, etc.). The division of the design set according to hybrid architecture is shown in Figure 3.15, which highlights how combustion engine displacement and hybridisation play a large role in dividing vehicle designs by energy consumption throughout the set. Appendix C contains a detailed description of the characteristics of several vehicle types in the design set.

The scenarios that are used to analyse the different technologies are depicted in Figure 3.16, which shows how the design set is divided across three distinct (but interrelated) geographic zones with their own policy measures, as well as three

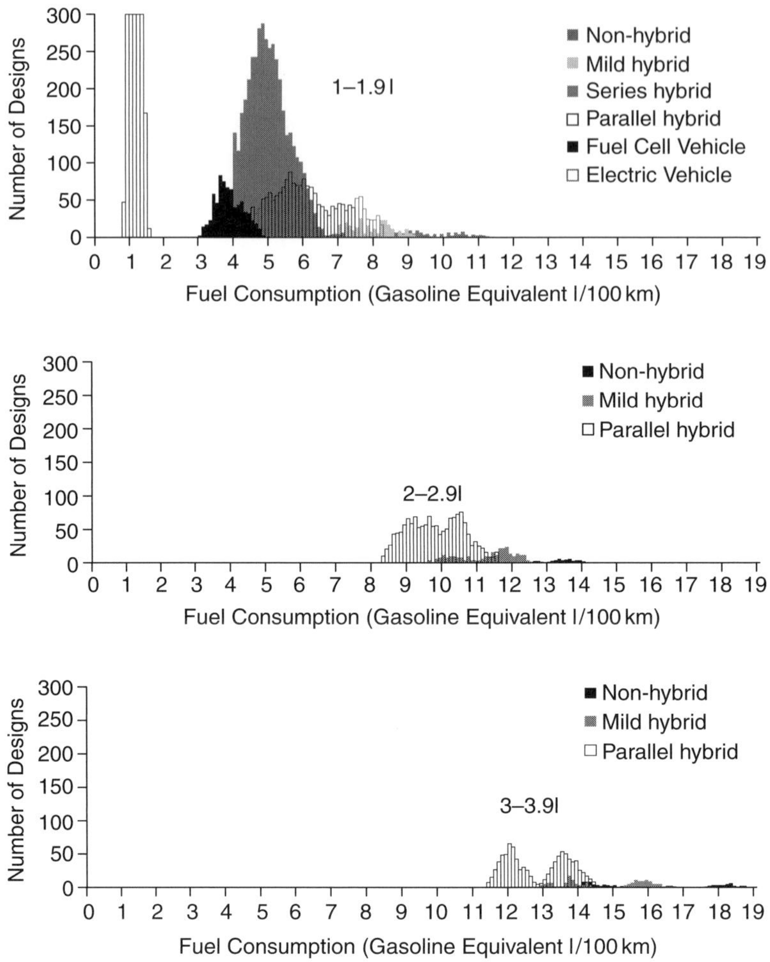

Figure 3.15 The distribution of fuel consumption of compact passenger vehicles in the design set. The consumption is primarily distributed according to hybrid architecture and displacement (see also colour plate 3).

points in time. The 2010 'current state' assumption is unusual in such analysis, but provides interesting insight into the effect that taking current 'off the shelf' technology prices (i.e. with no volume production assumptions taken) has on the various consumer criteria. In this chapter, most results are based on North American 2010 volume production (US2010v) and 2035 (US2035ren) prices with dominant future renewable energy production assumptions.

The trade-off analysis is divided into hybrid architecture/powertrain, materials, and fuel options, although the results are drawn from the same design set results. When analysing the results, it is important to consider that there are many variables driving criteria indicators, and the influence of a single variable may require effort to isolate.

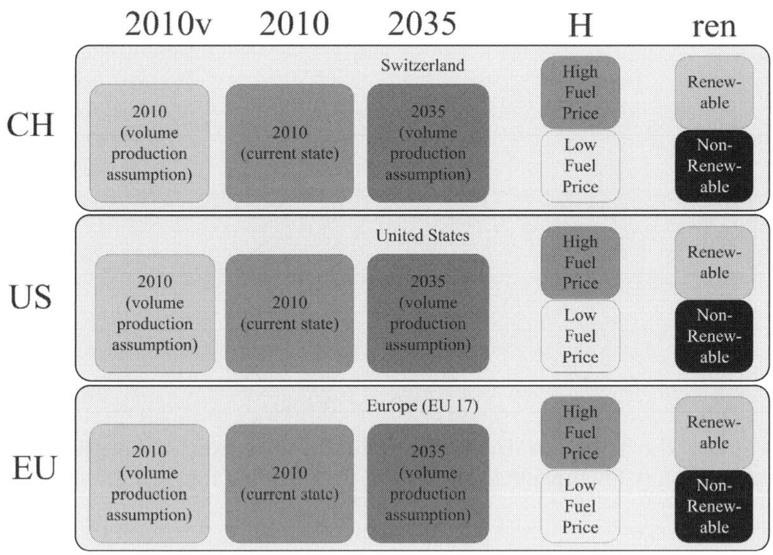

Figure 3.16 Various geographical regions and time horizons analysed.

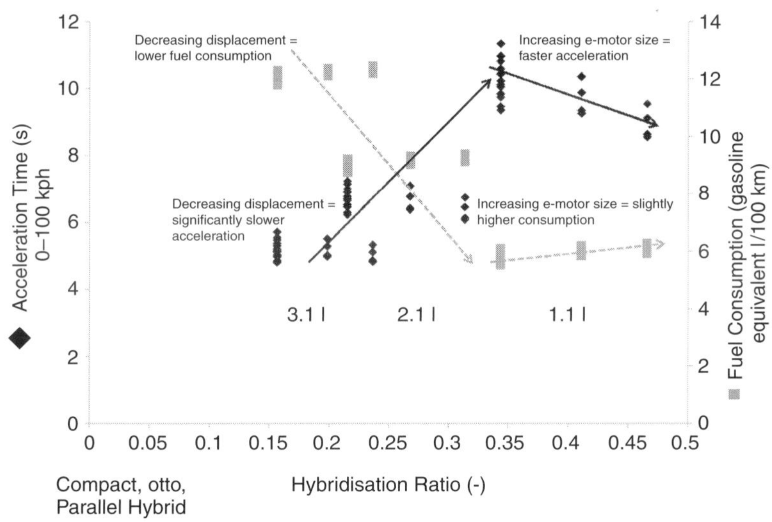

Figure 3.17 The performance versus fuel consumption trade-off is mitigated through electrification. A larger hybridisation ratio is the automatic consequence of increasing battery/motor size with fixed combustion engine displacement. This increases acceleration performance more than increasing fuel consumption.

3.5.1 Powertrain technology

Arguably one of the most important consumer trade-offs is that between performance (represented in Figure 3.17 by the time needed to accelerate from 0–100 kph) and fuel consumption. Adding hybrid electric components increases performance while simultaneously reducing fuel consumption.

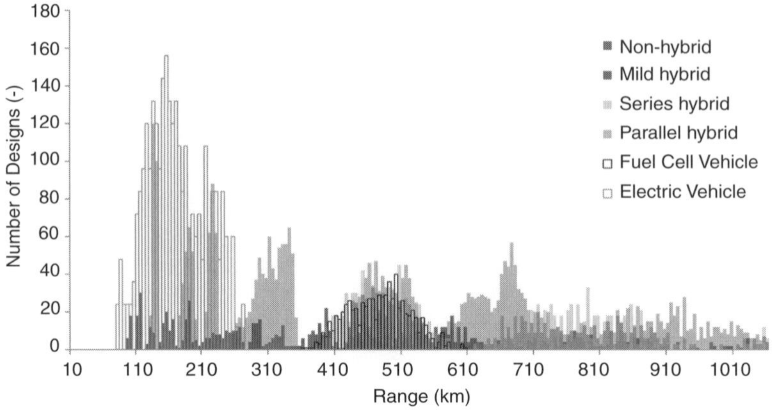

Figure 3.18 Distribution of driving range in the design set. Although all-electric vehicles emit very few life cycle emissions, their driving range is limited (see also colour plate 4).

In fact, the degree to which performance is increased by hybridisation is greater than can be achieved by increasing engine displacement (on a per kW basis). The hybridisation ratio (*HR*) is defined as the peak power (*P*) of the electric path divided by the total peak power of the vehicle drive systems as shown in Equation (3.7).

$$HR = \frac{\max(P_{motor})}{\max(P_{motor} + P_{engine})} \tag{3.7}$$

Adding hybrid components comes at a cost, which must of course be considered in the trade-off analysis. The Section 3.5.2 addresses the costs of lightweighting and hybridisation technology.

All-electric vehicles (EVs) have very positive values for many indicators. They are cheap to own and operate, have low direct emissions, potentially very low life cycle emissions, and can provide satisfactory driving performance. Their major shortcoming is shown in Figure 3.18, and that is their limited driving range.

All-electric and FCVs have comparable CO_2 emissions when electricity from solar or wind generation is assumed to be the primary energy, but fuel cells presently enable a driving range between filling of greater than 500 km, and are therefore more acceptable to consumers than the range of electric vehicles which is typically closer to 200 km.

Using electricity from the US grid for electric vehicles and hydrogen from central SMR, the CO_2 emission versus performance trade-off is dominated by parallel hybrids (along the Pareto frontier) as shown in Figure 3.19. These vehicles have the advantage of significantly more power for acceleration, as well as reduced fuel

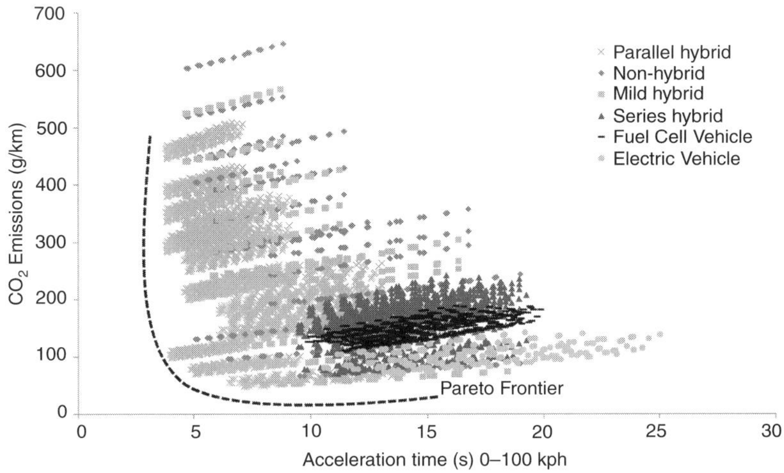

Figure 3.19 Parallel hybrids dominate the performance versus life cycle CO_2 emissions trade-off. Note that the high life cycle CO_2 emissions for conventional, mild, and parallel hybrid powertrains come from the fact that the design set is composed according to heuristics, and not sales weighted, which would tend to mean smaller-displacement vehicles would bring the average down. The average CO_2 emissions for a 2010 vehicle in the European fleet are closer to 180 g/km (see also colour plate 5).

consumption through idle stopping and engine operating point optimisation. This result agrees well with the previous discussion of hybridisation ratio advantages, and demonstrates how adding electric components adds power for increased performance while reducing fuel consumption. The added powertrain efficiency and performance does result in an increased manufacturing and purchase cost, which is explored in more detail in the multi-criteria analysis.

3.5.2 Lightweighting technology

The results presented in this section continue with a discussion of hybrid architecture as well as introducing the lightweighting technology options, which include aluminium, high-strength steel, and composite materials (Bjelkengren, 2008). The real-world weight versus fuel consumption trade-off is shown in Figure 3.20. Here, the fuel consumption of real conventional, hybrid, and all-electric vehicles is plotted against vehicle kerb weight and the relationship between weight and fuel consumption is explicitly given for each class of powertrain. It is clear that the sensitivity to weight is lower for hybrids, and even lower for electric vehicles.

This trend is also exhibited by the simulation results in Figure 3.21, which shows the average increase in fuel consumption across each powertrain type from adding 100 kg. This result is explained by the higher average efficiency of the hybrid and electric powertrain, partially achieved by the ability of those vehicles to recapture braking energy. The exact dynamics of this effect are in fact relatively complex, and

often higher than fleet averages due to the fact that very large displacement Otto powertrains as well as small electric vehicles are equally represented in the set. In this manner, the comparison of technology can be accomplished without bias imposed by purchase statistics. The importance of fuel choice can be seen in Figures 3.23 to 3.26 where life cycle CO_2 and NO_x emissions, respectively, are shown for different vehicle powertrain configurations over the US2010v and US2035ren input assumptions. Compressed hydrogen (C. Hydrogen) combustion yields the highest carbon emissions using hydrogen from SMR, while hydrogen used in FCVs yields the second lowest value after natural gas-powered series hybrids. Biodiesel performs well amongst the combustion options (total emissions shown with lighter line), with CO_2 and NO_x emissions very close to those of compressed natural gas vehicles. The poor performance of ethanol is attributed to its lower energy density and its unfavourable NO_x emissions during production.

It is clear from examining Figures 3.23 through 3.26 that the use phase dominates the CO_2 emissions for all vehicles except FCVs. Ethanol production sequesters CO_2, but because of its higher tank-to-wheel consumption it has overall higher CO_2 emissions. FCVs perform best in the set for all future US2035ren fuel options due to their lower fuel consumption and only slightly higher manufacturing emissions. It is very clear that the method of producing hydrogen and electricity is an important factor in how attractive fuel cell and electric vehicles are. With non-renewable 2010 production techniques, advanced powertrains do not produce significantly fewer emissions than conventional transportation fuels in conventional powertrains. Electric vehicles shift energy conversion losses from the vehicle to the power plant. Assuming that the refining losses from crude oil to gas or diesel are about 23%, that power line transmission and distribution losses are about 7%, and that power plant conversion efficiencies range from about 33% to 55%, then the maximum reduction in primary energy demand due to shifting from oil to electric vehicles will not be 75%, but rather about 37% to 62%.

3.5.4 Multi-criteria decision analysis

For this work the 'pairwise-outperformance approach' (POA) to MCDA was adapted from the EU NEEDS project (Makowski *et al.*, 2009), and was used to rank technologies according to stakeholder preferences. The POA is characterised by Equations (3.8) through (3.11), where technologies i and j are described by the criteria indicator vector r, and ranked according to the stakeholder weighting vector w, with their relative performance being scaled by the scaling function β to improve algorithm performance.

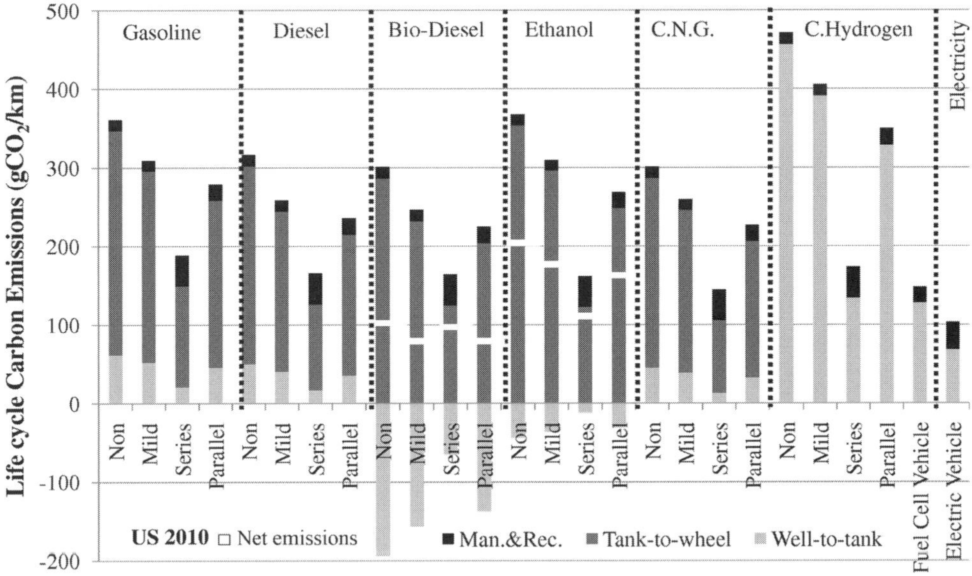

Figure 3.23 Life cycle CO_2 emissions for various hybrid architecture and fuel combinations using 2010 US assumptions. This shows that hydrogen from central steam methane reforming in a fuel cell vehicle can yield very low emissions, comparable to those of a compressed natural gas-operated series hybrid (biofuels total emissions shown with lighter lines). Note that the emissions are design set averages and therefore cannot be directly compared to fleet averages.

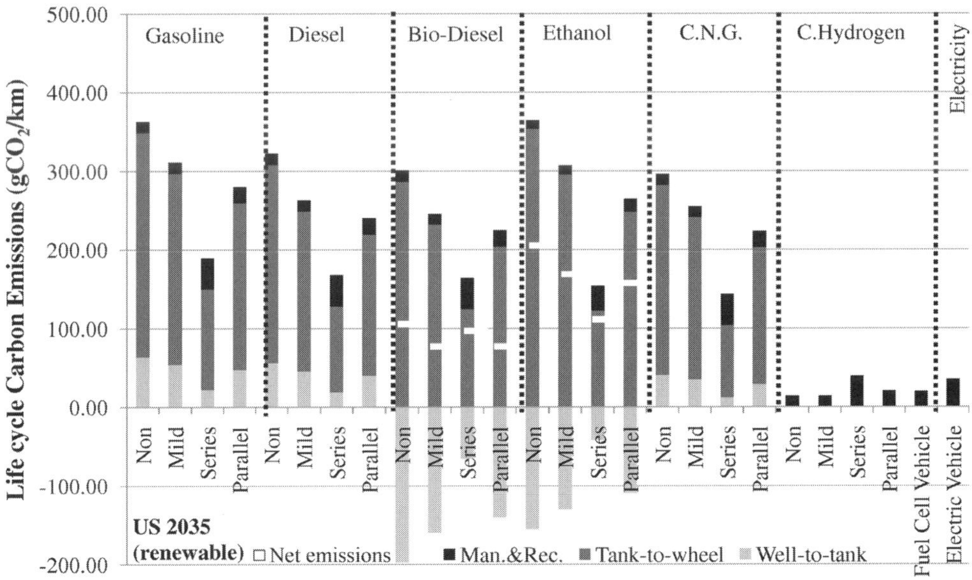

Figure 3.24 Life cycle carbon dioxide emissions for various hybrid architecture and fuel combinations using 2035 US renewable primary energy assumptions. This shows how using renewable or nuclear generated electricity for producing hydrogen or charging batteries can significantly improve vehicle emissions (biofuels total emissions shown with lighter line). Note that the emissions are design set averages and therefore cannot be directly compared to fleet averages.

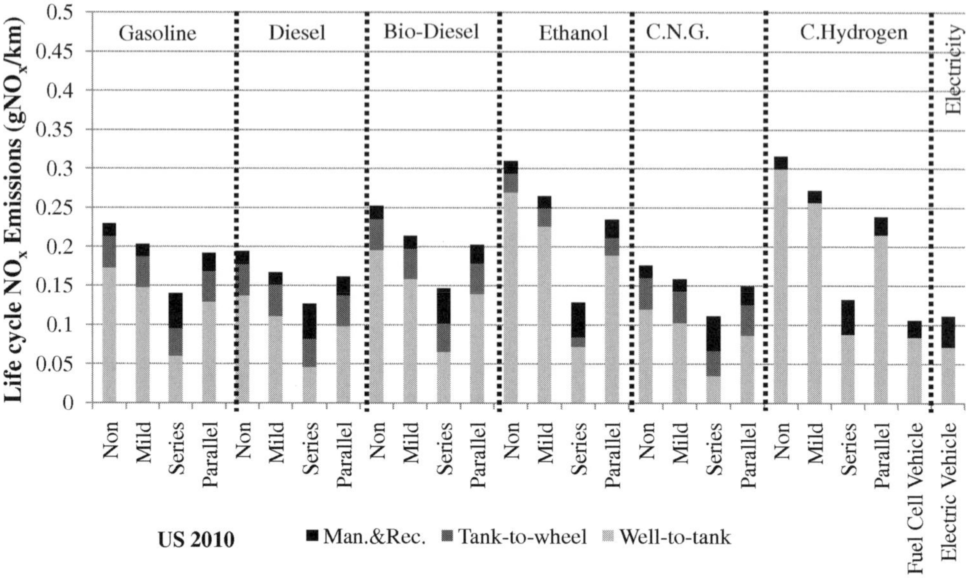

Figure 3.25 Life cycle nitrogen oxide emissions for various hybrid architecture and fuel combinations using 2010 US assumptions. Diesel vehicles score well due to low fuel consumption and advanced emissions control systems, whereas ethanol vehicles score poorly due to high fuel consumption. Fuel cell vehicles produce the lowest NO_x emissions on a per-kilometre basis. Note that the emissions are design set averages and therefore cannot be directly compared to fleet averages.

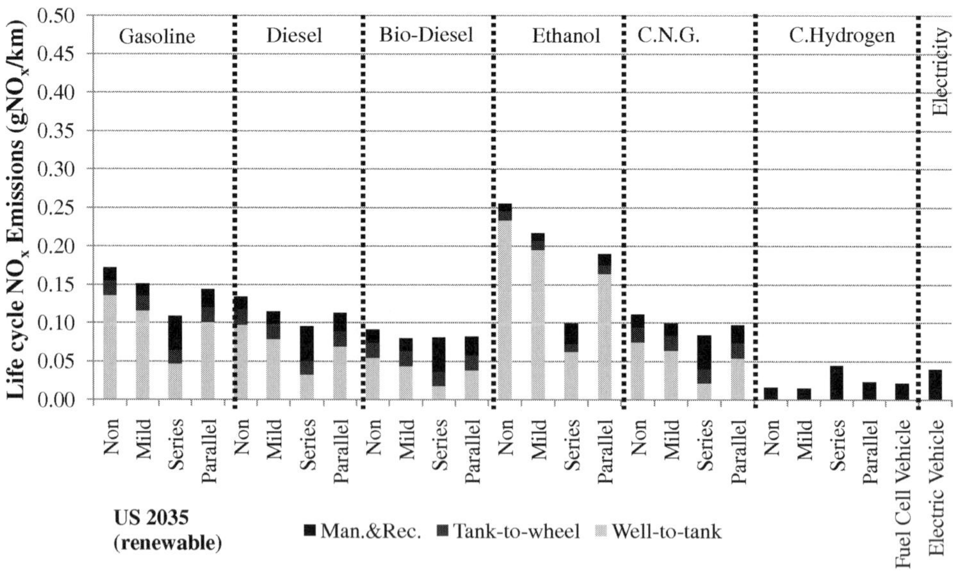

Figure 3.26 Life cycle nitrogen oxide emissions for future hybrid architecture and fuel combinations using 2035 US renewable primary energy assumptions. Hydrogen fuel has dramatically lower emissions. Note that the emissions are design set averages and therefore cannot be directly compared to fleet averages.

$$dc_{ijk} = w_k \cdot (r_{ik} - r_{jk}) \cdot \beta(r_{ik}), \tag{3.8}$$

$$dc_{jik} = w_k \cdot (r_{jk} - r_{ik}) \cdot \beta(r_{jk}), \tag{3.9}$$

$$\beta(x) = \alpha^{-x} (\alpha = 10; 0 \le x \le 1), \tag{3.10}$$

$$d_{ij} = \sum\nolimits_{k=1}^{n} (dc_{ijk} - dc_{jik}), \tag{3.11}$$

If $d_{ij} > 0$ then vehicle design i is preferred to alternative j. The HDA, drivetrain simulation, and the multi-criteria analysis were all performed in the MATLAB environment using vectorised code structures to minimise execution time.

An online survey generated 70 complete responses (an overall 40% participation rate), a summary of which can be found in Appendix D. A cluster analysis was performed on the set of complete survey responses in order to cluster the respondents into three stakeholder groups with partitioning using the k-means method. The best metrics for distinguishing between groups was vehicle ownership and daily kilometres travelled, with North American respondents travelling an average of 13 km more than European respondents per day. Figure 3.27 shows how individual responses regarding daily travel, tendency to plan finances, and desire for good acceleration performance are grouped around three centroids. The ‘Frequent’ driving group tended to own their own vehicles, commute an average of 100 km daily, and valued speed and financial planning more than the ‘Seldom’ driving group, who tended not to own their own vehicles, travelled by car less than 7 km per day on average, and neither planned their finances nor needed fast vehicles. Members of the ‘Occasional’ driving group travelled an average of 41 km by car per day, also tended to own their vehicles, and represented a slightly younger age demographic.

The eight vehicle characteristics shown at the circumference of Figure 3.28 were chosen as being representative of the most important consumer criteria in the questionnaire, and served as inputs to the MCDA algorithm: two from each of the performance, utility, environment, and cost categories. The criteria preferences for each of the three groups are very similar, which supports the finding that the most statistically significant grouping factors were whether a respondent owned a car and how much they travelled by car. It should be noted that, while the survey did not explicitly ask respondents about their preference for low ‘total cost of ownership’, the responses for ‘operating cost’ were deemed an appropriate surrogate for this criteria. Additionally, CO_2 emissions were taken as a proxy for ‘GHG emissions’ criteria preferences (GHG emissions) and NO_x vehicle performance indicators were a proxy for ‘health-related emissions’ throughout the MCDA.

The result of performing MCDA using criteria weights from the three stakeholder groups for both 2010v and 2035ren vehicle technologies are shown in Figures 3.29 and 3.30. The top five vehicles chosen in the near term by all three groups were

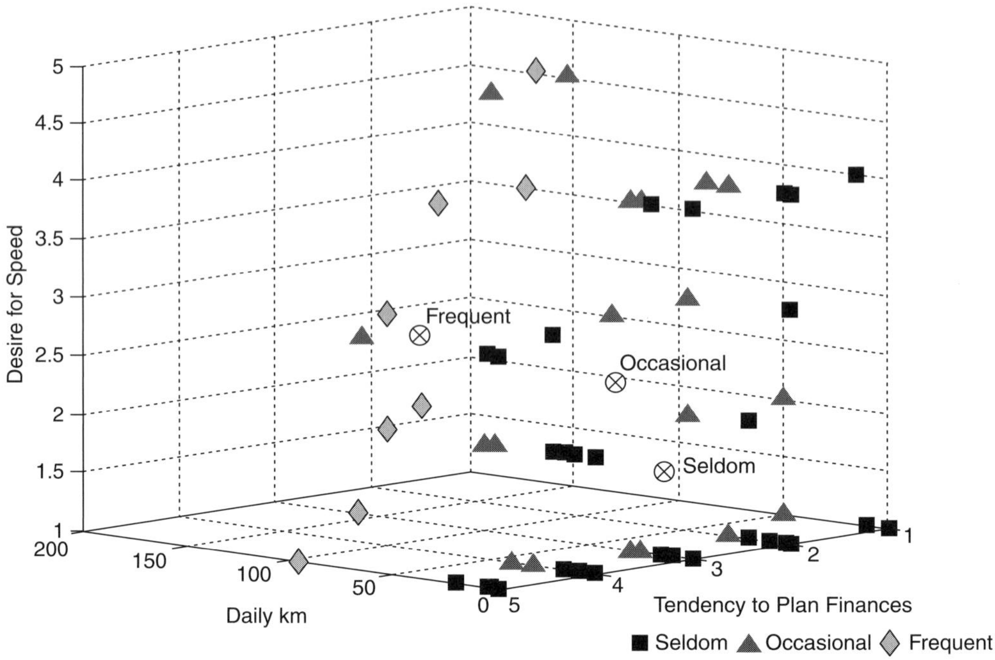

Figure 3.27 Survey responses clustered around stakeholder groups according to daily travel (km), tendency to plan finances (survey rank 1–5), and desire for acceleration performance (survey rank 1–5). The analysis shows a correlation between those who seldom drive tending not to value good acceleration performance, and frequent drivers, who are more likely to value detailed financial planning (see also colour plate 2).

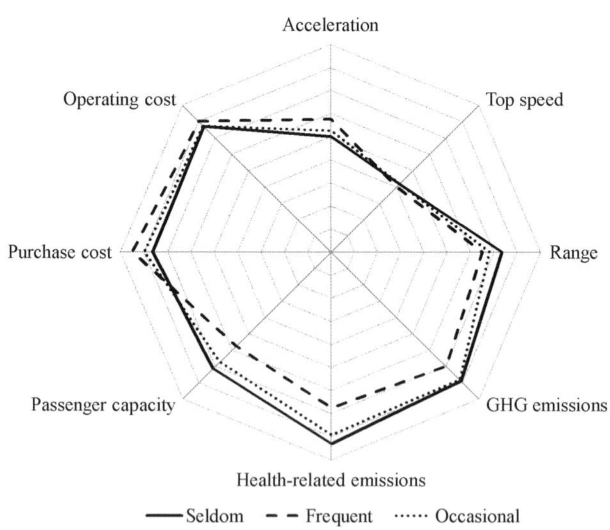

Figure 3.28 Normalised averages of clustered stakeholder criteria weightings, from survey responses.

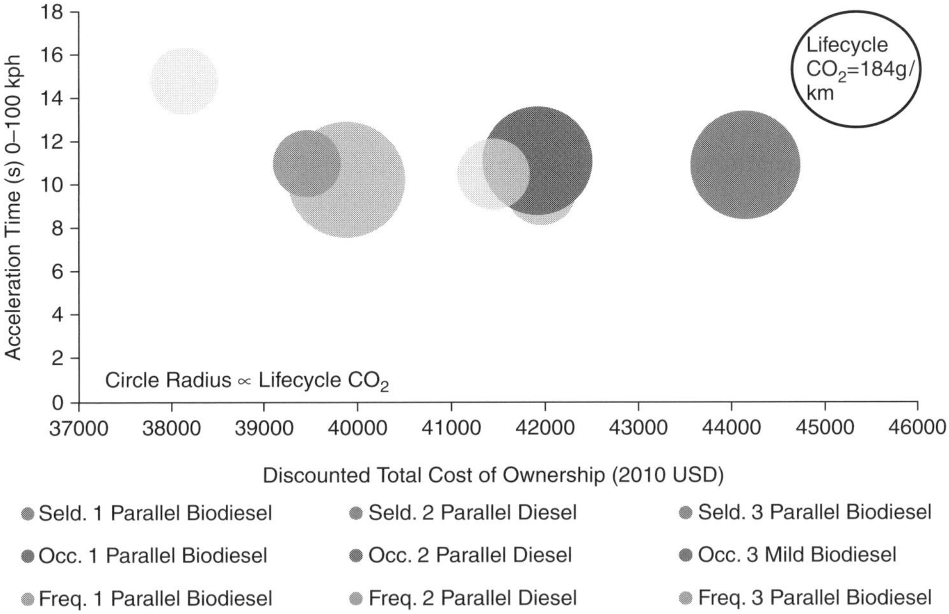

Figure 3.29 Multi-criteria decision analysis results for survey respondents using US 2010v technology characteristics. The CO_2 emissions reference of 184 g/km was selected because this represents the largest emissions for the present case, corresponding to the parallel gasoline vehicle (see also colour plate 6).

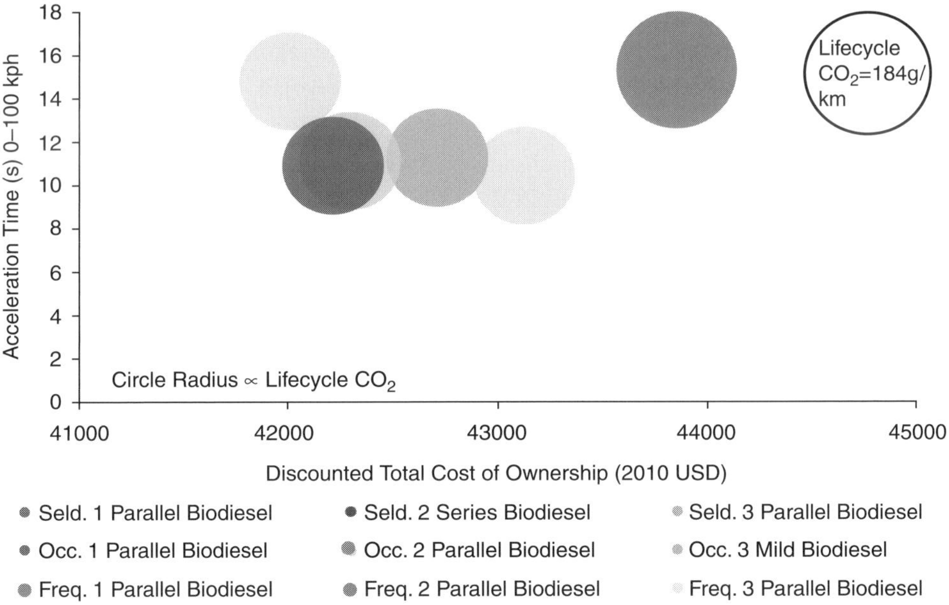

Figure 3.30 Multi-criteria decision analysis results for survey respondents using US 2035 renewable technology characteristics. The CO_2 emissions reference of 184 g/km was selected because this represents the largest emissions for the present case, corresponding to the parallel gasoline vehicle (see also colour plate 7).

very similar, mostly due to how small the differences in clustered criteria preferences were between stakeholders. It is interesting that diesel/biodiesel parallel hybrids were selected most often, and that, in the near term, parallel hybrids were preferred to mild hybrids, which in turn were preferred to no hybridisation. This reflects the trend (at least in Europe) toward diesel vehicles, as well as the current tendency toward parallel hybrid architectures before plug-in series architectures. The future options do not diverge significantly from the near-term choices, likely because of the high weight that survey respondents placed on cost criteria.

3.5.5. Robustness of technology options using multi-criteria decision analysis

A sensitivity analysis was performed in order to better understand the results of the stakeholder survey MCDA. The first series of tests were performed to determine the influence that each individual stakeholder criterion has on the top design choice. Table 3.3 shows the results of performing eight 'polarised' runs where each criterion weight was sequentially set to 100%, while all other criterion weights were set to 0% for current (2010) and future (2035ren) technologies. Hybrid vehicles were selected when top speed criteria were maximised, which is consistent with the performance advantages offered by electrified drivetrains. The results for autonomy are not surprising: diesel series hybrids have the highest on-board energy content, and are among the most efficient vehicles. That the lowest total cost of ownership is achieved by biodiesel vehicles also comes as no surprise, given the competitive price of fuel and low life cycle CO_2 emissions results for current and future low-displacement B-100 options. The choice of hydrogen-fuelled vehicles as the top-performing design on the basis of CO_2 and NO_x emissions for future sets is also logical, assuming that the hydrogen would be produced from solar photovoltaic (PV) electricity with low associated life cycle NO_x emissions.

To further examine the sensitivity of the optimal design to changes in criteria preferences, two criteria weightings were increased as two criteria weightings were decreased, while the remaining four were held constant at a nominal value of 0.2. The current technology vehicle design set US2010v was used for all of the following analyses. Figure 3.31 shows how the vehicle performance is as sensitive to environmental as it is to performance preference, with biodiesel options being selected most often due to their low life cycle CO_2 emissions and relatively good performance in parallel form. As environmental preference increases, engine displacement decreases.

Considering environmental and cost preferences, Figure 3.32 shows that, as the preference for low cost increases with a constant, low environmental preference, the life cycle CO_2 emissions are a maximum when an Otto-parallel design is selected.

Table 3.3. *Resulting technology chosen by 'polarised design choices' setting the criteria in question to 100% while all others to 0%.*

	Powertrain	Disp. (l)	Fuel	Battery	Battery Power	Battery Energy
	With current technology (US2010v)					
Acceleration	Parallel	3.1 l	Gasoline	LiON	70 kW	8 kWh
Top speed	Parallel	3.1 l	Gasoline	NiMH	70 kW	6 kWh
Autonomy	Series	1.1 l	Diesel	LiON	90 kW	25 kWh
CO_2 emissions	Parallel	1.0 l	B–100	LiON	50 kW	2 kWh
NO_x emissions	EV	–	Electricity	LiON	60 kW	16 kWh
Passenger volume	None	1.9 l	Gasoline	–	–	–
Purchase cost	None	1.9 l	E–85	–	–	–
Total cost	None	1.7 l	B–100	–	–	–
	With future technology (US2035ren)					
	Powertrain	Disp. (l)	Fuel	Battery	Battery Power	Battery Energy
Acceleration	Parallel	3.1 l	Gasoline	LiON	70 kW	8 kWh
Top speed	Parallel	3.1 l	Gasoline	NiMH	70 kW	6 kWh
Autonomy	Series	1.1 l	Diesel	LiON	90 kW	25 kWh
CO_2 emissions	Mild	1.9 l	Hydrogen	NiMH	3 kW	1.4 kWh
NO_x emissions	Mild	1.9 l	Hydrogen	NiMH	3 kW	1.4 kWh
Passenger volume	None	1.9 l	Gasoline	–	–	–
Purchase cost	None	1.9 l	E-85	–	–	–
Total cost	Mild	1.7 l	B-100	NiMH	3 kW	1.4 kWh

For low environmental and cost criteria weights, diesel parallel models are selected. For almost all cost and environment preferences, however, biodiesel hybrids are selected, signifying their robustness across these stakeholder preferences.

So far this sensitivity analysis has highlighted how certain fuel, powertrain, and hybridisation technologies are robust over a varied set of stakeholder preferences. Care should be taken in interpreting these results, however, because of the method used of holding four specific criteria weightings constant while varying the remaining four to perform the sensitivity analysis. This may lead to missing certain designs which are robust over other sets of stakeholder preferences. To maximise the utility of MCDA results, individual stakeholders should be allowed the opportunity to iteratively set and test their preferences and interpret the results. To simulate this process, a series of 256 preference weighting sets were generated using the eight criteria, each set to a 'high' and 'low' value of 1 and 0.1, respectively. Using different input assumptions for fuel and technology prices, as well as primary energy

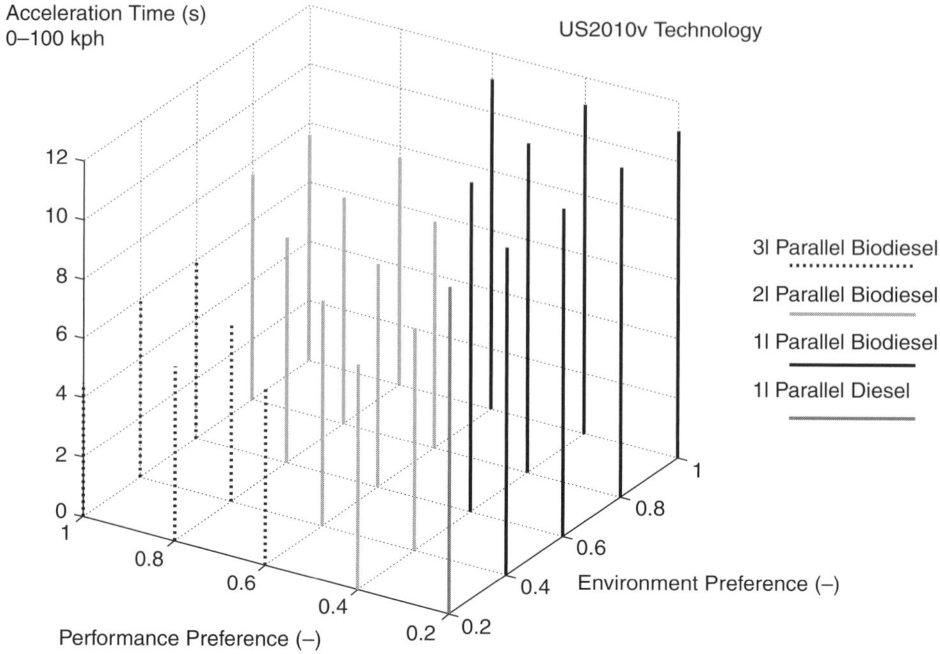

Figure 3.31 The sensitivity of design choice to environmental and performance preference, quantified by vehicle acceleration time. Acceleration is equally sensitive to performance and environmental preference, with biodiesel dominating the selected technology based on low life cycle CO_2 emissions (see also colour plate 8).

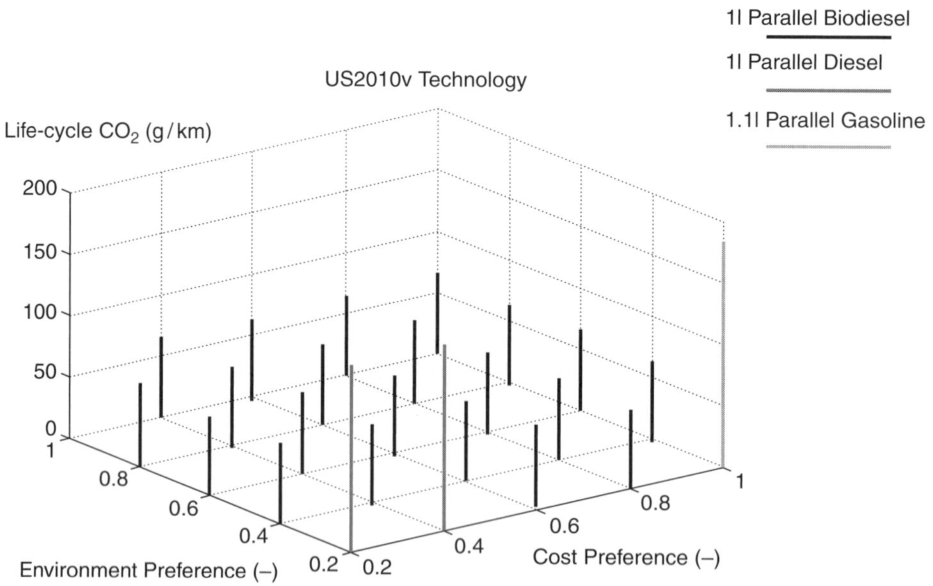

Figure 3.32 Sensitivity of design choice to environment and cost preference, quantified by life cycle CO_2 emissions. Biodiesel options dominate for high environmental preference weightings; Otto powertrains dominate where cost preference is high (see also colour plate 9).

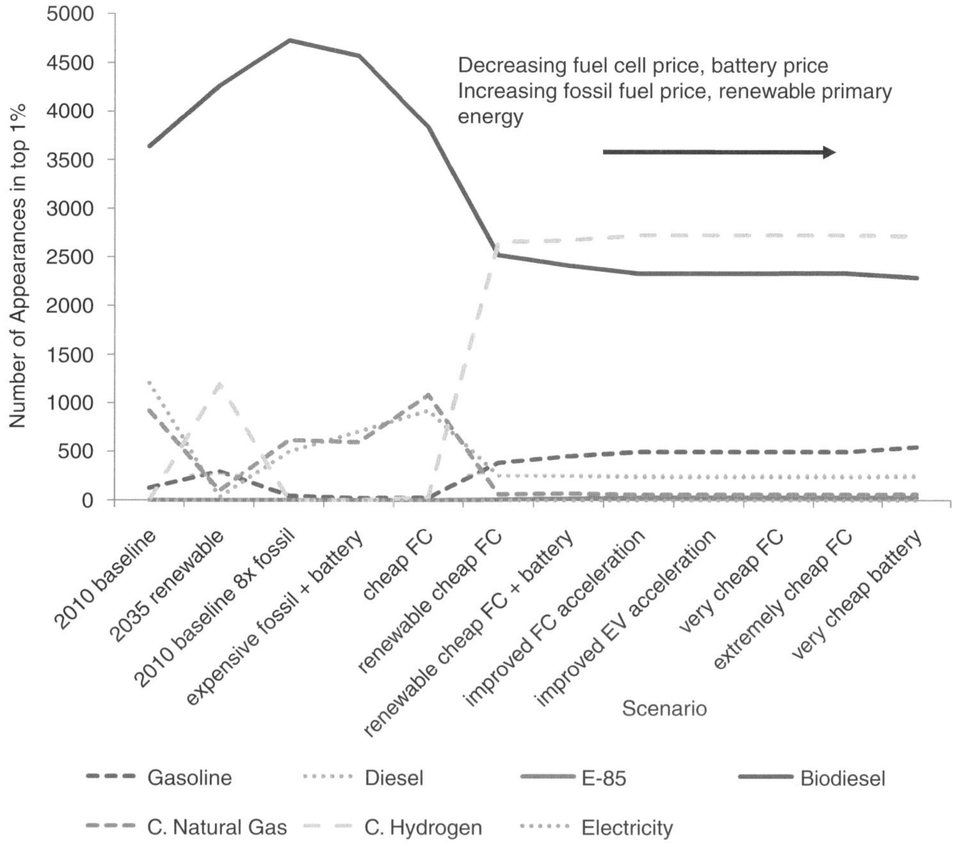

Figure 3.33 Number of times a fuel appears in the top 1% of technologies selected by all possible combinations of two-level criteria weightings. The trend toward hydrogen combustion (and, to a lesser extent, fuel cells) can be seen as fossil fuel price increases and primary energy CO_2 emissions decreases.

source, MCDA was performed on each of the input assumptions. The trend over these input scenarios for the fuel most often found in the top 1% of all designs is shown in Figure 3.33. The trend toward hydrogen combustion vehicles can be clearly seen as advanced technology decreases in price, primary energy CO_2 emissions decrease, and fossil fuel price increases. What is also notable is that battery electric vehicles are still not selected, even when battery cost is reduced to extremely optimistic levels. This can be explained by the fact that, even when cost and emissions criteria are fully satisfied, the range offered by electric vehicles still remains below the consumer acceptability threshold for this type of MCDA sensitivity testing. This does not imply that preference sets for which electric vehicles are selected do not exist, as demonstrated in Table 3.3 where an electric vehicle is chosen for low NO_x preferences.

3.6 Summary and conclusions

An attempt at answering the questions posed in Section 3.1.2 regarding the effects of introducing advanced powertrains, lightweighting, and alternative fuels was made in this chapter. It is hoped that the overarching goal of assessing the impact of hydrogen and fuel cell transportation technology on stakeholder criteria was also achieved. A summary of the conclusions reached throughout this chapter are that:

1. Heuristic design techniques allow passenger vehicle technologies to be combined in a self-consistent manner that enables a large number of technology options to be simulated and analysed.
2. Optimising the vehicle's hybrid control strategy using DDP methods ensures that all technology will be compared on a fair basis without variability introduced by using non-optimal rule-based hybrid control algorithms.
3. The sensitivity of vehicles with electric drives to weight reduction is lower (in some cases by a factor of two) than that of conventional, non-hybrid, vehicles.
4. Hence, lightweighting materials should be applied preferentially to non-hybrids, where they reduce overall emissions, increase performance, and reduce the total cost of vehicle ownership.
5. Hybridisation allows downsizing of combustion engines while maintaining performance, which reduces fuel consumption and thereby mitigates one of the most important consumer trade-offs. Although hybridisation of ICE-powered vehicles increases initial purchase costs, the resulting reduction in fuel consumption is generally sufficient to reduce the total cost of vehicle ownership.
6. Hybrid technology has the potential to reduce energy consumption by 10% for mild to 60% for plug-in series hybrids. Lightweighting the chassis can reduce energy consumption by 0.2–5.0% with constant vehicle size and powertrain characteristics (i.e. no engine downsizing is assumed). Switching fuel can reduce on-road energy consumption from 5% (for ethanol/gasoline) to 75% (when electricity is used as a transport energy vector). These efficiency gains may be diminished or magnified when these technologies are combined.
7. Switching to hydrogen or electricity from renewable or nuclear primary energy as a transportation fuel reduces most life cycle emissions, and all but eliminates tailpipe emissions. This results in some life cycle emissions, such as NO_x and volatile organic compounds, being drastically reduced.
8. The cost of fuel cell and battery electric powertrains must be offset by the fuel savings in a realistic time frame. The incremental cost of these advanced powertrains with 2010 technology is only paid back over 12 years with an eight-fold increase in current fossil fuel price, and is hence prohibitively expensive.

9. Stored energy must be produced from low-CO_2 (e.g. renewable or nuclear) sources. If electricity is taken from the 2010 US grid, electric vehicles emit an average of 210 g CO_2/km, compared with FCVs which emit an average of 170 g CO_2/km using hydrogen from SMR.
10. Survey results suggest that consumers are willing to pay a premium for the improved performance and reduced emissions of parallel/mild hybrids, and that biodiesel is a very attractive fuel for present as well as future consumers.
11. All-electric vehicles are typically only interesting if the energy density of battery systems improves to accommodate a longer driving range.
12. When environmental preferences are highly weighted, hydrogen powered vehicles are preferred if the fuel is produced renewably.

Although uncertainty regarding future technology development exists, this chapter has explored trade offs and presented MCDA results that quantify the strengths and weaknesses of the available options to allow informed decisions to be made. In general, hydrogen and FCV options are attractive if hydrogen is produced renewably, and if stakeholder emphasis placed on environmental preferences. Additionally, scarcity of supply leading to oil price increases as well as government subsidy distribution between fossil and renewable energies have not been treated extensively in this chapter. These aspects are very important, and will be treated in Chapters 5 and 6.

The MCDA methods described here are available online at http://multicriteria-analysis.com/TtH and allow the modification of input assumptions such as fuel costs, powertrain characteristics, etc., for a customised analysis of advanced vehicle technologies.

3.7 Acknowledgements

This work was supported by the Swiss Competence Centre Energy and Mobility (CCEM-CH), in collaboration with MIT and the industrial partners Ford and ENI under the framework of the Alliance for Global Sustainability.

4

Hydrogen emissions to the atmosphere from industry and transportation

STEVEN W. BOND, ROBERT ALVAREZ, STEFAN REIMANN, MARTIN WEILENMANN, AND BRIGITTE BUCHMANN

4.1 Introduction

Hydrogen has the potential to become a sustainable energy carrier of the future, particularly for transportation. Until recently, H_2 emissions to the atmosphere from anthropogenic sources have been largely disregarded and viewed in large part as a product of incomplete fossil fuel combustion. As H_2 production and the fraction of energy-based H_2 applications increase, a potential accumulation of H_2 in the atmosphere from direct emissions, losses and leakage, and changes in chemical processes in both the troposphere and the stratosphere could result. Vehicle exhaust gas is currently an important source of anthropogenic H_2 to the atmosphere, and losses of H_2 from production, distribution, storage, and other end-use systems could become increasingly significant contributors. With changing vehicle fleet composition and increasing H_2 demand and production for industrial and direct energy-based end-uses, an assessment of well-to-wheel H_2 emissions, including H_2 emissions from current and emerging vehicle technologies, will help assess the magnitude of future technological anthropogenic H_2 emissions to the atmosphere.

4.1.1 Objectives and scope

Apart from the challenging technical questions associated with the production, transport, and storage of H_2, future anthropogenic H_2 emissions are an area of increased interest due to their prominence in the global H_2 budget (see Figure 4.1). Currently, large uncertainty is associated with the contribution of H_2 from technological processes, primarily from fossil fuels (white arrow, Figure 4.1). A detailed

Transition to Hydrogen: Pathways Toward Clean Transportation, ed. E. Wilhelm and A. Wokaun. Published by Cambridge University Press.

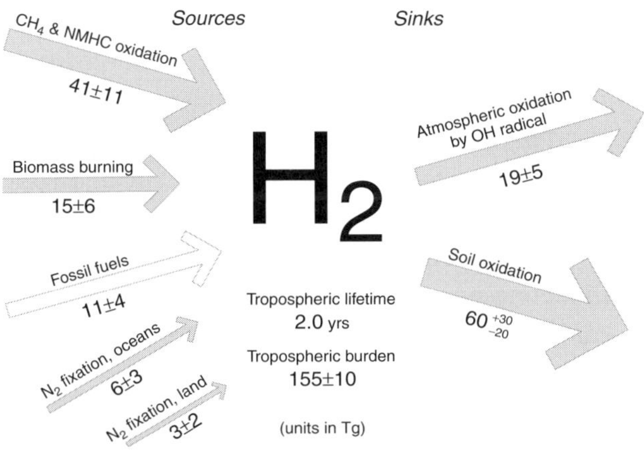

Figure 4.1 Global atmospheric H_2 budget. Sources are detailed on the left, sinks on the right. Anthropogenic emissions from fossil fuel-based processes (white arrow) are the focus of this chapter. Adapted from Ehhalt and Rohrer (2009).

analysis of H_2 emissions from the major anthropogenic processes that contribute to the present and future global H_2 budget has been performed in this work and serves to better quantify the contribution from this sector.

In a collaboration between the Laboratory for Air Pollution/Environmental Technology and the Laboratory for Internal Combustion Engines at Empa, H_2 emissions in the exhaust from various vehicle technologies were measured in order to help quantify the contribution of H_2 to the atmosphere from direct emissions, and to map the processes involved in H_2 evolution from current ICEV technologies. In addition, emission factors have been calculated to help formulate global H_2 scenarios from road-based transportation considering variable vehicle technologies, emission standards, geography, and time periods. The main objective of this chapter is to apply this information to estimate future global technological anthropogenic H_2 emissions considering changing vehicle fleet composition and global H_2 production. This task is approached through scenarios on different time scales under the assumption that H_2-based mobile applications and H_2 production will increase throughout the twenty-first century.

Two separate scenarios (see Chapter 6) are assessed that relate emissions of H_2 from road-based transportation with emissions derived from industrial and direct energy-based processes. In the baseline (business-as-usual) scenario, CO_2 emissions are left to increase unabated without the intervention of any specific climate policies aimed at addressing global climate change. Under this scenario, H_2 production for energy services increases only very modestly throughout the twenty-first century, and FCVs do not penetrate the market at all. Under the 450 ppm CO_2 climate policy scenario, specific mechanisms are applied that are designed to stabilise the atmospheric CO_2 mixing ratio at 450 ppm by the end of the twenty-first century. Under this scenario, H_2 production for direct energy services increases significantly over the period of study, notably with the penetration of FCVs in the second half of the century and an exponential increase in H_2 production for primary energy consumption. Under both scenarios, the changing H_2 emissions from transportation are combined with projected losses from H_2 production, distribution, storage, and other end-uses in order to provide an overview of projected overall technological anthropogenic H_2 emissions to the atmosphere throughout the twenty-first century. Scenarios have been formulated for 2010, 2020, 2050, and 2100.

4.2 State of research

4.2.1 Hydrogen in the atmosphere

After methane (CH_4), H_2 is the second most abundant trace gas in the atmosphere. H_2 can be considered an indirect GHG because it can lead to the production of tropospheric ozone (O_3), and it participates in a competing reaction with CH_4 for the hydroxyl radical (OH) (Derwent *et al.*, 2001; Prather, 2003), which is the primary atmospheric 'cleansing' agent. OH initiates the removal of both H_2 and CH_4 from the atmosphere. It is therefore plausible that increasing H_2 mixing ratios in the atmosphere could increase the lifetime and thus the radiative forcing of CH_4, since the OH radical may react increasingly with H_2 rather than with CH_4 (Schultz *et al.*, 2003). Through its indirect radiative effects and relationship with other direct GHGs, climate implications have therefore been coupled with tropospheric H_2 (Forster *et al.*, 2007). Furthermore, it has been suggested that increasing H_2 emissions at the surface could result in increasing H_2 in the stratosphere[1], which in turn could lead to an increase in stratospheric water vapour, with potentially dire consequences for the ozone layer (Tromp *et al.*, 2003). This assertion is, however, debatable, as Ehhalt and Rohrer (2009) have pointed out that the weak vertical gradient of hydrogen in the upper troposphere[2] and lower stratosphere means that virtually no export of H_2 into the stratosphere will occur. Rohs *et al.* (2006) have also reported that the contribution of stratospheric H_2 to the water vapour trend is minor. Any losses of H_2 to the stratosphere would, however, likely occur as convective injections in the tropical regions. The focus here will be on ground-based emissions.

One of the earliest estimations of H_2 in the troposphere comes from measurements made in 1923, where Paneth (1937) estimated the atmospheric H_2 mixing

[1] The stratosphere is the atmospheric layer from approximately 10 to 50 km above the earth's surface.
[2] The troposphere is the atmospheric layer up to approximately 10 km above the earth's surface.

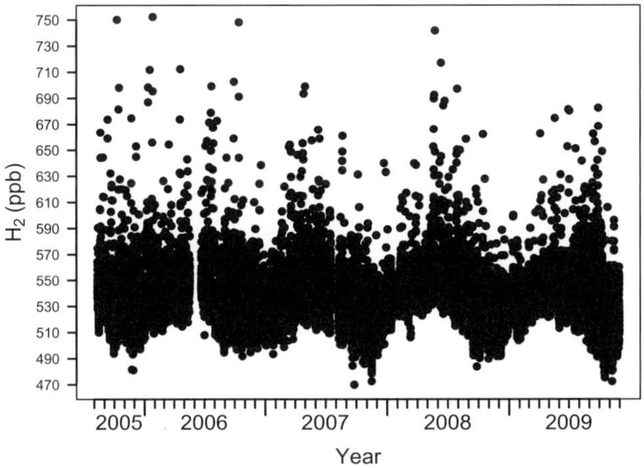

Figure 4.2 Atmospheric H_2 measured at the high altitude site Jungfraujoch, Switzerland. The dataset covers the period August 2005–November 2009. The black band represents background levels, while pollution and depletion events are seen bordering either side of the background measurements. Adapted from Bond *et al.* (2011b).

ratio to be approximately 0.5 parts per million (ppm). This estimation is remarkably similar to today's much more accurate measurements of approximately 530 parts per billion (ppb; ≈ 0.53 ppm) globally (Novelli *et al.*, 1999), and would suggest that the overall atmospheric mixing ratio of H_2 has not changed significantly over the past century, despite the annually increasing number of motor vehicles on the global scale. The lack of a long-term trend in background tropospheric H_2 in the last decades in Europe (Grant *et al.*, 2010) supports this finding. In fact, measurements at remote tropospheric sites should function as effective tools to monitor the behaviour of H_2 during the transition to more H_2-intensive applications (e.g. through long-term trend monitoring). As an example, Figure 4.2 depicts atmospheric H_2 mixing ratios at the remote high-Alpine site Jungfraujoch, Switzerland, (adapted from Bond *et al.*, 2011b).

The sink strength in the global H_2 budget is dominated by oxidative processes within the soil involving enzymes called hydrogenases. This term has been estimated at representing between ~75–90% of the global H_2 sink (Conrad and Seiler, 1985; Hauglustaine and Ehhalt, 2002; Rahn *et al.*, 2003; Sanderson *et al.*, 2003; Rhee *et al.*, 2006; Price *et al.*, 2007). Given the first-order reaction kinetics of H_2 oxidation by soil processes (Conrad and Seiler, 1981), it is plausible that an increase in atmospheric H_2 mixing ratios would be offset by enhanced H_2 uptake by soils. This assumption is backed by the striking similarity in the results of measurements of mean tropospheric H_2 throughout the twentieth century, despite global increases in motor vehicles with inefficient emission control systems compared with today's standards. Nevertheless, an upper-end threshold boundary beyond which this assumption may not hold is equally plausible.

The global source strength of H_2 from fossil fuels has been estimated at between 11–20 Tg/year (Novelli *et al.*, 1999; Sanderson *et al.*, 2003; Rhee *et al.*, 2006; Xiao *et al.*, 2007; Ehhalt and Rohrer, 2009), a significant contribution to the entire source term of atmospheric H_2 (see Figure 4.1). The contribution from traffic exhaust represents the major source of H_2 from fossil fuel-based processes (Schmidt, 1974; Novelli *et al.*, 1999; Barnes *et al.*, 2003; Ehhalt and Rohrer, 2009). Novelli *et al.* (1999) reported a range of H_2 emissions strictly from transportation at 5–20 Tg/year, while Vollmer *et al.* (2007) and Ehhalt and Rohrer (2009) reported estimates at the lower end of that range at 5.4 ± 3.2 and 9 ± 3 Tg/year, respectively. Other processes, such as the reformation of hydrocarbons, currently represent a comparatively minor fossil fuel-based source of H_2 to the atmosphere (Ehhalt and Rohrer, 2009).

4.2.2 Hydrogen emissions in internal combustion engine exhaust: the role of air/fuel ratios and lambda

H_2 emissions in ICEV exhaust result from incomplete fuel combustion in the absence of oxygen (O_2). Carbon monoxide (CO) emissions evolve under these same conditions, which creates a strong positive correlation between H_2 and CO in direct vehicle exhaust measurements. The air/fuel ratio — the ratio of the mass of air supplied to the engine to the mass of fuel supplied to the engine over the same time period — plays an important role. Lambda (λ) is a measure of the extent of the excursion of the air/fuel ratio relative to stoichiometry[3]. A λ value of 1 is at stoichiometry, whereas a value above 1 indicates an excess of air (lean-burn (oxidising) condition), and a value below 1 indicates the opposite (rich-burn (reducing) condition). Modern engines control λ at 1 using lambda sensors mounted both up- and downstream of the catalyst. With current technologies, gasoline engines for passenger vehicles are typically operated under stoichiometric conditions whenever possible, during which the air/fuel ratio is optimal to ensure complete fuel combustion in both the engine and catalyst. Exceptions are situations requiring extended periods of high power, during which rich stoichiometry is selected to cool the engine and the catalyst, short periods where the engine control unit (ECU) cannot perfectly maintain lambda at 1, or when rich conditions are chosen to clear

[3] The stoichiometric air/fuel ratio for a particular fuel type is the air/fuel ratio that ensures complete combustion of all carbon and hydrogen without any remaining oxygen.

out O_2 from the catalyst after fuel cut-off phases. Diesel vehicles always operate under 'fuel-lean' oxidising conditions, where air/fuel ratios are consistently above stoichiometric conditions, leading to much lower mixing ratios (and often untraceable levels) of H_2 in exchange for higher levels of nitrogen oxides (NO_x, nitric oxide (NO) + nitrogen dioxide (NO_2)).

4.2.3 Three-way catalytic converters

Increasingly stringent legislation on emissions of CO, NO_x, and hydrocarbons in response to air pollution and the rising awareness of adverse effects on human health, anthropogenic infrastructure, and biota has led to the advent of three-way catalytic converters (TWCs), which simultaneously promote the oxidation of CO and hydrocarbons, along with the reduction of NO_x. Although TWCs are not always mandated by legislation, it has become impossible to reduce emissions below legislated limits without them.

Overall, TWCs remove H_2 from raw vehicle exhaust. H_2 emissions from vehicles without TWCs are much higher than from those equipped with exhaust after-treatment systems[4]. However, today's TWCs are also designed to promote water-gas shift and steam reforming reactions (Auckenthaler, 2005), which ultimately results in the production of some H_2 through the destruction of hydrocarbons, CO, and water (H_2O) under fuel-rich conditions. In steam reforming reactions, hydrocarbons are oxidised by H_2O to produce CO and H_2. Steam reforming reactions are endothermic (i.e. they require the input of heat and proceed at high temperatures). The general equation for the steam reforming of hydrocarbons is:

$$C_nH_{2n+2} + nH_2O \rightarrow nCO + (2n+1)H_2 \qquad (4.1)$$

(Enthalpy change (ΔH) positive)

The water-gas shift reaction is exothermic (i.e. gives off heat and proceeds at lower temperatures), consuming CO and producing H_2 in the process:

$$CO + H_2O \rightarrow CO_2 + H_2 \qquad (4.2)$$

(ΔH negative)

Ageing of a catalytic converter can have a significant effect on the level of H_2 emissions over the lifetime of a TWC. For example, Auckenthaler (2005) observed much higher water-gas shift and steam reforming activity with a fresh TWC than with an aged device.

[4] It is important to note that not only do TWCs remove H_2 and other substances from vehicle exhaust, but current vehicle technologies also often have more sophisticated and precise onboard computer systems that consistently regulate the air/fuel ratio at the stoichiometric point. The result is much lower emissions compared with older vehicle technologies (or vehicles without a TWC).

4.2.4 Hydrogen emissions from test bench, chassis dynamometer, and real-world studies involving internal combustion engine vehicles

The processes of H_2 production and emission have been assessed in a limited number of cases through studies involving vehicles simulating real-world driving conditions. A wide range of H_2 emission factors for a series of gasoline-powered Euro-3 automobiles tested on a chassis dynamometer were observed (Heeb *et al.*, 2006). The range was largely attributed to the various driving conditions characteristic of the different cycles and sub-cycles. H_2 emission factors were found to be highly dependent on speed and acceleration patterns. The mean H_2 emission factor (E_{H2}) across a range of urban, rural, and highway sub-cycles was calculated to be 59.4 mg/km. A separate chassis dynamometer study involving gasoline-powered Euro-4 automobiles and the same driving conditions resulted in a mean E_{H2} of 26.5 mg/km (Bond *et al.*, 2010), illustrating the systematic development of TWCs and exhaust after-treatment systems as emissions limits tighten. Newer vehicles are also equipped with better λ control systems, which provide more consistent regulation around the stoichiometric air/fuel ratio, and thus fewer fuel-rich excursions and lower resultant emissions.

Tunnel studies eliminate the influence from other local source and sink factors, and can thus provide robust information about H_2 emissions from real-world vehicle fleets. In a tunnel study in Switzerland, an E_{H2} of 67 mg/km for a gasoline-based traffic fleet comprising a wide range of emission standard technologies was reported (Vollmer *et al.*, 2007). A much smaller E_{H2} was found for heavy-duty vehicles than for LDVs, a trait pinned to the much lower H_2 emissions from diesel compared with gasoline technologies.

In a test bench study involving emission measurements from an LDV engine (Vollmer *et al.*, 2010), pre-TWC measurements displayed predictable concentrations of both H_2 and CO in the exhaust gas under variable air/fuel ratios (see Figure 4.3). Because of this correlation, and the much more well-constrained CO contribution from vehicle exhaust, the molar ratio of H_2/CO has been considered a central topic in previous studies involving older vehicle technologies for establishing the H_2 contribution from fossil fuel combustion processes, primarily transportation (Schmidt, 1974; Novelli *et al.*, 1999; Barnes *et al.*, 2003; Steinbacher *et al.*, 2007; Vollmer *et al.*, 2007; Aalto *et al.*, 2009; Hammer *et al.*, 2009; Yver *et al.*, 2009).

4.2.5 Direct emissions of hydrogen to the atmosphere from production, distribution, storage, and non-transport end-use systems

Direct emissions of H_2 to the atmosphere from non-transport end-use systems also contribute to the overall source from anthropogenic processes. Currently, the

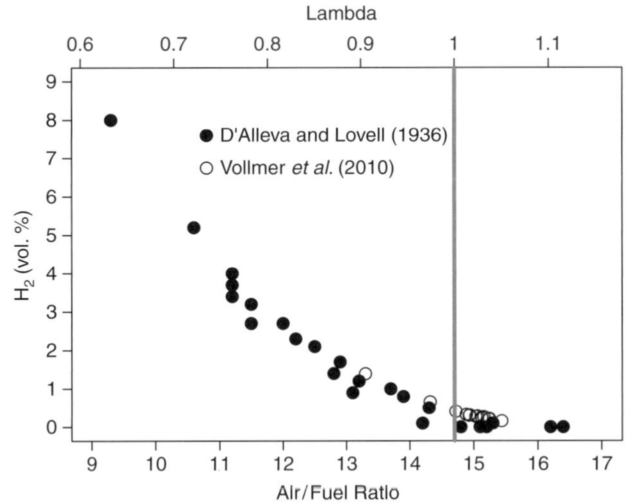

Figure 4.3 H_2 mixing ratios in raw gasoline exhaust gas. Measurements cover an air/fuel ratio ranging from approximately 9–17. The approximate corresponding λ values are shown on the upper x-axis. The filled circles are from D'Alleva and Lovell (1936), the open circles from Vollmer *et al.* (2010). The dark grey vertical line represents stoichiometric (λ = 1) conditions.

majority of these losses are likely derived from fossil fuels during the production of H_2 for industrial processes (e.g. fossil fuel reforming). Captive uses of H_2 are those where hydrogen is produced and consumed directly on-site. These are primarily the petroleum refining, ammonia (NH_3) fertiliser production, and methanol (CH_3OH) production industries, which, combined, account for ~92% of global H_2 production (CEH, 2007). Of this H_2 production, approximately 95% is produced from fossil fuels (Ogden, 2004). Future losses and leakage of H_2 will not only depend on methods of production, but also on strategies for the distribution, storage, and end-uses of H_2. It is widely accepted that emissions from cryogenic liquid H_2 dominate those of H_2 stored in the compressed gaseous form. An estimated range of H_2 loss rates due to 'boil off' or evaporation from liquid H_2 would be on the order of 0.1–10% over the entire production, distribution, and end-use chain (Zittel and Altmann, 1996; Schlapbach and Züttel, 2001; Burns *et al.*, 2002; Kammen and Lipman, 2003; Schultz *et al.*, 2003; de Wit and Faaig, 2007; Sherif *et al.*, 2007; R.J. Natkin, personal communication, 2008). This range is based on leakage from mature delivery and end-use systems, including pipelines, storage systems, compressors, pumps, and vehicles. Losses associated with high-pressure gaseous H_2 are lower than those associated with liquid H_2. The estimated loss of compressed gaseous H_2 from mature systems is marginal up to approximately 4% over the entire production, distribution, and end-use chain (Zittel and Altmann, 1996; Kammen and Lipman, 2003; Lovins, 2003; Schultz *et al.*, 2003; R.J. Natkin, personal communication, 2008; F. Balzer, personal communication, 2008).

Although a large-scale H_2-intensive economy would involve significantly greater H_2 production volumes for direct energy services, losses of H_2 to the atmosphere are not unanimously expected to increase considerably, despite the diffusive nature of H_2. Instead, it could be anticipated that losses actually decline as the maturity of the hydrogen chain develops (Bond *et al.*, 2011a). Loss rates estimated by industry tend to be lower, which may indeed be based on more realistic and reliable information from those involved in the industry, where safety, ecological, and monetary concerns highlight the agenda. The higher-end loss rates are often based on results from model studies (Tromp *et al.*, 2003). The high-end loss rate scenarios, on the one hand, provide important reminders about the proper diligence required during the transition toward an increasingly H_2-intensive economy, while on the other hand, they are potentially overestimated in today's technology-driven society.

4.3 Methodology for the assessment of hydrogen emissions

To characterise H_2 emissions from current technology vehicles, the exhaust gas from a number of vehicles was measured. Vehicle types included light-duty gasoline and diesel vehicles, two-wheelers (motorcycles and scooters), and natural gas automobiles. All four-wheeled vehicles were subject to the constraints of the Euro-4 emission standards, while the two-wheelers complied with the two-wheeler, Euro-3 emission standards. All vehicle measurements were conducted on a chassis dynamometer, which allows for the simulation of dynamic driving conditions under controlled laboratory settings. For details of the experimental setup, see Vasic and Weilenmann (2006) and Alvarez *et al.* (2009). A total of six cycles were tested for the four-wheeled vehicles and four cycles were tested for the two-wheeled vehicles in order to adequately capture the real-world characteristics of urban, rural, and highway driving conditions. Each cycle comprises three separate sub-cycles. For details of the individual driving cycles, see Bond *et al.* (2010).

All direct online H_2 measurements were conducted with an H-Sense process mass spectrometer (V&F Analyse- und Messtechnik GmbH), which operates according to the electron ionisation mass spectrometry measurement principle. See Heeb *et al.* (2006) and Bond *et al.* (2010) for further details of the instrument.

Emission factors are widely used to quantify vehicle emissions and are often expressed as the mass of a pollutant per unit time or vehicle distance travelled. The calculation of vehicle emission factors is not only important for monitoring regulated and other pollutants of interest in vehicle exhaust, but also serves as a tool for quantifying and upscaling emissions to constrain specific source terms of atmospheric budgets. Hydrogen emission factors (E_{H2}) were calculated through the combination of molar H_2 mixing ratios in the exhaust gas and the exhaust mass flow. A detailed description of the calculation of E_{H2} from vehicles measured directly on the chassis dynamometer can be found in Bond *et al.* (2010). For vehicle technologies not directly tested on the chassis dynamometer, emission factors were calculated from other independent measurements and assumptions made about the individual technologies. For details of the derivation of E_{H2} for these vehicle technologies, see Bond *et al.* (2011a).

The emissions factors (E_{H2}) for the various vehicle types were applied to globally upscale emissions from transportation (Section 4.4.2, Table 4.1). As vehicle technologies, global kilometres travelled, technologies designed to address vehicle exhaust emissions, and emissions/emission standards for different world regions are different and change with time, a number of assumptions were formulated to develop an accurate account of global H_2 emissions over the subject time horizon. In formulating these assumptions, effort has been made throughout this chapter to coordinate and reconcile details with other chapters, particularly Chapter 6 for fleet progression and Chapter 3 for vehicle technology. The main assumptions are as follows:

1. The percentage share of each LDV technology for 2010, 2020, 2050, and 2100 are based on results of least-cost optimisation modelling, using the PSI's Global Multi-regional MARKAL (GMM) model. This is a bottom-up model of the global energy system, with detailed representation of energy and passenger transport technologies, energy resources, and demands. The model is applied to study scenarios of global energy and transport system development – in this case, a baseline (business-as-usual) scenario and a scenario consistent with achieving a stable atmospheric CO_2 mixing ratio of 450 ppm. See Chapter 6 for details of the model and list of LDV technologies considered in this study.
2. Projections of global LDV kilometres separated by world region in 2010 and 2020 are based on model results from Fulton and Eads (2004). This second model (the Sustainable Mobility Project (SMP) transport model) captures the variability surrounding global emissions based on region, emissions/emission standards, and vehicle technologies supplementary to LDVs. In addition to LDVs, two-wheelers, medium-sized trucks, large buses, and mini-buses are considered, to be as inclusive as possible with vehicle technologies responsible for significant H_2 emissions. Heavy-duty trucks are all assumed to be diesel-based, from which H_2 emissions are assumed to contribute only marginally on a global scale (Bond *et al.*, 2011a). These additional criteria allowed us to complement the GMM and more comprehensively map the complex variability inherent to current and near-term emissions conditions.
3. Beyond 2020 (i.e. for 2050 and 2100 scenarios), projections of global LDV kilometres and emissions are exclusively according to the GMM. The separation by world region (as described in assumption 2) was not applied because

Table 4.1. *H_2 emission factors for all vehicle technologies and world regions considered in this study (Bond et al., 2011a). The letters for each time period represent the E_{H2} that was applied for each world region. 'all' signifies that all world regions were subject to that particular E_{H2}. All E_{H2} are reported in mg/km. Letters for each time period refer to: (a) Organisation for Economic Co-Operation and Development (OECD) North America, (b) OECD Europe, (c) OECD Pacific (Japan, Korea, Australia, New Zealand), (d) Former Soviet Union, (e) Eastern Europe, (f) China, (g) other Asia, (h) India, (i) Middle East, (j) Latin America, and (k) Africa.*

Vehicle Technology Class	E_{H2}	2010	2020	2050	2100
Gasoline automobiles					
1. pre-Euro-1	1130	i,k	–	–	–
2. Euro-1	283	d,e,f,g,h,j	–	–	–
3. Euro-2	108	–	i,k	–	–
4. Euro-3	59.4	a,b,c	d,e,f,g,h,j	–	–
5. Euro-4	26.5	–	–	–	–
6. post Euro-4	26.5	–	a,b,c	all	all
Gasoline two-wheelers					
1. Based on E_{CO} of 20 g/km	793	d,e,f,g,h,i,j,k	–	–	–
2. Based on E_{CO} of 15 g/km	595	a,b,c	–	–	–
3. Based on E_{CO} of 7.5 g/km	297	–	d,e,f,g,h,i,j,k	–	–
4. Based on E_{CO} of 6.575 g/km	261	–	b	–	–
5. Based on E_{CO} of 6.467 g/km	256	–	a	–	–
6. Based on E_{CO} of 4.628 g/km	184	–	c	–	–
7. Euro-3	141	–	–	–	–
8. Post-Euro-3	26.5	–	–	all	all
Diesel vehicles					
Assumed to be same as diesel Euro-4 vehicles based on oxidising operating principle of diesel engines	0.1	all	all	all	all
Gasoline medium-sized trucks/mini-buses					
1. Based on E_{CO} of 35 g/km	1390	d,e,f,g,h,i,j,k	–	–	–
2. Based on E_{CO} of 15 g/km	595	–	d,e,f,g,h,i,j,k	–	–

Table 4.1 (*cont.*)

Vehicle Technology Class	E_{H2}	2010	2020	2050	2100
3. Based on E_{CO} of 7.5 g/km	297	a,b,c	–	–	–
4. Based on E_{CO} of 5.625 g/km	223	–	a,b,c	–	–
5. Post-Euro-4	26.5	–	–	all	all
Gasoline large buses					
1. Estimated, based on E_{CO} of 46.7 g/km	1850	d,e,f,g,h,i,j,k		–	–
2. Estimated, based on E_{CO} of 20 g/km	793	–	d,e,f,g,h,i,j,k	–	–
3. Estimated, based on E_{CO} of 10 g/km	397	a,b,c		–	–
4. estimated, based on E_{CO} of 7.5 g/km	297	–	a,b,c	–	–
5. post-Euro-4	26.5	–	–	all	all
Natural gas vehicles	73.5	a,b,c	a,b,c	all	all
Biofuels vehicles	13.3	a,b,c	a,b,c	all	all
Gasoline HEVs	23.1	a,b,c	a,b,c	all	all
Natural gas hybrid vehicles	58.8	a,b,c	a,b,c	all	all
Biofuels hybrid vehicles	10.6	a,b,c	a,b,c	all	all
Plug-in hybrid vehicles	14.8	a,b,c	a,b,c	all	all
H_2 hybrid vehicles	33.1	–	a,b,c	all	all
H_2 fuel cell vehicles	vary	–	–	all	all

all vehicles of similar technologies were assumed to have same E_{H2} in 2050 and 2100, irrespective of region.

4. Projections of global kilometres for medium-sized trucks, large buses, and mini-buses for specific regions and time periods were derived from Fulton and Eads (2004), and based on kilometres driven by region for 2010 and 2020, total global kilometres in 2050 and the forecast according to regression for global kilometre projections to 2100.
5. Projections of global two-wheeler kilometres for specific regions and time periods were derived from Fulton and Eads (2004), and based on kilometres driven by region in 2000, projected model results based on two-wheeler ownership by region to 2050, and the forecast according to the regression for kilometre projections to 2100.
6. The percentage share of fuel types for conventional vehicle types (gasoline and diesel) are based on projections by Fulton and Eads (2004). This percentage share of gasoline- and diesel-based transportation applies to the oil products and synfuels ICEVs category described in Chapter 6.
7. Medium-sized trucks, large buses, and mini-buses were all assumed to be fuelled by either gasoline or diesel over the course of the study period.
8. Technologies other than conventional gasoline (including two-wheelers) and diesel vehicles (i.e. natural gas, biofuels, and all hybrid vehicles) were assumed to be incorporated only into developed-country vehicle fleets for 2010 and 2020.
9. A 10-year time lag for each world region was applied according to predicted emission standards or emission factors (E_{H2}) for LDVs from Fulton and Eads (2004). For example, LDV E_{H2} derived for 2010 are based on the emission standards or assumed emissions predicted for each region as of 2000. This was done to account for older vehicles still present within the fleet following the introduction of a new emission standard. This was done only for the years 2010 and 2020 (see assumption 11).
10. For large buses, medium-sized trucks, and mini-buses, emission factors were derived based on the assumed average emissions for each vehicle stock from Fulton and Eads (2004). In contrast to the LDVs (assumption 9), average emissions are provided for each time period rather than adopted emissions standards for these vehicle technologies. Thus, for these vehicle technologies, a time lag was not applied since an average emission for each vehicle technology and time period was provided. This was done only for the years 2010 and 2020 (see assumption 11).
11. Beyond 2020 (i.e. for 2050 and 2100) similar vehicle technologies were assumed to have the same E_{H2}, irrespective of world region. For example, conventional gasoline LDVs were assumed to have the same E_{H2} in Europe as in China.

12. The calculation of emission factors from vehicles measured on a chassis dynamometer can be found in Bond *et al.* (2010). The derivation of other E_{H2} can be found in Bond *et al.* (2011a).
13. E_{H2} from technologies other than conventional gasoline and two-wheeler vehicles were assumed to be constant over time.
14. E_{H2} from diesel vehicles were assumed to be constant with time based on the consistently oxidising operating point of diesel engines (Bond *et al.*, 2010).

In addition to these assumptions made for the global upscaling of emissions from transportation, leakage rates of H_2 to the atmosphere from production, distribution, storage, and non-transportation-based end-use systems (Section 4.4.3) were deduced based on assumptions and information from personal communication from industrial and commercial sources, along with estimations from the literature (Zittel and Altmann, 1996; Schlapbach and Züttel, 2001; Burns *et al.*, 2002; Kammen and Lipman, 2003; Lovins, 2003; Schultz *et al.*, 2003; Tromp *et al.*, 2003; Warwick *et al.*, 2004; de Wit and Faaig, 2007; Sherif *et al.*, 2007 Balzer, personal communication, 2008; Natkin, personal communication, 2008). H_2 emissions from other fossil fuel-based processes such as industrial and residential heating systems are not accounted for in this study. However, H_2 emissions from such systems have been found to expel only marginal H_2 to the atmosphere (M.K. Vollmer, personal communication, 2010), suggesting that H_2 emissions from other industrial processes not related to intentional (e.g. the petroleum refining, NH_3 production, and CH_3OH production industries) H_2 production may also be minor (compared to CO), particularly as fossil fuel consumption begins to decline, cleanup technologies improve, and renewable primary energy resources gain in market share.

4.4 Results and discussion

4.4.1 Hydrogen emissions in vehicle exhaust gas: a qualitative survey

Hydrogen emissions from today's vehicles powered by ICEs are dependent on several factors, including:

- the presence/absence of a TWC;
- the engine state (engine speed and torque), which is defined by the vehicle state (e.g. speed, acceleration);
- the quality of lambda control;
- vehicle and fuel type; and
- engine and catalyst temperature at start up.

Ambient conditions (e.g. temperature, humidity) are also factors contributing to H_2 in vehicle exhaust. For example, cold-start emissions increase at lower ambient temperatures (Weilenmann *et al.*, 2009). However, specific results involving variable ambient conditions are not included here.

For vehicles that operate under typically fuel-rich conditions, the presence or absence and the state (fresh or aged) of the TWC dictates, to a large extent, the timing and magnitude of H_2 emissions in vehicle exhaust. An example illustrating the predictability of H_2 concentrations in raw gasoline exhaust is presented in Figure 4.3. D'Alleva and Lovell (1936), in their early twentieth-century study of the relationship of exhaust gas composition to the air/fuel ratio, revealed H_2 concentrations in raw exhaust that decreased in a slight non-linear fashion as the air/fuel ratio increased. In a recent study involving pre- and post-catalytic converter measurements of gasoline exhaust at various engine operating points, a similar relationship was confirmed (Vollmer *et al.*, 2010). The consistency in the results from both studies illustrates the predictability of the combustion principle and the dependence of H_2 emissions on the air/fuel ratio under fuel-rich conditions when not subject to the complexities of a TWC.

The state at which the engine is operated (engine speed and torque) is an important factor contributing to H_2 emissions in vehicle exhaust. This observation is depicted in Figure 4.4a for a gasoline Euro-4 automobile over the course of the Common

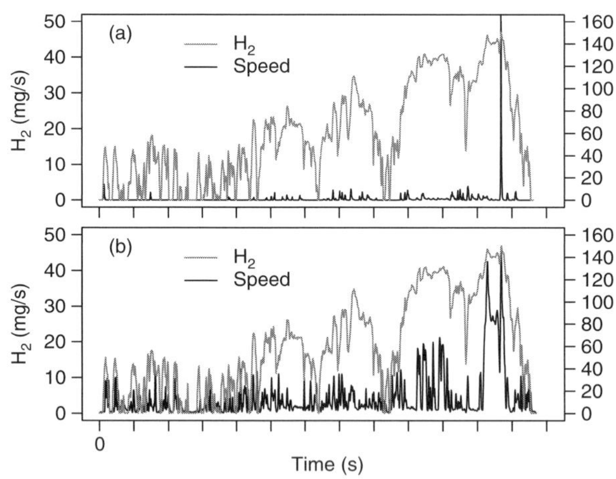

Figure 4.4 H_2 emissions from a gasoline Euro-4 automobile (a) and a gasoline Euro-3 motorcycle (b). The high emissions for the automobile at the end of the cycle are due to the engine control unit switching to a fuel-rich (low lambda) regime under high engine speed and load conditions. Emissions from the automobile are much lower throughout the cycle due to the more sophisticated lambda control system compared with the motorcycle, which displays consistently higher emissions under the same driving conditions, irrespective of the engine state.

Artemis Driving Cycle (CADC). The greatest emissions occur under conditions of high engine speed and torque. These high rpm (revolutions per minute) and load situations occur at greater speeds and during periods of acceleration, or when a vehicle requires significant power to climb a hill, for example. Fuel-rich conditions also equate to more power. Thus, although high H_2 emissions are typically most prominent as speeds approach the upper end of the speed range for a highway-based chassis dynamometer sub-cycle, high H_2 emissions would also be observed if slope profiles were likewise incorporated into the driving cycles. In order to prevent the engine from overheating under such conditions, some ECUs are set to convert to a fuel-rich regime as the excess evaporated fuel cools the engine and TWC.

However, emissions from current ICEVs are not simply a result of the engine state, but also of the quality of lambda (λ) control. H_2 emissions are elevated in cycles with higher proportions of fuel-rich phases, and fuel-rich phases are more frequent with less sophisticated λ control systems. This is evident in Figure 4.4a with emissions only markedly elevated as the automobile accelerates at already high vehicle speed. Hydrogen emissions from two-wheelers tend to be substantially greater than from gasoline-powered automobiles, largely due to the less developed ECUs for motorcycles, resulting in less precise λ control and higher emissions, despite identical driving conditions. Figure 4.4b illustrates this phenomenon for a Euro-3 motorcycle.

Vehicle and fuel type are further parameters that largely dictate the magnitude of H_2 in ICEV emissions. Figure 4.5 depicts H_2 emissions from a motorcycle (A), a natural gas automobile (B), a gasoline automobile (C), and a diesel automobile (D). Emissions are depicted as cumulative emissions in milligrams (mg) over the CADC driving cycle. The highest H_2 emissions from vehicles tested in this study

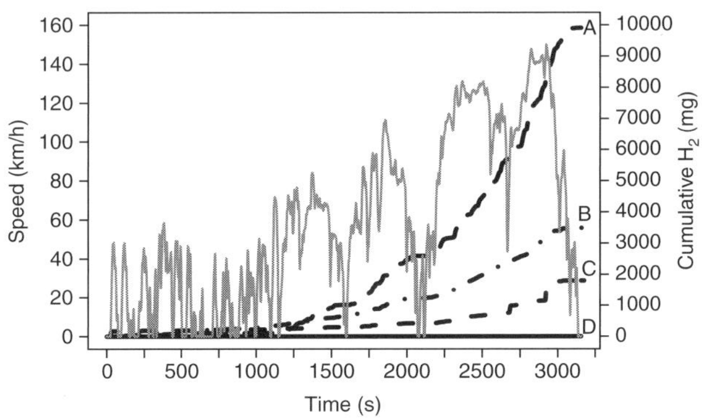

Figure 4.5 Cumulative H_2 emissions for a motorcycle (A), natural gas automobile (B), gasoline automobile (C), and diesel automobile (D). Emissions (in mg) are depicted by the variable black lines. The speed profile for the CADC driving cycle is depicted by the solid line.

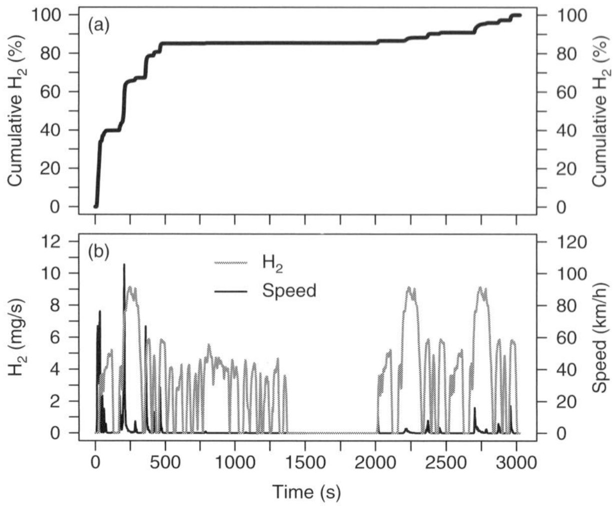

Figure 4.6 Importance of a cold start for a gasoline Euro-4 automobile. (a) Approximately 40% of total emissions are due to the cold start, with an additional 25% of the total occurring during the first major acceleration event. (b) H_2 emissions (mg/s) and the speed profile of the United States Federal Test Procedure FTP-75 are depicted by the black and grey lines, respectively.

resulted from two-wheelers, followed by natural gas vehicles, gasoline automobiles, and diesel vehicles[5]. H_2 emissions from gasoline-powered vehicles are much higher than those from diesel-powered vehicles, and in many instances H_2 is not detectable at all over the course of a typical driving cycle conducted by a diesel vehicle. This is expected due to the constant oxidising combustion characteristic of diesel vehicles. H_2 emissions from natural gas vehicles are generally higher than for gasoline-based vehicles because the hydrogen-carbon ratio in natural gas is approximately double that of gasoline. Furthermore, emissions from natural gas vehicles also have higher water vapour content, which facilitates a greater degree of water-gas shift activity.

The temperature of the engine and TWC are further important parameters dictating emissions of H_2 from a vehicle. Figure 4.6 illustrates the importance of cold-start emissions from a gasoline-powered Euro-4 automobile. For a ride of 23.5 km in this example, approximately 40% of total H_2 emissions occur at ignition of the cold engine, with an additional 25% of total emissions resulting from the first major acceleration event (Figure 4.6a). The rate of emissions (mg/s) and

[5] The purpose of the cumulative emissions depicted in Figure 4.5 is to qualitatively illustrate how vehicle and fuel type may affect H_2 emissions, and not to make any quantitative statements regarding the average of all vehicles powered by such fuel types (apart from the diesel vehicle, D). However, the total cumulative emissions from the vehicles depicted in Figure 4.5 do reflect the hierarchy of mean H_2 emissions factors of the measured vehicles (Bond *et al.*, 2010).

the speed profile of the United States Federal Test Procedure FTP-75 are depicted by the black and grey lines in Figure 4.6b, respectively. Emissions are significantly higher during the cold start (Weilenmann *et al.*, 2005). When prolonged, fuel-rich conditions do not dominate the cumulative emissions (e.g. during urban or rural driving conditions), cold-start emissions generally contribute the greatest cumulative H_2 emissions. An initially cold catalyst needs to attain a minimum temperature before reaching an operationally high conversion rate. The light-off temperature is often used to describe the temperature at which the catalyst becomes more than 50% effective with respect to CO oxidation (Heywood, 1988). This temperature is above 200°C (Favez *et al.*, 2009). In addition, most gasoline cars are run with a rich air/fuel ratio at start up because this guarantees the smooth running of the cold engine where gasoline condenses on the cold walls. The ECU usually then converts to $\lambda = 1$ at the catalyst light-off temperature. Cold-start emissions generally affect all vehicle and fuel types measured in this study, although their relative magnitude in terms of total emissions is different for each vehicle, fuel type, and cycle driven. An advantage of natural gas vehicles over conventional gasoline-powered vehicles is the reduced cold-start and low temperature emissions (Thurnheer *et al.*, 2009).

4.4.2 Hydrogen emission factors and global scaling of transportation emissions

In this section, E_{H2} from the vehicle technologies relevant to this study are used to upscale H_2 emissions and project global emission scenarios from road-based transportation. The ongoing tightening of regulations has resulted in the systematic improvements of engines, TWCs, and engine controls to reduce the number and length of fuel-rich events and, hence, emissions of CO and hydrocarbons. In parallel, H_2 emissions have also decreased in newer generations of cars. This is apparent in the E_{H2} found by Heeb *et al.* (2006) and Vollmer *et al.* (2007) – studies involving slightly older gasoline vehicle technologies – compared with the lower E_{H2} of gasoline Euro-4 automobiles (Bond *et al.*, 2010). Table 4.1 provides an overview of all the E_{H2} for vehicles considered in this chapter.

The emission factors listed in Table 4.1 were applied in combination with vehicle kilometres driven by technology and world region (see Table 4.1 caption) in order to globally upscale H_2 emissions from road-based transportation (Bond *et al.*, 2011a). Table 4.2 lists results of the global projections (in Tg/year). Current conditions represent H_2 emissions based on global kilometres travelled and vehicle fleet composition according to the baseline scenario (i.e. no effective climate policy implementation) since it was assumed that the efforts to mitigate climate change to date have had an inconsequential effect on a global scale. For 2020, 2050, and 2100, results have been applied for both the baseline scenario and the 450 ppm

Table 4.2. *Global projections of H_2 emissions (Tg/year) from road-based transportation under both the baseline and the 450 ppm CO_2 climate policy scenarios.*

Year	Baseline scenario	450 ppm CO_2 climate scenario
2010	4.51	–
2020	2.33	2.33
2050	1.08	1.05[a]
2100	1.20	0.77[a]

[a] Assuming no net H_2 emissions from fuel cell vehicles.

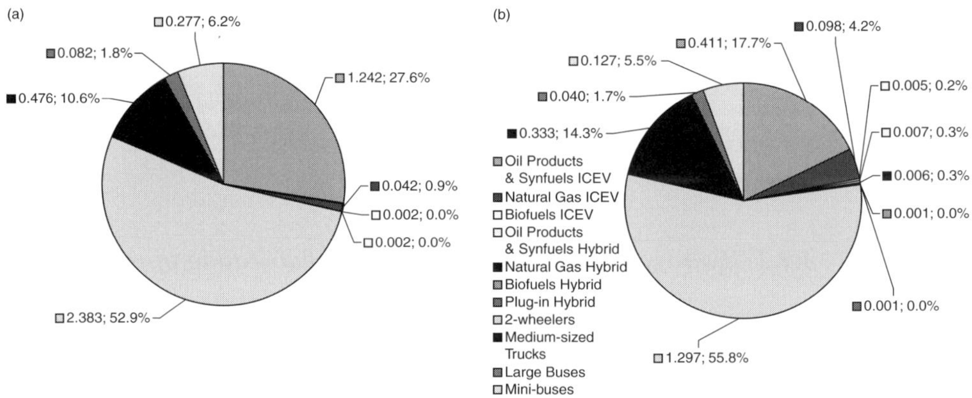

Figure 4.7 H_2 emissions from each vehicle technology in 2010 (a) and 2020 (b) separated by absolute emissions (in Tg) and percentage of total emissions (see also colour plate 10).

CO_2 climate policy scenario. For the breakdown of global LDV kilometres for the baseline and 450 ppm CO_2 climate policy scenarios, see Chapter 6, Figures 6.8 and 6.9, respectively.

Global H_2 emissions from transportation are projected at 4.51 Tg for 2010. Emissions are dominated by two-wheelers (~53%), along with conventional (predominantly gasoline) ICEVs (~27.5%), and the 'other' category, consisting of medium-sized trucks, large buses, and mini-buses (~18.5%), the emissions from which are also dominated by the gasoline-powered vehicles. Natural gas ICEVs account for ~1% of total emissions (and ~3% of LDV emissions), while biofuels and gasoline hybrid-electric vehicles (HEVs) each make up <0.1%. Figure 4.7 depicts the H_2 emissions from each of the contributing vehicle technologies in 2010 (a) and 2020 under the 450 ppm CO_2 climate policy scenario (b), respectively. Figure 4.8 separates the vehicle technologies further into emissions by world region for each technology.

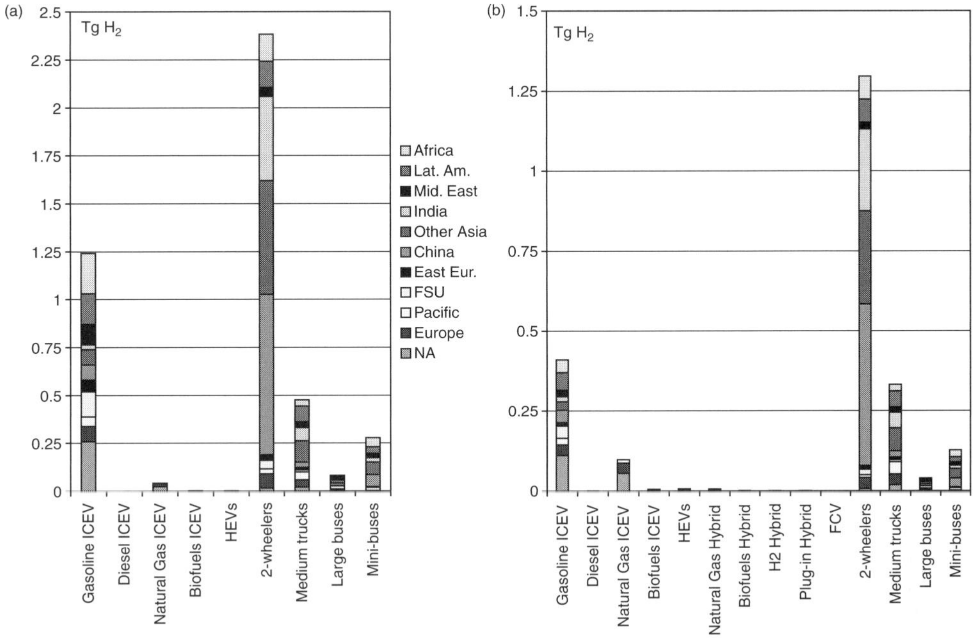

Figure 4.8 H_2 emissions from each vehicle technology in 2010 (a) and 2020 (b) separated by the absolute contribution (in Tg) from each world region (note the difference in scales) (see also colour plate 11).

Global H_2 emissions projections for 2020 result in 2.33 Tg under both scenarios due to very similar vehicle fleet compositions. These projections are approximately half of that projected for 2010, primarily due to tightening emission standards and corresponding improvements in exhaust after-treatment systems, despite the greater number of global vehicle kilometres travelled (VKT). The rate of decrease from 2010–2020 (~6.5%/year) is broadly in line with that projected by Vollmer *et al.* (2007) for the period 2004–2015, based on the measured H_2/CO ratio (7.3%/year). In 2020, the same general picture is apparent with two-wheelers (~56%), conventional ICEVs (~18%), and the group of other vehicles (~21.5%) making up the vast majority of total H_2 emissions. Natural gas ICEVs account for ~4% of total H_2 emissions from transportation (18.5% of LDV emissions), while the group with biofuels vehicles, gasoline HEVs, natural gas hybrids, biofuel hybrids, and plug-in hybrids accounts for a combined 0.5% of the total.

In 2050, emissions are expected to be the same within each vehicle technology class, irrespective of world region (see Table 4.1). FCVs account for 8.3% of the global LDV vehicle kilometres under the 450 ppm climate policy; FCVs do not penetrate the global vehicle fleet at all under the baseline scenario. An E_{H2} of ~43 mg/km has been reported for a cryogenic liquid H_2 ICEV (Wallner *et al.*, 2008), but to our knowledge, published E_{H2} data for FCVs do not exist. Here, we have selected two criteria for FCV emissions when assessing emissions from transportation in 2050 and 2100: (1) no net H_2 emissions from FCVs; and (2) FCV E_{H2} that would maintain overall H_2 emissions from transportation within the realm of current and near-term estimates. The reason for the selection of these FCV E_{H2} is: (1) to illustrate overall road-based transport emissions based on optimal fuel cell efficiency and complete recycling of purged H_2, for example, through the development of catalytic burners in the fuel cell exhaust systems, and (2) to illustrate upper-end boundaries using current emissions as an acceptable guide, as it is currently unknown to what extent an increasing anthropogenic H_2 source would be balanced by increasing removal of atmospheric H_2 through soil oxidation, for example. As a third independent criterion in 2100, we have selected a small set of FCV E_{H2} (265 mg/km, 200 mg/km, and 70 mg/km) to provide a quantitative picture of overall transport-based H_2 emissions based on various mean FCV E_{H2}. These FCV E_{H2} have not been entirely arbitrarily chosen; the selected highest E_{H2} (265 mg/km) is based on what we consider to be a high-end FCV E_{H2}, which is in accordance with our knowledge of emissions from prototype FCVs. A reduction in emissions of 50% for production FCVs by 2020 is conceivable (C. Bach, personal communication, 2010). For an overview of emissions scenarios where set low- and high-end FCV E_{H2} have been selected, see Bond *et al.* (2011a).

Discounting the H_2 contribution from FCVs for now, baseline conditions result in 1.08 Tg H_2 from transportation in 2050, while 450 ppm CO_2 policy conditions result in 1.05 Tg H_2. Results for the 450 ppm CO_2 policy scenario can therefore be interpreted as road-based emissions if FCVs were to achieve no net H_2 emissions. Reasons for the perpetual decrease in H_2 emissions are, again, emissions regulations stringency and improvements in exhaust after-treatment systems, particularly now in the world's less developed regions, where a previous vehicle technology time lag existed, and where the highest emissions are typically found in the near term. By 2050, the shift from conventional gasoline vehicles to emerging technologies such as biofuels and hybrid vehicles is marked, and this development plays an important role in the decrease in global H_2 emissions. The introduction of the Euro-4 and Euro-5 emission standards for two-wheelers in the coming decade should reduce two-wheeler emissions significantly, conceivably to levels of current technology automobiles (ACEM, 2010). This is a primary reason why H_2 emissions are expected to decrease considerably throughout the twenty-first century, despite perpetually increasing kilometres travelled globally by two-wheelers.

Although FCVs contribute less than 10% of global vehicle kilometres in 2050, it is illustrative to consider how emissions from FCVs may relate to current and near-term emissions scenarios. Based on global vehicle kilometres under the 450 ppm CO_2 climate policy scenario and vehicle E_{H2} projected for 2050, we estimate that a range of E_{H2} for FCVs of 600–1625 mg/km would result in similar global H_2 emissions from transportation as conditions from 2010–2020 (2.33–4.51 Tg). In other words, barring a major breakdown in the development of fuel cell efficiency during the first half of this century, emissions from transportation in 2050 are expected to remain well below those currently observed. To put this last statement into a quantitative perspective, a FCV E_{H2} of ~16 mg/km under the 450 ppm CO_2 climate policy scenario in 2050 would result in similar levels of H_2 from transportation that is estimated under the baseline scenario (~1.08 Tg).

By 2100, FCVs are expected to make up an estimated 80% of global LDV VKT under the 450 ppm CO_2 climate policy scenario (see Chapter 6); again, FCVs are not expected to penetrate the global vehicle fleet under the baseline scenario. H_2 emissions from transportation under the baseline scenario are estimated at ~1.20 Tg. Assuming no net H_2 emissions from FCVs under the 450 ppm target scenario, road-based H_2 emissions are estimated at ~0.77 Tg. Two-wheelers still represent approximately 33% of total road-based H_2 emissions in 2100 under baseline conditions, but make up only ~16% of two-wheeler emissions from 2010 in absolute terms. Under the 450 ppm CO_2 policy scenario, an E_{H2} range of 55–135 mg/km from FCVs is estimated to result in H_2 emissions from transportation remaining within the realm of current and near-term (2020) global H_2 emissions from transportation (2.33–4.51 Tg). It is clear that the E_{H2} for FCVs would have to be significantly lower in 2100 than in 2050 due to the sheer number of vehicle kilometres driven by FCVs. In 2050 (under the 450 ppm policy scenario), FCVs will travel approximately 2.1 trillion kilometres globally, while in 2100, FCVs will travel approximately 27.5 trillion kilometres.

As a comparison for global H_2 emissions from transportation in 2100, hypothetical E_{H2} for FCVs of 265, 200, and 70 mg/km would result in total H_2 emissions from road-based transportation of approximately 8.1, 6.3, and 2.7 Tg – values 1.8, 1.4, and 0.6 times 2010 levels, respectively. Figure 4.9 portrays this comparison graphically. The first column represents 2010 baseline conditions. The four additional columns (2100a, b, c, and d) represent 2100 projections for H_2 emissions from road-based transportation based on the 450 ppm CO_2 policy scenario for various FCV E_{H2}. The first two columns (2100a and b) represent total road-based emissions considering FCV E_{H2} of 270 and 200 mg/km, respectively. The third column (2100c) represents total transportation-based emissions with a mean FCV E_{H2} of 135 mg/km. The comparison with emissions in 2010 shows that this FCV E_{H2} in 2100 will maintain total transportation-based H_2 emissions similar to current levels, despite increasing global kilometres travelled throughout the twenty-first century. The final column (2100d) represents total transportation-based emissions according to an FCV E_{H2} of 70 mg/km. In this case, total emissions are considerably lower than 2010 emissions. In all cases, it is apparent that H_2 emissions from

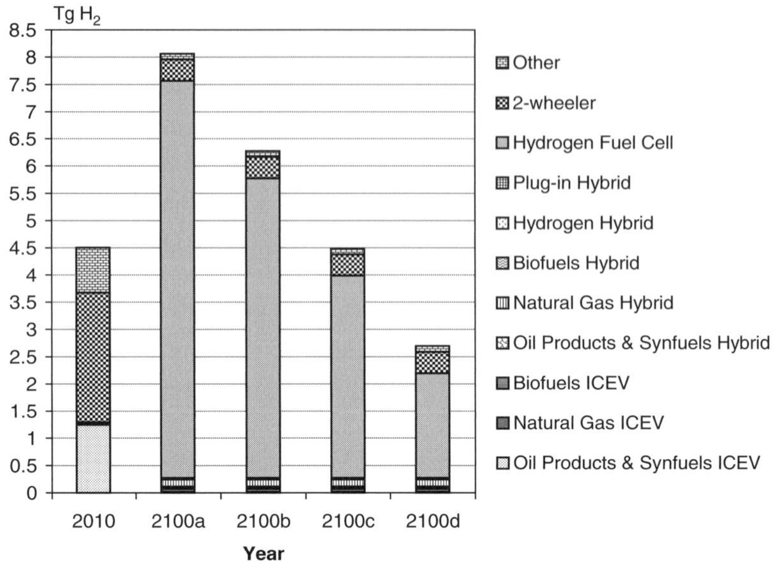

Figure 4.9 H_2 emissions in 2010 and 2100 separated by the absolute contribution from each vehicle technology (in Tg) and illustrating scenarios based on variable H_2 emission factors from fuel cell vehicles. The first two columns for 2100 (a and b) represent total road-based transportation H_2 emissions based on an FCV E_{H2} of 265 and 200 mg/km, respectively. Column 2100c represents total H_2 emissions based on an FCV E_{H2} of 135 mg/km. With this FCV E_{H2} in 2100, total road-based H_2 emissions would be similar to total emissions in 2010. Column 2100d represents total H_2 emissions based on an FCV E_{H2} of 70 mg/km.

FCVs in 2100 make up a considerable fraction of total emissions because FCVs account for about 80% of global LDV kilometres under the 450 ppm CO_2 climate policy scenario.

In summary, H_2 emissions from road-based transportation in 2010 and 2020 are dominated primarily by two-wheelers, gasoline-powered ICE LDVs, and a group of 'other' gasoline-powered vehicles consisting of medium-sized trucks, large buses, and mini-buses. The primary reason for the drastic decrease in emissions between 2010 and 2050 is the cleanup of emissions from two-wheelers, and to a lesser extent, those from gasoline-powered ICEVs and 'other' category of vehicles. Thus, transport-based H_2 emissions in 2050 are expected to be significantly lower than in both 2010 and 2020. As an example, the E_{H2} of 265 mg/km used as a worst case scenario in 2100 would result in total H_2 emissions from transportation of 1.6 Tg in 2050, approximately one-third of 2010 transport-based emissions estimates. In 2100, FCVs will have far more influence on transport-based H_2 emissions than in 2050 due to the 13-fold increase in global FCV kilometres. Thus, smaller ranges of FCV E_{H2} will have greater effects on total emissions.

It has been demonstrated that fuel cells have a variable purge requirement in order to prevent the accumulation of inert gases, water, and impurities such as CO and CO_2 in the anode channels (Ahluwalia and Wang, 2008; Owejan *et al.*, 2009a, b). Purging at regular intervals occurs in order to maintain efficiency and stable cell performance (Yu *et al.*, 2009), and since H_2 is delivered via the anode channel, H_2 is simultaneously purged during these events. An important consideration might therefore be how much effort will be invested in optimising the trade-off between fuel cell efficiency and loss of H_2 to the atmosphere due to the purge requirement of fuel cells. Such an optimum has been demonstrated by Ahluwalia and Wang (2007), which could well ensure that future H_2 losses to the atmosphere from marketed FCVs are minimised. As noted by Owejan *et al.*, 2009a, b), water management is one of the key engineering challenges facing the commercialisation of proton-exchange membrane fuel cells, suggesting that such areas are already being contemplated. Further, the development of catalytic burners within the fuel cell exhaust systems is another initiative that would both reduce emissions of H_2 to the atmosphere and improve fuel economy for FCVs.

4.4.3 Current and future hydrogen emissions from hydrogen production, distribution, storage, and non-transportation end-use systems

An important category of anthropogenic H_2 emissions is that of leakage or other fugitive emissions from production, distribution, storage, and non-transportation-based end-uses. This category includes all direct energy and industrial end-use sectors (Bond *et al.*, 2011a), except transportation. In 2010, global production of H_2 for all end-uses has been estimated at 47.0 Tg from Bond *et al.* (2011a). Currently, H_2 production for industrial end-uses vastly outweighs that for direct energy-based end-uses. In the United States, for example, ~99% of H_2 production is used in the chemical and petrochemical industries, primarily the petroleum-refining and NH_3 fertiliser production industries (EIA, 2008b). Although H_2 production in these areas is projected to increase in the coming decades, due largely to increasing gasoline and diesel consumption, the rate at which H_2 production increases is expected to slow as global oil supply declines and more advanced alternative energy technologies continue to gain market share (see Chapter 6).

Peak oil production is likely to occur in the first half of the twenty-first century (Aleklett *et al.*, 2010). However, global H_2 production in the refining industry will likely continue to increase even after the peak in oil supply, due to the decreasing quality petroleum feedstocks (e.g. oil sands), coupled with stricter environmental regulations (S. Schlag, personal communication, 2010). Based on global H_2 production in the petroleum refining industry of 18.6 Tg in 2010 (Bond *et al.*, 2011a),

Table 4.3. *H_2 production scenarios for industrial- and energy-based end-uses for the years 2010, 2020, 2050, and 2100. Figures have been provided for both the baseline and 450 ppm CO_2 climate policy scenarios. Units are in Tg/year.*

	Baseline scenario		450 ppm CO_2 climate policy scenario	
Year	Industrial	Direct energy	Industrial	Direct energy
2010	47.0	<0.1	–	–
2020	69.4	0.1	69.9	3.1
2050	137.5	0.8	132.3	92.6
2100	263.5	5.1	254.9	1436.6

we consider global oil supply figures from Gül (2008) as a guide for projections of global H_2 production in this industry throughout the twenty-first century. In parallel, we apply a continuous 1% annual growth rate in H_2 production in the refining industry to account for the growing requirements of H_2 per unit of refinery throughput. This results in H_2 production of 34.1 and 46.0 Tg under the baseline scenario for 2050 and 2100, respectively, and 28.9 and 37.3 Tg under the 450 ppm CO_2 climate policy scenario in 2050 and 2100, respectively.

For the other industrial end-uses (primarily NH_3 and CH_3OH production), we consider global economic growth rates as a guide for H_2 demand in these industries. These growth rates are based on the B2 scenario of the Intergovernmental Panel on Climate Change's (IPCC) Special Report on Emissions Scenarios (SRES; IPCC, 2000), which is the same report from which long-term economic growth in Chapter 6 is derived (Gül, 2008). For the period 2020–2050, we apply an annual growth rate of 2.8%, which is the same average growth rate projected for the period 1990–2050. For 2050–2100, we apply an annual growth rate of 1.5%, which is considered an average value for the global economic growth projected for the period 1990–2100. This results in H_2 production from the other significant industrial end-uses (i.e. industrial end-uses other than petroleum refining) of 103.4 and 217.6 Tg in 2050 and 2100, respectively. We do not assume production capacity for these end-uses to differ between climate policies.

For overall H_2 production for all industrial end-uses, we first adopt the global H_2 production and consumption projection of 69.4 Tg (baseline) and 69.9 Tg (450 ppm) for 2020 from Bond *et al.* (2011a). Based on the above projections for total industrial H_2 production (i.e. petroleum refining and other industrial end-uses), total H_2 production for industrial end-uses is projected at 137.5 and 263.5 Tg under the baseline scenario in 2050 and 2100, respectively, while production is projected at 132.3 and 254.9 Tg under the 450 ppm CO_2 climate policy scenario in 2050 and 2100, respectively (see Table 4.3).

H_2 production for direct energy-based end-uses (e.g. transportation, commercial and residential heat and power) has been projected with the GMM to 2100 (see Chapter 6); production is expected to increase considerably throughout the twenty-first century – an annual growth rate of 14.5% from 2010–2100 under the 450 ppm CO_2 climate policy (considering baseline production of H_2 for direct energy services in 2010). Table 4.3 provides results of upscaled industrial and direct energy-based H_2 production. Baseline and 450 ppm CO_2 climate scenario projections are included for 2010, 2020, 2050, and 2100. Although industrial H_2 production will still be greater in 2050, direct energy-based H_2 production would be more than five times higher than industrial H_2 production in 2100 under the 450 ppm CO_2 climate policy. Indeed, climate policies influence H_2 production for direct energy services throughout the twenty-first century, and the more stringent the policy with respect to atmospheric CO_2 concentrations, the more H_2 is produced (see Chapter 6).

From Table 4.3, it is apparent that direct energy-based H_2 production remains only a small fraction of total H_2 production through to 2100 under the baseline scenario. Under the 450 ppm CO_2 climate policy scenario, direct energy-based H_2 production in 2020 also makes up only a small fraction of total H_2 production (~4%). However by 2050, the fraction of H_2 produced for direct energy-based applications increases considerably to ~41% of the total, while in 2100, this figure reaches 85% of total H_2 production.

Table 4.4 outlines the total global production capacity of H_2 in 2010, 2020, 2050, and 2100, along with a range of loss rates of H_2 to the atmosphere from production, delivery, storage, and end-use systems.

If we consider a current loss rate range of 1–4% over the entire production, distribution, storage, and end-use chain (with 4% considered to be an upper-end figure), between 0.5 and 1.9 Tg H_2 will be emitted to the atmosphere in 2010 under the baseline scenario. However, Zittel and Altmann (1996) and Schultz *et al.* (2003) have reported that losses of gaseous H_2 from the existing H_2 distribution grid in Germany are in the order of 0.1%, suggesting that future systems should attain lower loss rates. This argument is driven primarily by financial, safety, and environmental concerns. For example, Air Liquide Deutschland GmbH is responsible for the largest hydrogen pipeline network in Germany. According to one of their engineers, the goal for H_2 production, delivery, and storage is to achieve zero losses. In reality, however, a 0.1% leakage rate to the atmosphere is more realistic. A rate of 1% would result in massive annual monetary losses (F. Balzer, personal communication, 2008), an incentive for large industrial gas suppliers to minimise leakage of H_2 to the atmosphere. By 2050, under the 450 ppm CO_2 climate policy, as the fraction of global H_2 production for direct energy services begins to gain a foothold, a loss rate range of 0.1–2% is assumed to be representative over the entire

Table 4.4. *Overall global anthropogenic H_2 emissions to the atmosphere from transportation and H_2 production, delivery, storage, and other end-use systems. Global projections of H_2 losses (Tg) under a range of loss rates (%) have also been included. Total production and losses comprise H_2 production for both industrial- and energy-based end-uses. See Table 4.3 for the breakdown of total production between industrial- and energy-based end-uses. Projections are separated according to the baseline scenario and the 450 ppm CO_2 climate policy scenario for the years 2010, 2020, 2050, and 2100. The first column of the overall anthropogenic emissions for 2050 and 2100 under the 450 ppm CO_2 climate policy scenario represent overall emissions, should no net emissions from FCVs result; the second column when a fuel cell vehicle (FCV) E_{H2} would lead to total emissions from transportation similar to 2010 (~4.5 Tg). These FCV E_{H2} correspond to ~1625 mg/km in 2050 and ~135 mg/km in 2100.*

	Baseline scenario					450 ppm CO_2 climate policy scenario					
Year	Production capacity (Tg)	Loss rate (%)	Total loss (Tg)	Emissions from transport (Tg)	Overall anthropogenic emissions (Tg)	Production capacity (Tg)	Loss rate (%)	Total loss (Tg)	Emissions from transport (Tg)	Overall anthropogenic emissions (Tg)	
2010	47.0	1.00	0.47	4.51	4.98	–	–	–	–	–	
		2.00	0.94		5.45		–	–		–	
		3.00	1.41		5.92		–	–		–	
		4.00	1.88		6.39		–	–		–	
2020	69.5	0.50	0.35	2.33	2.68	73.0	0.50	0.37	2.33	2.70	
		1.00	0.70		3.03		1.00	0.73		3.06	
		2.00	1.39		3.72		2.00	1.46		3.79	
		4.00	2.78		5.11		4.00	2.92		5.25	
2050	138.3	0.10	0.14	1.08	1.22	224.9	0.10	0.22	1.05	1.27	4.73
		0.20	0.28		1.36		0.20	0.45		1.50	4.96
		0.50	0.69		1.77		0.50	1.12		2.17	5.63
		2.00	2.77		3.85		2.00	4.50		5.55	9.01
2100	268.6	0.01	0.03	1.20	1.23	1691.5	0.01	0.17	0.77	0.94	4.68
		0.05	0.13		1.33		0.05	0.85		1.62	5.36
		0.10	0.27		1.47		0.10	1.69		2.46	6.20
		0.50	1.34		2.54		0.50	8.46		9.23	12.97

production, distribution, storage, and end-use phases (excluding transportation) of the H_2 life cycle. Under such conditions, losses of H_2 to the atmosphere would be in the order of 0.2–4.5 Tg. Direct energy-based H_2 applications would be responsible for ~41% of these losses.

By 2100, approximately 1400 Tg H_2 will be produced for direct energy services, equivalent to ~170 EJ and 10% of primary energy consumption (total primary energy consumption of ~1690 EJ) under the 450 ppm CO_2 climate scenario. With H_2 providing such a significant fraction of primary energy consumption, fully mature systems should be advanced enough to attain a maximum 0.5% loss rate of H_2 due to safety, monetary, and ecological concerns. Leakage rates of 0.5% over the entire chain would result in emissions to the atmosphere in the order of approximately 8.5 Tg/year, with direct energy applications accounting for 7.2 Tg. Most hydrogen applications are fuelled by gaseous H_2, however, from which losses to the atmosphere are significantly less than those from liquid H_2 (Lovins, 2003). If we consider a loss rate of 0.1% over the entire production and end-use chain with the annual H_2 production of 1690 Tg/year for all end-uses, an annual leakage of 1.7 Tg H_2 would result, with direct energy-based applications accounting for 1.4 Tg. In other words, by 2100 with a well-to-wheel H_2 loss rate of 0.1%, an economy reliant on H_2 for approximately 10% of primary energy consumption would release a comparable amount of H_2 into the atmosphere as estimations of today's anthropogenic, non-motor vehicle source of 0.5–2.0 Tg (Table 4.4). By 2100, lower loss rates are conceivable, and loss rates of 0.01% and 0.05% have been considered for illustrative purposes. Should such leakage rates be achieved by 2100, well-to-wheel losses (excluding emissions from transportation) would be in the order of 0.17 Tg and 0.85 Tg, respectively, at the lower end of current H_2 emissions from the non-motor vehicle source.

It is apparent in Table 4.4 that, in the majority of cases, total losses remain within the realm of H_2 production from the current non-motor vehicle source. Exceptions are the 4% loss rate in 2020 for both the baseline and 450 ppm CO_2 climate policy scenarios, the 2% loss rate in 2050 for both scenarios, and the 0.5% loss rate for the 450 ppm CO_2 climate scenario in 2100. It should be noted, however, that these are all considered worst case loss rates for these time periods, and that industrial end-uses are also responsible for large fractions of the losses – 95.7–99.9% in 2020, 58.8–99.4% in 2050, and 15.1% of losses under the 450 ppm CO_2 climate policy in 2100.

4.4.4 Current and future overall anthropogenic hydrogen emissions from technological sources

In this section, overall anthropogenic H_2 emissions from technological sources (taken as the sum of road-based transportation emissions (Section 4.4.2) and emissions from hydrogen production, distribution, storage, and other non-transport, end-use systems (Section 4.4.3)) are combined and scaled-up to provide scenarios of overall direct anthropogenic H_2 emissions to the atmosphere under 2010, 2020, 2050, and 2100 conditions. Table 4.4 outlines the results.

In 2010, based on a loss rate range of 1–4%, overall baseline anthropogenic H_2 emissions are estimated at 5.0–6.4 Tg. This range is at the lower end of overall fossil fuel-based anthropogenic H_2 emissions from other studies (Novelli *et al.*, 1999; Hauglustaine and Ehhalt, 2002; Sanderson *et al.*, 2003; Price *et al.*, 2007; Xiao *et al.*, 2007). These lower-end estimations are possibly due in part to slight overestimations in other studies, as many of those estimates are based on scaling atmospheric H_2 to the much more widely known CO inventories, due to the known positive correlation between H_2 and CO in the exhaust gas of motorised vehicles (the most significant source of atmospheric H_2 from fossil fuels). In support of this presumption, it has been shown that some fossil fuel combustion processes emit much smaller concentrations of H_2 relative to CO compared with the exhaust gas of motor vehicles. For example, the flue gas of industrial and residential heating systems has been found to be much higher in CO than H_2 (M.K. Vollmer, personal communication, 2010).

In 2020, with a loss rate of 0.5–2%, emissions would be lower than those estimated for current overall anthropogenic emissions. However, based on an assumed high-end loss rate of 4% under both the baseline and 450 ppm CO_2 climate policy scenarios in 2020, overall anthropogenic emissions would be within the range estimated for 2010.

The combined anthropogenic emissions column for both 2050 and 2100 has been split for the 450ppm CO_2 climate policy scenario. The first column represents combined emissions without accounting for H_2 emissions from FCVs (i.e. no net H_2 emissions from FCVs, see Section 4.4.2). The second column represents combined emissions based on FCVs contributing to overall emissions, such that emissions from transportation remain similar to the 2010 projection of 4.51 Tg. As detailed in Section 4.4.2, these E_{H2} for FCVs correspond to approximately 1625 mg/km and 135 mg/km for 2050 and 2100, respectively.

Considering only the 450 ppm CO_2 climate policy scenario in 2050 – under the baseline scenario, virtually all H_2 production is destined for industrial end-uses – loss rates of 0.1% and 0.2% would result in overall anthropogenic H_2 emissions to the atmosphere of approximately 4.7 and 5.0 Tg, respectively, if emissions from FCVs are limited to ~1625 mg/km. Both of these estimations would be at the lower end of the range of current baseline anthropogenic H_2 emissions estimations. A 0.5% and worst case 2% loss rate at such a production capacity (224.9 Tg) would lead to anthropogenic H_2 emissions of ~5.6 Tg and ~9.0 Tg, respectively, should FCV emissions be held to a level that would lead to H_2 emissions from transportation comparable to today. Under these conditions, overall emissions based on the 2% loss rate would exceed current overall anthropogenic H_2 emissions estimations, whereas the 0.5% loss rate would result in similar overall emissions compared with today. Of course, H_2 emissions from FCVs will almost certainly be substantially lower than 1625 mg/km for reasons detailed at the end of Section 4.4.2, suggesting that overall emissions from anthropogenic processes in 2050 will be largely dictated by leakage and losses and not transportation, and remain significantly lower than current estimates.

In 2100 under the 450 ppm CO_2 policy scenario and based on a leakage and loss rate of 0.5% and an FCV E_{H2} of 135 mg/km, overall anthropogenic H_2 emissions would be roughly a factor of 2–2.5 higher than current emissions. However, based on a loss rate range of 0.01–0.1% and an FCV E_{H2} of 135 mg/km, overall anthropogenic H_2 emissions to the atmosphere are projected to be within the range of 2010 estimates. In other words, under the 450 ppm CO_2 climate policy scenario, and based on a well-to-wheel (excluding transportation) leakage range of 0.01–0.1% in 2100, overall anthropogenic H_2 emissions to the atmosphere should be, at worst, comparable with today's anthropogenic source of H_2 from fossil fuels should E_{H2} for FCVs not exceed ~135 mg/km.

4.5 Summary and conclusions

Anthropogenic H_2 emissions from fossil fuels originate primarily from the exhaust gas of motorised transportation and through losses and leakage from H_2 production, storage, distribution, and end-use systems, with the former currently contributing the largest fraction to the overall source term of the global hydrogen budget. The majority of current H_2 emissions from transportation result from the incomplete combustion of fossil fuels and depend on several factors, including the presence or absence of a TWC, engine speed and load, the quality of lambda control, vehicle and fuel type, engine and catalyst temperature at start up, and ambient conditions.

Current H_2 emissions calculated from road-based transportation (~4.51 Tg) are at the lower end of the figures provided in previous literature (~5–20 Tg). Emissions from road-based transportation are expected to perpetually decrease throughout the twenty-first century as emission standards tighten, exhaust after-treatment systems develop and improve, and alternative vehicle technologies increase their market share in relation to conventional gasoline vehicles. Throughout the first half of the century, two-wheeler emissions are expected to dominate emissions from transportation, contributing ~53% in 2010 and ~56% in 2020 to total H_2 emissions. Other, primarily gasoline-based, technologies account for the bulk of the remaining fractions. Non-conventional vehicles such as natural gas ICEVs, biofuels, and hybrid vehicles contribute to overall H_2 emissions only very marginally in the first half of the century, and although their relative fraction and absolute number of global kilometres travelled increases, emissions from non-conventional vehicles remain small and never approach those of the most significant current technologies (i.e. four-wheeled gasoline-powered ICEVs and two-wheelers).

In the second half of the century, emissions from those technologies important throughout the first half of the century are expected to decrease considerably. In 2050, as FCVs are estimated to account for ~8% of global vehicle kilometres under a climate policy aimed at stabilising atmospheric CO_2 concentrations at 450 ppm, a seemingly immense FCV E_{H2} of ~1625 mg/km is estimated to maintain total H_2 emissions from transportation at similar levels to those in 2010. In 2100, an FCV E_{H2} of ~135 mg/km is estimated to result in total emissions from transportation comparable to road-based transport emissions in 2010. This lower E_{H2} (more than an order of magnitude compared with 2050) is largely due to the 13-fold increase in global FCV kilometres driven between 2050 and 2100. For reference, FCV E_{H2} of 265, 200, and 70 mg/km in 2100 are estimated to result in H_2 emissions approximately 1.8, 1.4, and 0.6 times 2010 emissions from transportation, respectively. Should H_2 emissions from FCVs remain below 135 mg/km throughout the century, total emissions from road-based transportation are projected to remain lower or comparable to today's estimated road-based transport H_2 emissions. Further research into the direct H_2 emissions from FCVs would be beneficial, particularly into optimising the trade-off of fuel cell efficiency and purge losses of H_2.

Losses and leakage of H_2 to the atmosphere from production, storage, distribution, and end-use systems other than transportation should be driven throughout the coming century primarily by safety, ecological, and monetary concerns. Particularly as H_2 production for direct energy-based end-uses begins to increase, efforts invested in controlling such losses should intensify. Through 2020, combined anthropogenic H_2 emissions from transportation and well-to-wheel leakage based on a 0.5–4% leakage rate should remain within the realm of current overall emissions. Virtually all non-transport-based emissions would originate from industrial end-uses (primarily petroleum refining and ammonia production).

Based on a leakage rate range of 0.1–2% in 2050, overall H_2 emissions should remain either lower or within the range projected for 2010 conditions. This is even the case for all loss rates where an FCV E_{H2} of 1625 mg/km has been applied. An exception is the 2% loss rate under the 450 ppm CO_2 climate policy scenario at this high FCV E_{H2}. In this case, overall anthropogenic emissions would be a factor of 1.4 to 1.8 higher than current overall emissions. However, considering the fact that average FCV emissions will almost certainly be substantially lower that 1625 mg/km, overall emissions in 2050 should remain lower or within the realm of 2010 estimates. Irrespective of transport-based emissions, emissions from direct energy-based applications would only be responsible for ~41% of non-transport emissions in 2050.

Considering a perhaps more representative leakage rate range of 0.01–0.1% from fully mature H_2 systems and a likely more realistic FCV E_{H2} of 135 mg/km in 2100 under the 450 ppm CO_2 scenario, overall emissions would remain within the range of estimates for 2010. With a leakage rate of 0.5% in 2100, overall H_2 emissions to the atmosphere would be a factor of 1.4 to 1.9 higher even if FCVs were to emit no H_2. In 2100, under the predicted scenarios, direct energy-based end-uses would account for ~85% of non-transport-related emissions. The keys here are minimising systems leakage and controlling emissions from FCVs.

In summary, climate objectives and vehicle fleet composition do not dictate to a large extent the overall emissions of H_2 to the atmosphere in the first half of the twenty-first century. Rather, global levels of anthropogenic H_2 in the atmosphere appear to be governed primarily by two-wheelers (at least until 2020), along with the variable loss rates from production, distribution, storage, and other non-transportation-based end-uses. Assuming an increase in FCVs after 2050, and particularly toward the end of the century, along with efforts invested in controlling leakage and other fugitive well-to-wheel emissions to the atmosphere, the mean E_{H2} for FCVs will largely influence overall anthropogenic H_2 emissions. Negative effects of a future H_2-intensive economy based on H_2 emissions to the atmosphere alone are difficult to envisage if overall emissions remain at similar levels to those currently observed. In a large-scale context, H_2 measurements at remote sites will contribute meaningfully to the monitoring of any fluctuations in the levels of atmospheric H_2.

4.6 Acknowledgements

Thank you to M.K. Vollmer and M. Steinbacher from the Laboratory for Air Pollution/Environmental Technology at Empa for their assistance, along with interesting and useful comments regarding the atmospheric aspects of the chapter. Thank you to C. Bach and P. Soltic from the Laboratory for Internal Combustion Engines at Empa for helpful insights and discussions regarding the technical automotive aspects of the chapter.

5

Regional fleet simulation

SILVIA ULLI-BEER, MATHIAS BOSSHARDT, AND ALEXANDER WOKAUN

5.1 Introduction

Deep cuts in global GHG emissions are required to keep the average increase in global temperature below 2°C (IPPC, 2007c; Sokolov *et al.*, 2009). A strong need for action concerning road traffic as a main originator of GHG emissions through their almost exclusive use of fossil energy carriers is acknowledged (Ribeiro *et al.*, 2007), as discussed in previous chapters. In this chapter, we analyse which technological drivetrain pathways and deployment strategies are required to meet the long-term, global challenge posed by climate change. We implicitly assume that the required technology change must be governed and managed by concerted decision making of entrepreneurial and political leaders. Many policy studies address the mid-term impact of incremental, energy-efficiency improvements of conventional ICE, or hybrid technologies and biofuels (Bandivadekar *et al.*, 2008b; Hankey and Marshall, 2009; Meyer and Wessely, 2009). However, a longer term view on fleet dynamics addressing not-yet-regulated policy measures is missing from the literature. Also, most of the fleet models do not discuss the impact of preference changes on the diffusion process of alternative drivetrain technologies and the effectiveness of corresponding policy measures (Greene *et al.*, 2007; McCollum and Yang, 2009; Thomas, 2009).

In this chapter, we present simulation experiments that illustrate the impact of vehicle technology purchase preferences. The system dynamics fleet model developed and used, therefore, offers a long-term view for the EU, addressing the question of what kind of drivetrain technologies have the potential to meet the scientifically indicated, GHG emission targets. Illustrative diffusion scenarios and the CO_2 impact of competing drivetrain technologies, including advanced ICE, hybrid technologies, and vehicles fuelled with gaseous fuel (i.e. LPG, CNG) as well as near zero-emission vehicles (i.e. renewable HFCV) are presented.

Transition to Hydrogen: Pathways Toward Clean Transportation, ed. E. Wilhelm and A. Wokaun. Published by Cambridge University Press.

According to complex systems theory, the development of mitigation and adaptation strategies should be guided by long-term system viability criteria. These should aim at building adequate adaptive capacities and supplement the cost-minimising implementation of a set of discrete policy or technology choices. Short-term adaptation actions may lead to immediate benefits but they would not necessarily mitigate the main vulnerabilities such as climate change and energy supply issues (see Nelson *et al.*, 2007). Therefore, the identification of game-changing technologies that offer good prospects to meet critical CO_2-emission reduction targets for the light-duty fleet on time is most important.

The system dynamics vehicle substitution model (VSM) provides insight into the dynamically interacting determinants of penetration rates and the competition between alternative drivetrain technologies. This has allowed the effect of policy changes in different scenarios to be investigated to address the critical question: what kind of technologies and fuels are required to meet critical climate policy targets?

The work supports the assumptions that nonlinear preference change processes, long fleet turnover rates, as well as industrial knowledge and capacity adaptation processes, make it necessary to deploy the near-term, fuel-saving potential of advanced ICEs and hybrid vehicles, as well as alternative fuels such as CNG enriched with biogas. Such a technology portfolio prepares a paradigm change within society and acts as a bridge toward clean and energy-efficient personal road transportation, based on radical advanced technologies, such as BEV fuelled by emission-free electricity, and FCV based on renewable hydrogen. Therefore, strongly committed efforts of all stakeholders at different levels within adequate organisational settings and institutions (e.g. public–private partnerships) are decisive. The dynamic decision support model is seen as a most helpful learning tool to enhance goal alignment between stakeholder groups, and test structural hypotheses concerning system behaviour in virtual laboratory experiments (Jackson, 2009). The model would yield its highest benefit as a steadily improved and extended experimental modelling setup, used by decision makers in industry and scientists to test and refine long-term strategies. It complements least cost planning model approaches (see Chapter 6) as well as engineering-based technology assessment models (see Chapters 2 and 3) by explaining how the technology options can be deployed.

5.1.1 Current situation in Europe and USA

The largest share of CO_2 emissions in Europe is emitted by energy plants, about 36% in the EU-15 countries (ACEA, 2007; Eurostat, 2008). However, road transportation is responsible for the second-largest part, 23% results from goods and passenger transportation in the EU-15. The numbers shift to smaller values for road transportation if all EU-25 countries are considered. However, this is mainly due to a smaller

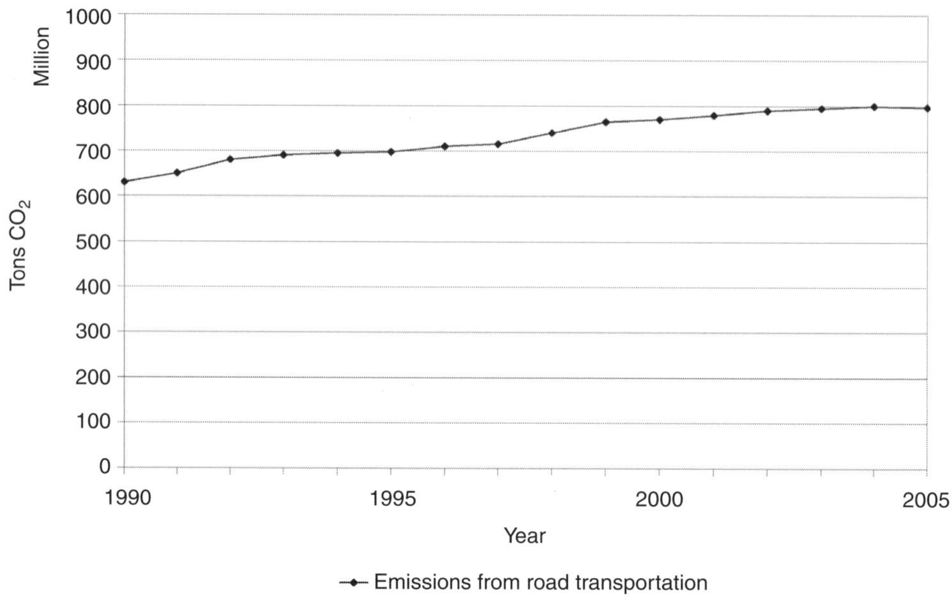

Figure 5.1 EU-15 CO_2 emissions from road transportation.

motorisation of the additional countries. Their relatively high growth potential will probably lead to a similar distribution in the future. The CO_2 emitted by passenger cars only varies for the different countries. In Germany, for example, it is rather high, with some 70% of all emissions coming from road transportation (Auto-Reporter.net, 2007) while the average in the EU-15 countries is smaller, at 50%. This results in an overall emission share of about 12%. Compared to other sectors, road transportation has shown an increasing trend over the last 15 years, which emphasises the need for action in this sector when it comes to emission reduction. Figure 5.1 shows the emission development from 1990 to 2005 for the EU-15 countries.

The new regulation on emission performance standards for new passenger cars (EC No. 443/2009) of the European Union (European Commission, 2009) is part of the EU's effort to reduce CO_2 emissions by 20% by 2020. Beyond this time frame, tougher reduction targets for CO_2 emissions based on the IPCC's recommendations are discussed. For our analysis we refer to a 50% reduction target for the light-duty fleet until 2050 by assumption.

To reach critical climate policy targets, both existing technologies must be improved, and alternative fuel technologies promoted.[1] Tapping the full potential

[1] In this chapter alternative fuel or drivetrain technologies are distinguished from conventional drivetrain technologies, based on internal combustion engines fuelled with petrol or diesel. The possibility of mixing or replacing fossil fuels with their biofuel counterparts is not seen as an alternative propulsion system in the model presented in the following chapters.

of existing drivetrain systems is an important step, but the success of fuel efficiency improvement largely depends on the *emphasis on reducing fuel consumption* (ERFC). In the past, about 50% of the benefits achieved were not used to reduce consumption (and, therefore, emissions) but to improve performance (Bandivadekar and Heywood, 2004). Cars became heavier with increased acceleration while reducing consumption only moderately versus the potential. Furthermore, the costs of reducing CO_2 are placed solely upon the automotive industry. Promoting alternative technologies brings a better cost share between consumers and government, as well as fuel and automotive industry (Evans *et al.*, 2007). The introduction of alternative propulsion technologies is also regarded as a necessary step for sustainable development in terms of further reducing fuel use and emissions. Although its effect in the start-up phase of market penetration might still be smaller than that of mere technology improvement in the next one or two decades, it is crucial for a higher, long-term reduction potential. The cost per saved ton of CO_2 might therefore become lower for alternative technologies in the future.

The ongoing improvement of ICEs in the past has already led to a continuous reduction in fuel consumption per driven kilometre and, therefore, also in CO_2 emissions. However, growing vehicle stocks and partly increasing VKT have overridden these benefits in so-called rebound effects (Sorrell *et al.*, 2009). Although many passenger car fleets in Europe seem to experience saturation, there are countries, such as Poland, that still have a strong growth potential due to their relatively low motorisation rate. This growth trend and the limited potential of reducing fuel consumption of mainstream technology indicate that incremental technology improvement may not be sufficient to reach emission reduction targets in the long run. Therefore market introduction and penetration of alternative drivetrain technologies may become crucial.

The situation for the United States is slightly different concerning the achievable improvement potential of the current technology trajectory. Historically, the US passenger car fleet has had a greater emphasis on increasing performance and weight than in the case of the EU fleet. Subsequently, the US has not achieved the same improvement path in fuel consumption as the EU passenger car fleet. In Europe, the diesel share has exceeded 50%, and vehicle size and weight are some two-thirds of US vehicle values.

Hence, the tank-to-wheel emission values of the EU-15 passenger car fleet average have reached a comparatively lower level, with 160 g CO_2 equivalent/km as compared to the US passenger car fleet average, with 255 g CO_2 equivalent/km around 2006 (An *et al.*, 2007). In the MIT report 'On the road in 2035' (Bandivadekar *et al.*, 2008b), the authors identified a potential for 30–50% reduction in US fleet fuel consumption by consistently directing the technological advances toward reducing vehicle fuel consumption by a combination of improved gasoline and diesel engines and transmission as well as gasoline hybrids. However, for meeting long-term climate targets, in the United States as well as Europe, further advanced vehicle technologies, such as battery plug-in electric vehicles, pure battery electric vehicles, and hydrogen-based fuel cell cars, need to play a crucial role.

'On the road in 2035' includes an important reference fleet model developed at MIT. The MIT LDV fleet model was used to simulate representative scenarios on fuel use and GHG emissions until 2035, based on assumed market penetration rates and different values of emphasis on reducing fuel consumption by advanced powertrain technologies such as turbo gasoline engines, low-emission diesel vehicles, gasoline hybrids and plug-in hybrids. The scenario results suggest that, to reduce fuel consumption by a factor of two in the new vehicle sales mix in the United States, major changes would be required. 'To meet the target two-thirds of the efficiency improvements must be used to reduce fuel consumption rather than emphasising performance, alongside more than 20% vehicle weight reduction, and an 80% market share of advanced powertrains' (Bandivadekar *et al.*, 2008b: ES-7). The MIT research team also investigated some scenarios for selected EU countries with their fleet model directing attention to the future ratio of diesel to gasoline. Their results suggest that gasoline hybrid vehicles could have a stronger impact on reducing fuel consumption than enforcing the penetration of diesel vehicles. They point out that reducing fleet fuel use by more than 10% up to 2035 may be a major challenge. Also, they expect that the proposed 2012/15 CO_2/km GHG emission targets (130 g/km and 120 g/km with biofuels) of the European Commission (2007, 2009) may not be reached on time.

5.1.2 Objectives and scope

The adoption process of one or more new drivetrain technologies throughout society represents an innovation diffusion process (Rogers, 2003). The aim of this work is twofold: first, we introduce an adapted methodology of describing and explaining the diffusion and substitution rates of new drivetrain technologies and the impact of a critical mass with a simulation model; second, we investigate the effectiveness of different policy packages. This work is relevant for stakeholders who are interested in the diffusion potential of new drivetrain technologies, for instance, car manufacturers, fuel suppliers, or governmental authorities at different levels. Learning about the critical quantities or significant leverage points in the market of alternative cars is fundamentally important for policy implementation, in order to overcome the current lock-in situation of the internal combustion regime.

The promotion of market introduction of alternative drivetrain technologies raises important questions, such as what sort of policy packages are the most effective and can help to reach policy targets, as well as what are the main barriers to the successful market penetration of alternative vehicles. The research questions are formulated in a broader way to leave room to interpret the results in different country-specific contexts:

1. When do stable states and self-sustaining diffusion processes emerge in the system?
2. What are important levers in the system?
3. What policies help reach emission reduction targets; and on what time scale?

It is difficult to anticipate scale and timing of emerging technologies (Bandivadekar *et al.*, 2008b). Estimation of market penetration rates without accounting for any feedback processes in the system provides a preview on the development of future fuel use and resulting CO_2 emissions under certain conditions. Dynamic technology diffusion modelling can additionally contribute to understanding the behaviour generated by fundamental causal relations concerning market penetration of new technologies. This knowledge forms the basis of long-term policy making for sustainable development. The two main challenges for technology diffusion in the case of alternative drivetrain vehicles are the *replacement of existing technologies* and the *'chicken-or-egg' problem of vehicles and infrastructure*.

The first is actually not only about technology diffusion, but also technology replacement. The new alternatives are intended to replace a well-established technology of high performance. The model must therefore address the question of changing consumer preferences, against established behavioural norms and technological paradigms (Ulli-Beer *et al.*, 2010). The role of soft factors, such as the attractiveness of a clean and efficient technology, based on its perception, must also be taken into account. This aspect will be particularly emphasised in this chapter.

The second challenge is the chicken-or-egg problem between infrastructure and vehicles: on the one hand, consumer preferences and the resulting market potential determine the investment in fuelling stations or the development of new vehicle types supplied. On the other hand, consumers see fuel availability as well as a broad spectrum of vehicle types offered as a precondition for buying vehicles. The VSM presented maps the causal structure of such problems and identifies leverages to intervene in the system.

5.1.3 Model purpose and boundaries

The purpose of the VSM is to explain the chronological behaviour of the diffusion rates of competing drivetrain technologies in the passenger car market.[2] Its main function therefore has two components: linking the technologies in the market to

[2] *Passenger car* comprises vehicles for passenger transport with a maximum gross vehicle weight of 3500 kg, corresponding to the European light-duty vehicle class.

describe the competition between them by means of technology substitution, and explaining the relation between attributes, consumer preferences, and diffusion rates using the postulated feedback structure. In other words, the VSM must be able to cope with different drivetrain technologies, each one characterised by several attributes. The easiest way to map this system is a multi-dimensional matrix. In the model it is done using different technology layers, which are referred to as technology platforms, referring to the columns of the matrix, for instance. A platform comprises all aspects of a drivetrain technology deduced from its attributes (Struben, 2004).[3] The causal relations between attributes remain the same for each platform. The competition between platforms is introduced by the consumer decision, determining the market shares of all technologies. The attributes going into the decision function are allocated to three sectors representing the endogenous stakeholders: consumers, fuelling station owners,[4] and retailers.[5] The endogenous stakeholders experience basic conditions such as technology availability, performance, fuel cost, and car purchase price. Their options basically consist of whether to adopt new drivetrain technologies at the moment when each consumer decides whether or not to buy a car of a certain drivetrain technology. Their actions have an influence on the decisions of other stakeholders within their own as well as in other sectors. This defines our first, inner model boundary for the VSM, which is shown in Figure 5.2. As shown in this box diagram, there are important stakeholders outside the first boundary which are not part of the VSM, but play an important role in relation to the whole system. The two stakeholder groups must be distinguished from each other when defining the boundaries of the model. The three different sorts of exogenous policy-making stakeholders in the system determine the basic conditions of the market: car manufacturers, fuel suppliers, and municipal and governmental authorities. In order to simulate realistic behaviour, the exogenous stakeholders build a framework to make different scenarios reasonable and put the model parameter values in a major context. Their influence is considered by exogenous parameters of policy measures such as subsidies, tax reductions, or area access permits, referred to in the following as 'policy variables'.

5.2 The vehicle substitution model

Modelling the future development of individual transport is a challenging task, involving technological, economical, ecological, and social issues. An important

[3] In Chapter 3, the alternative heuristic design approach to vehicle technology assessment is discussed.
[4] The term 'owner' in this context refers to stakeholders, or decision makers, respectively, representing the fuelling infrastructure, although not every fuelling station is operated by its owner but by employed managers.
[5] This term comprises all decision-making stakeholders representing the car import, retail, and service sector.

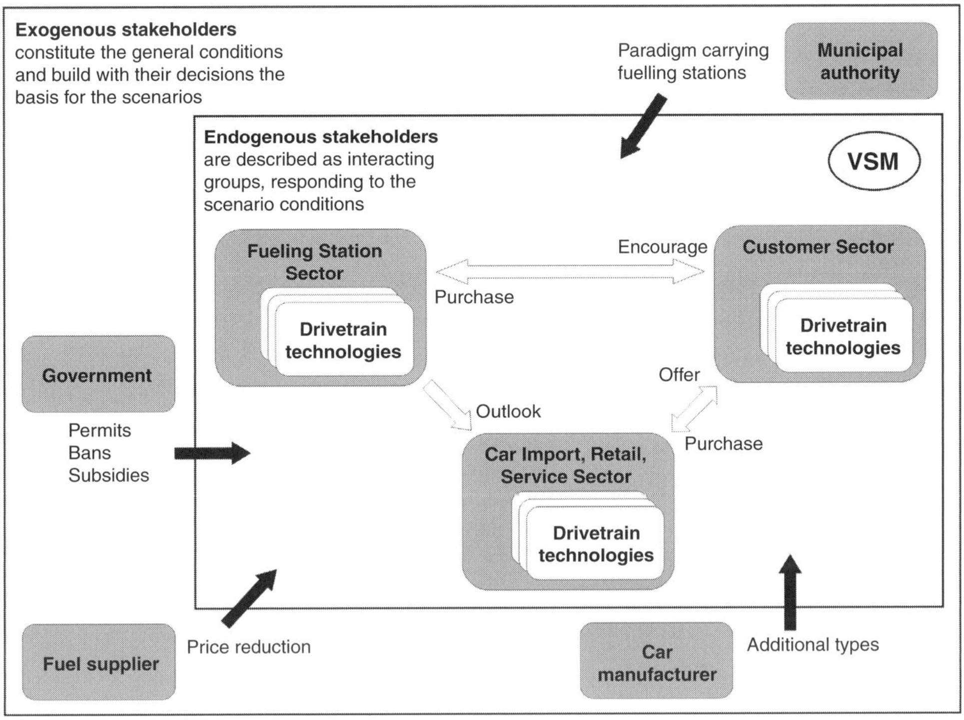

Figure 5.2 Model boundaries for the vehicle substitution model.

part is the market penetration and diffusion of drivetrain technologies, which has been addressed in the past by several models focusing on diverse aspects with different approaches and methods, such as probabilistic models with Weibull (Zachariadis *et al.*, 1995) or logit decision functions (Brownstone *et al.*, 2000; Greene, 2001; de Haan *et al.*, 2009), as well as system dynamics (Schneider *et al.*, 2004; Janssen *et al.*, 2006; Struben and Sterman, 2006). System dynamics, the modelling approach used for this work, is a methodology used to study complex feedback structures developed in the 1950s by Jay W. Forrester, an engineer at the MIT (Forrester, 1969).

5.2.1 Data and variables

There is still a high uncertainty concerning the time frame for alternative vehicles to reach substantial market share. With data from the Swiss Federal Statistical Office, the average age of scrapped passenger cars can be calculated as roughly 12 years for 2004 (BFS, 2004), with an upward trend, following the trend in the United States (Davis and Diegel, 2007). The same result holds for Germany (DVV, 2006), and generally for wider Europe. Using adopter categories, according to the theory

Table 5.1. *Main model variables. Underlined variable names represent stocks, while italic variable names are rates; the others are either multipliers or exogenous-given parameters or lookup functions (see Figure 5.4).*

Endogenous	Exogenous	Excluded
Light-duty vehicles	Median lifetime	Fleet ageing
Sales	New sales	Vehicle classes
Discards	Availability	
Social behavioural norm	Fuel availability satisfaction	
Comparative attractiveness	Spectrum satisfaction	
Adopter potential	Economical attractiveness	
Forecasted car sales	Inherent attractiveness	
Fuelling stations	Purchase price	
Fuelling station construction	Fuel price	
Fuelling station abolishment	Mileage	
Fuelling station load	Fuel economy	
Fuelling station growth	Population	
Fuelling station profitability	GDP	
Types offered		
Additional types		
Desired imports		
Consumer satisfaction		

of Rogers (2003), the time frame for a vehicle fleet turnover turns out to be much longer that one would intuitively expect. Market shares or sales numbers of a technology may be high, while the resulting fleet share is delayed for each group by the length of a vehicle's service life. We have previously shown (Ulli-Beer *et al.*, 2008) that even with a best-case diffusion curve guided by social behavioural norm incentives, it would take 36 years for a new technology to reach 50% fleet share, assuming a vehicle lifetime of 12 years. First empirical data on the diffusion of sport utility vehicles (SUVs) and hybrids give evidence that this is realistic. A further increase in average vehicle service life would extend the time frame even further.

Variables representing quantities of the VSM are subdivided into three groups: endogenous, exogenous, and excluded. Table 5.1 gives an overview of the main model variables (see also Figure 5.4).

The first column includes the dynamically tied up variables which are fully determined by the simulation run. Exogenous variables are given parameters from outside the model, in most cases time-dependent series.[6] Variables that represent

[6] Data acquisition proved to be difficult, since there is no collected edition on all data related to the transportation sector for Europe. The most important source for such information is EUROSTAT, providing free online access to their database. However, the countries' statistical agencies (who provide the data to EUROSTAT) collect their data at different intervals and with varying quality and completeness. This causes gaps or even a complete lack of data series in some countries.

Table 5.2. *Countries under consideration with their fleet and population shares (Eurostat, 2005).*

Country	% pop. (EU25)	Vehicles (million)	% of fleet(EU25)	Share of diesel sales	Unique features
Germany	17.8%	44.6	21.2%	42%	Largest fleet
France	13.6%	30.0	14.3%	69%	Diesel share
Great Britain	13.0%	26.5	12.6%	37%	Equal/high fuel prices
Italy	12.7%	33.6	16.0%	58%	CNG/LPG history
Spain	9.4%	18.8	9.0%	68%	Diesel share
Poland	8.2%	11.0	5.2%	17%	55% w/out car, LPG history
Total	74.7%	164.5	78.3%		
Sweden	1.9%	4.1	2.0%	34%	Ethanol
Switzerland	–	3.8	–	32%	Diesel expensive

aspects not considered by the model are listed as excluded. The main model quantities which are treated as aggregated quantities, represented by physical stocks, are LDVs, fuelling stations, and vehicle types offered. Vehicle sales and discards, fuelling station construction and abolishment, as well as additional types are the corresponding flow rates. All other endogenous variables are causal constructs.

The model has been evaluated using detailed input data for the eight different countries listed in Table 5.2. Germany has been chosen because it is the largest country in the EU-25, on a fleet stock as well as population basis. Therefore it has a strong influence on the overall trends of the EU-25 CO_2 emissions. France is a typical representative of a country where diesel ICEs are currently booming. An increase in the diesel share is observed in many European countries. Great Britain was chosen because of its fuel price characteristics. Diesel prices have approached those of gasoline over recent years and will perhaps become more expensive in the future. Italy is an interesting case with a relatively high installed base of liquid petroleum gas vehicles (LPGVs) and natural gas vehicle (NGVs). Spain has – similar to France – a high diesel sales share. Poland has a high growth potential since more than 50% of all households do not yet own a car. Sweden is an innovative country in terms of energy policy. After the introduction of a CO_2-related annual road tax on passenger cars, the diesel and ethanol share of new registrations increased (Pock, 2007). Finally, Switzerland shows an increasing diesel share despite prices being higher for diesel than gasoline. The first six countries together (excepting Sweden and Switzerland) are considered as representative of the diversity of the fleet characteristics and their overall vehicle share (78%) of the EU-25 as indicated in Table 5.2.

To get a picture of the CO_2 emission trends in the EU-25, all member states were clustered under the seven members mentioned above (excepting Switzerland). The result of the representative country of each cluster was scaled up according to the number of vehicles in the whole group. The results presented give a rough impression of possible trends under a few different scenarios.

Here, an overview on the key assumptions and input data is given. Most parameters used in the model are represented (as far as possible) by country-specific average numbers on an annual basis. When no information was available, vehicle-specific parameters for each drivetrain category are taken from the Volkswagen Golf IV in its different variations as a typical representative (see CONCAWE *et al.*, 2007).

The key assumptions are:[7]

- *Time horizon* The time horizon is set from 1995 to 2100, as about 100 years. Real-world data are used as inputs and to calibrate the model until 2005–8. After this, the discussed data are always simulation data.
- *Inherent attractiveness* The inherent attractiveness plays a major role in model calibration. It represents the promise that a certain technology makes to the society. A value of 0.5, for example, is seen as the 50th percentile of people who would classify a technology as persuasive based solely on its inherent appeal. For values as of 2008, it is assumed to reach a value of 0.2 for alternative vehicles, i.e. for gasoline and diesel HEVs, NGVs, and LPGVs. For Sweden LPGVs are replaced by ethanol flexible fuel vehicles (FFVs). Because of their environmental aspects and upcoming policy incentives in several countries, they are likely to be seen as more attractive than conventional ICE vehicles, the inherent attractiveness of which is set to 0.1.
- *Vehicle purchase price* For the vehicle purchase prices, the following order is assumed: gasoline ICE $<$ diesel ICE $<$ NGV $\leq$ LPG $<$ gasoline HEV $<$ diesel HEV. Bandivadekar *et al.* (2008b) argue that diesel vehicles will increase in price compared to gasoline vehicles because the after-treatment systems needed to meet future emission standards may be costly. Gas vehicles need additional components, but will be cheaper than vehicles with modern hybrid technology. The estimation is also based on German retail prices for the different versions of the VW Golf IV. In countries where data on prices were provided, it was preferred to this estimated price order.
- *Fuel prices* The basic assumption for fuel prices is a rise of 50% compared to 2007 until 2050 and then it is held constant until 2100. This is more pessimistic than the projections by the International Energy Agency (IEA, 2006).

[7] For the implemented parameter values, see Appendix A.

- *Fuel efficiency* The improvements of fuel efficiency reach about 30% for conventional ICEs and gaseous-fuelled vehicles, and about 50% for hybrids in 2045, compared to 2007. They are assumed to stagnate afterward (see IEA, 2006).
- *Vehicle service life* Following the trend in the United States, the vehicle service life in Europe will increase from its current 12 years to 14 years in 2045 (Davis and Diegel, 2007).
- *Vehicle kilometres travelled* The average VKT of diesel vehicles are higher than those of gasoline vehicles. However, they are assumed to level off as diesels increase their fleet share. The overall VKT is assumed to stabilise the 2008 levels.
- *Fleet growth* The net new sales are extrapolated based on the countries' trend over the past 15 years, assuming saturation over the next decades and a beginning decline toward mid-century.

5.2.2 Model structure

The stock and flow diagram presented in Figure 5.3 explains the interaction and the emergence of competition between different drivetrain technologies in the system. It also takes into account that the resulting market shares and vehicle stocks have an influence on technology-specific attributes, which, on their part again, determine the consumer decision, which is directly interpreted as sales or market shares. Figure 5.3 illustrates these relations, which are described later by means of the causal loop diagram in Figure 5.4.

The figure illustrates for two different drivetrain technologies the representative stock and flow structure of the model. The right-hand side of Figure 5.3 corresponds to the standard system dynamics notation, with twin line arrows representing physical flows between stocks. In this case, the stocks can be interpreted as the number of cars possessing one of two different drivetrain technologies, A and B. We assume that the total number of cars is at least constant. The outflow of the stocks is determined by the lifetime of its vehicles. Scrapped vehicles are substituted and do not diminish the fleet; either they are replaced by a car of the same drivetrain technology, leaving the stocks unchanged, or by its competition, resulting in an increase and decrease of the stocks.

The change of the total fleet size is determined by the net change rate, which can be positive (growing fleet) or negative (shrinking fleet). If net change rate is negative, scrapped vehicles are not replaced and the technology-specific fleet decreases. Vehicles that are replaced or that enter the fleet when growing are assigned to the stocks by the calculated market shares. The corresponding links are indicated in Figure 5.3 at the right-hand side by the thin black arrows. The values are based on the consumer decision, given by the adopter potential of each technology. The

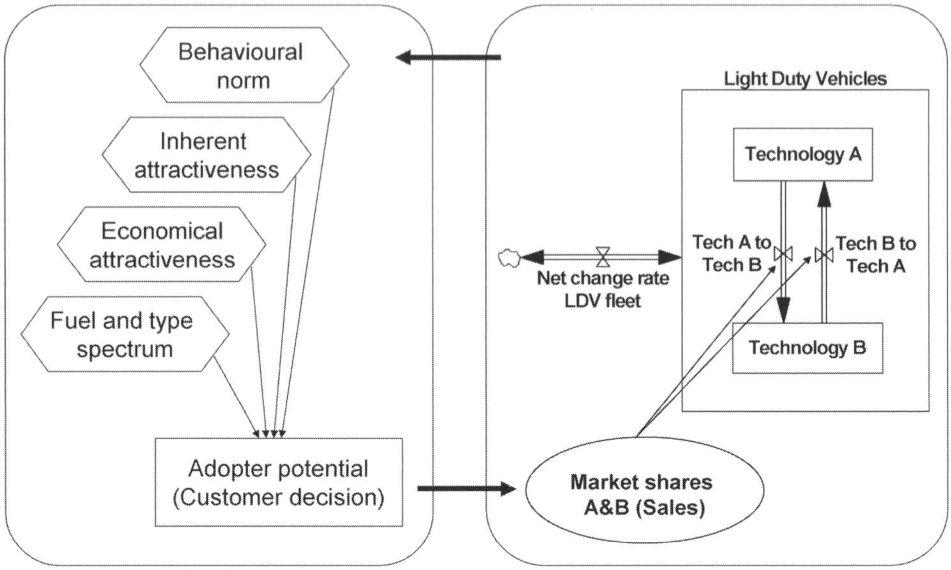

Figure 5.3 Representative stock and flow structure for two different drivetrain technologies.

latter is normalised to 100% as a result of comparing the potential of technologies A and B. The attributes behind the adopter potential are the social behavioural norm, the inherent and economic attractiveness, and the satisfaction concerning fuel availability and type spectrum offered, respectively. The stocks on their part have an influence on the behavioural norm as well as the fuel availability and type spectrum, which is not indicated in this graph. The symmetric body of the stock and flow structure on the right-hand side of Figure 5.3 allows each consumer to replace their car after scrapping with either of the drivetrain technologies considered, A or B in the picture. Options for policy making (such as subsidies or special driving permits) are implicitly manifest in variables such as purchase price or inherent attractiveness, for instance, and therefore influence technology substitution in the model. We assume that the competition of different drivetrain technologies is realised via their adopter potential, correlating them to each other. There is principally no limitation to the number of technologies considered, although there is certainly a limitation by the market itself.

The causal loop diagram (CLD) in Figure 5.4 illustrates the main endogenous dynamic processes guiding the consumers' purchase decisions, or the market shares. The structure presented in Figure 5.4 applies to all technology platforms, equally and simultaneously. Their specific differences are represented by the input parameter values, which are updated in each step of a model simulation. The basic structure is generic and can be applied to different countries and drivetrain technologies.

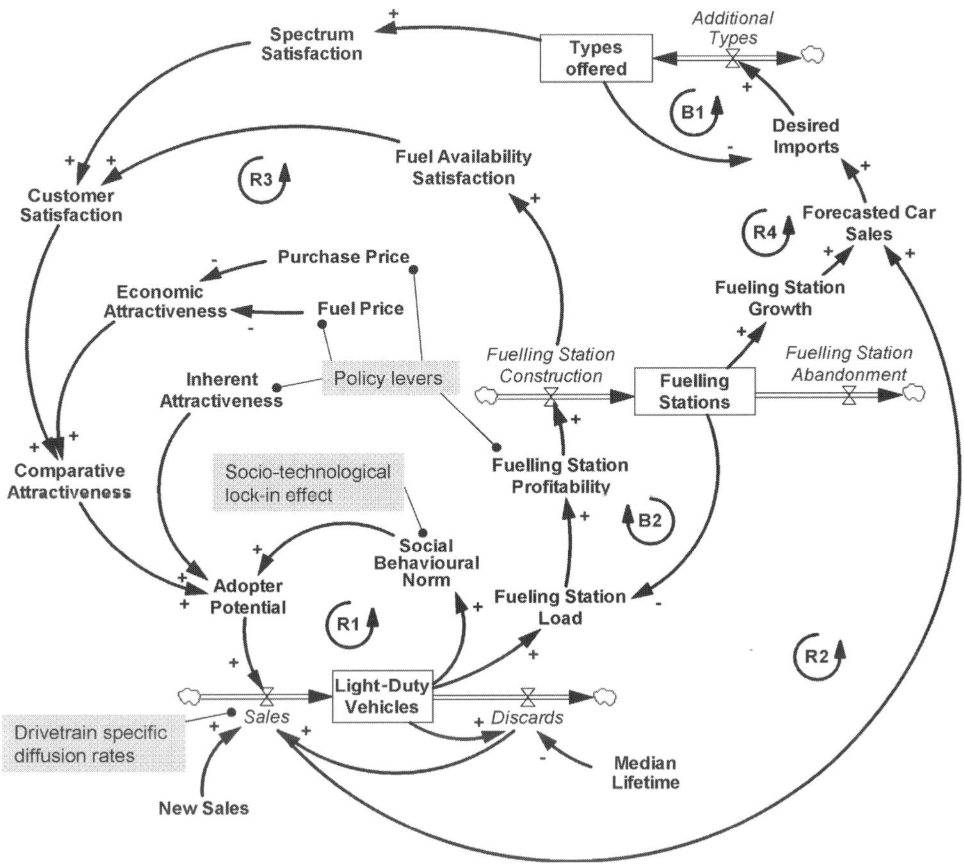

Figure 5.4 Causal loop diagram of the vehicle substitution model. The diagram shows the feedback loops within one single technology platform. Twin line arrows represent physical flows. Simple arrows denote causal links or information flows. The algebraic signs with the arrow heads indicate the type of causal link. '+' is interpreted as 'if the input variable increases, the target value is higher than it would have been otherwise'. The opposite interpretation holds for '-', 'if the input variable decreases, the target value is lower than it would have been otherwise'. An additional letter within a round arrow classifies each causal loop as reinforcing (R) or balancing (B).

The LDV stock and sales rate pictogram at the very bottom of the diagram links this view to the more detailed technology stock and flow diagram in Figure 5.3. The technology-specific, installed vehicle base is represented with a platform-specific stock; its inflow combines the new sales as well as the sales made at the expense of the alternative vehicle fleet stock. The outflow represents all vehicles leaving this technology-specific vehicle stock when being scrapped, even those that possibly are replaced with this technology and therefore re-enter the fleet stock through the sales. The other stock quantities in the system (fuelling stations and types offered, marked in Figure 5.4 by a surrounding rectangle) also represent a technology-specific platform.

The CLD illustrates four reinforcing and two balancing loops. The social behavioural norm (R1), the type spectrum (R2), and the fuelling station availability (R4) are three reinforcing loops emanating from the LDV stocks of each technology platform. They all connect to the adopter potential, where they are linked in the sense of technology competition as described in the previous section. The fourth reinforcing loop arises from coupling the car import and retail sector with the fuelling infrastructure sector by means of a market forecasting (R4) element (Janssen, 2004). The corresponding loop connects the number of fuelling stations to the forecasted car sales via the fuelling station growth ratio. These reinforcing loops must all be linked positively, meaning that growth in one loop enhances growth in the others and vice versa (Janssen, 2004). The two balancing loops arise from saturation effects. If more vehicle types are desired by consumers, the production or imports will increase, but a higher number of types supported leads to consumer satisfaction (B1). Analogously, a high fuelling station load (i.e. the number of cars per fuelling station) will increase their profitability and stimulate the construction of more stations. On the other hand, a higher supply of fuelling stations decreases the individual station's load and profitability (B4). Figure 5.4 also indicates that the inherent and economical attractiveness are not embedded dynamically. They are given by exogenous lookup functions, just as the new sales and the median lifetime are given, defining fleet growth and scrappage rates.

The emission module is not part of any feedback loop of the model. Its only purpose is to illustrate the impact of technological change in the vehicle fleet on CO_2 emissions. Applying a similarly sophisticated calculation as done by Zachariadis *et al.* (1995) considering the influence of a vehicle's age group on its emissions is not possible because of the higher aggregation of the vehicle fleet in the model. However, the average emissions factors have been calibrated based on the more detailed emission calculations. The vehicle stock of a technology platform is assigned only one specific emission factor, representing the averaged specific CO_2 emission, in grams of CO_2 per litre fuel, that is used while driving a vehicle, considered as a litre gasoline equivalent.

On the basis of the average VKT and the fuel consumption per vehicle, the fuel use of a car over one year is calculated. Together with the specific CO_2 emission mentioned above, the specific annual CO_2 emission of a single vehicle is determined. Considering the corresponding vehicle stock, this yields the overall annual CO_2 emissions for this technology platform. The equation presented by Zachariadis *et al.* (1995) reduces in this simple scheme to:

$$AnnualCO_2Emissions_i = VehicleKilometresTravelled_i \times SpecificFuelConsumption_i \times SpecificCO_2Emission_i \times LightDutyVehicles_i \quad (5.1)$$

5.3 Results

5.3.1 Policy and scenario specifications

The model is used to simulate the emerging technology diffusions to test different scenarios and five different policy packages. In addition to the baseline and the growing VKT scenarios, two preliminary worst-case experiments show what would happen if the historical initiated improvement efforts concerning improved energy efficiency and implementing alternative drivetrains and fuels do not take off. These scenarios and policy experiments are applied to the European car fleet. Illustrative, country-specific, diffusion simulation results are presented. The main research question guiding the simulation experiments is: what technologies and policies are needed to meet the postulated 50% CO_2 reduction target by 2050? The experiments are 'what if' analyses that show typical behaviour patterns and illustrative scenario results, but are not designed to predict specific penetration figures.

The scenario and policy analyses presented in Figure 5.5 are based on two levels of parameter settings. The first level, which we will refer to as scenarios, concerns different assumptions for some exogenous variables (such as VKT or fleet growth) that cannot be directly accessed by policy makers. These scenarios are subdivided into settings with different policy packages.

The key assumptions for the *baseline scenario* are those as described in the data and variable section for the fuel prices, fuel efficiency, vehicle service life, VKT, and fleet growth, as well as vehicle purchase price. Those also comprise a purchase price increase until 2045 of 6% for gasoline and diesel cars and a decrease of 1.5% for hybrids, because of expected exhaust after-treatment measures and technological improvements, respectively (Bandivadekar *et al.*, 2008b). The inherent attractiveness is assumed to reach 20% for alternative vehicles mainly due to their environmental and political aspects, whereas it is set to 10% for conventional ICE vehicles.

The *growing VKT scenario* assumes (in comparison to the baseline scenario) an increase in the number of VKT. As in the baseline scenario, the VKT of diesel and gasoline vehicles level off, but they experience a rise due to a generally growing demand for individual transport. The scenario anticipates an increase of 10% up to 2045 compared to 2008. The same assumptions for the fleet development are applied as in the baseline scenario, that is, an extrapolation of country-specific trends toward saturation in the long run.

The third and final scenario is the *constant fleet scenario*. In comparison to the baseline, it explores the situation where the vehicle fleets do not grow any further. The number of vehicles is held constant after 2008. An exemption to this is Poland, because its fleet is very likely to continue growing in the context of its economic development. The VKT are also held constant, as in the baseline scenario.

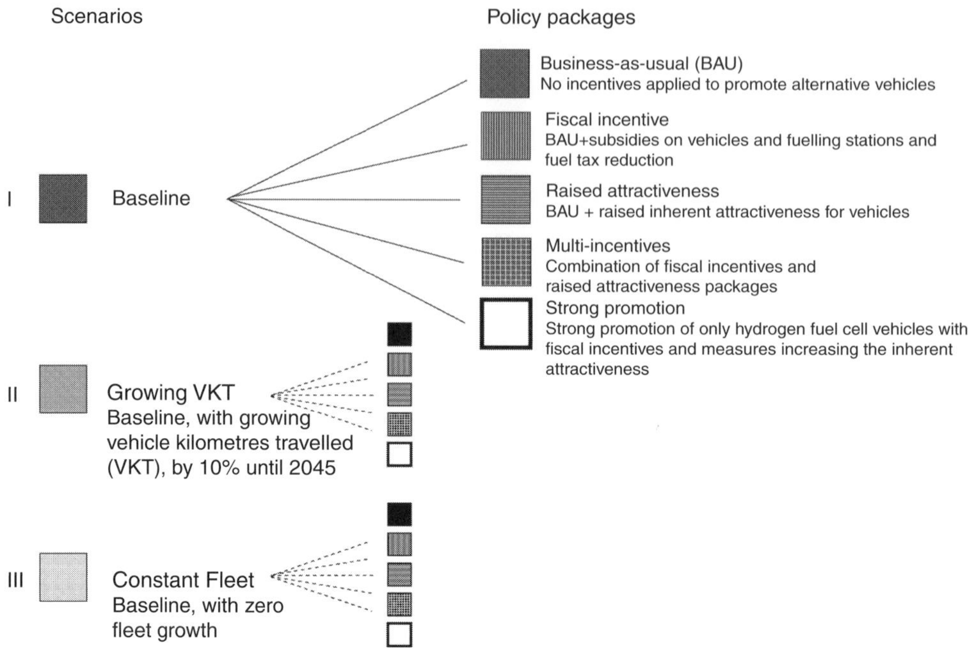

Figure 5.5 Illustration of the combination of scenarios and policy packages. The figure indicates the greyscale (scenarios) and hatching scheme (policy packages) that is used for the graphical representation of the emission results. The greyscale refers to the scenarios baseline, growing vehicle kilometres travelled (VKT) and constant fleet. Hatching represents different policy packages that are applied in the context of a scenario.

In the context of these three scenarios, different policy packages are evaluated that cover four fundamentally different strategies for the promotion of alternative vehicles. An overview is given in Table 5.3, leaving out the business-as-usual (BAU) package which actually does not consider any specific incentives. In the simulation experiments, the fiscal incentives package comprises the application of only fiscal instruments. Purchase price subsidies are granted equally on gasoline and diesel HEVs as well as NGVs and LPGs. The latter two are additionally promoted with fuelling station subsidies and fuel tax reductions. The raised attractiveness package excludes fiscal incentives and assumes a raised inherent attractiveness by a factor of four compared to the baseline assumption,[8] for all four of the drivetrain technologies mentioned. The multi-incentives package combines

[8] In the simulation model, the inherent attractiveness parameter of a vehicle technology can take a value between 1 and 100%. It represents all soft utility factors that are not monetised. In the real world, the inherent attractiveness of an energy-efficient vehicle is increased if the technological and environmental performance improves or if the vehicle users perceived utility is improved by policy measures (e.g. campaigns, introducing CO_2 emission labels, etc.). Please also see the discussion of the key assumption.

Table 5.3. *Applied incentives in the policy packages from 2008–2030.*

	Policy packages			
	Fiscal incentives	Raised attractiveness	Multi-incentives	Strong promotion
Vehicle price subsidy	2000 €[a]	–	2000 €[a]	5000 €[b]
Fuelling station	100 000 €[c]	–	100 000 €[c]	100 000 €[b]
Fuel tax reduction	0.3 €[c]	–	0.3 €[c]	0.5 €[b]
Inherent attractiveness	–	80%	80%	90%[b]

[a] Applied to gasoline HEV, diesel HEV, NGV, and LPG
[b] Applied to FCV or H_2, alternatively
[c] Applied to NGV

all incentives of the first two packages. Finally, the strong promotion package uses enhanced fiscal incentives as well as a higher inherent attractiveness for the promotion of HFCVs only.

The policy packages are standardised for all countries. Subsidies for vehicle purchase range from 7–12%, depending on the country, those for fuelling stations correspond to 40% of the original construction cost. The values for the inherent attractiveness reflect increased favourable infrastructure and policies for alternative vehicles. For the policy package strong promotion, the technology platform for natural gas vehicles is replaced by the technology platform for HFCVs, introducing an increase in parameters such as a higher fuelling station construction investment (€300 000) and vehicle price (10% higher than the most expensive technology in each country; see Appendix E). The replacement of natural gas was chosen because it still plays a minor role in most European countries today. The HFCV option stands for a game-changing technology that allows us to assess the impact of near zero-emissions vehicles. Therefore, it could also stand for a battery electric vehicle with a clean primary energy supply.

5.3.2 Impact of fuel efficiency improvement and technology diffusion

In this section, we present three different scenarios designed to explore both the effect of technology improvement and moderate technology diffusion. The first two experiments are not part of the scenarios and policy packages described above. They mainly demonstrate what would have happened if the past initiatives[9] were not implemented. The first case shows a model forecast result assuming that neither

[9] The voluntary agreements between the European Commission and the European Manufacturers Associations (ACEA), the Japan Automobile Manufacturers Association (JAMA), and the Korea Automobile Manufacturers Association (KAMA) to limit CO_2 emissions in vehicles sold in Europe.

technical improvement in terms of fuel efficiency takes place, nor can alternative vehicles significantly penetrate the market. The second case is limited to efficiency improvements, but still no alternative drivetrains are allowed to enter the fleet. These two no-compliance scenarios are finally compared to the baseline scenario with the BAU package, i.e. without any additional incentives applied but allowing for a very moderate diffusion of alternative vehicles.

In the first scenario, the current technical standards of gasoline and diesel ICE are frozen in 2007 and the fleet share of HEVs as well as NGVs and LPGVs is set to zero. In terms of emissions, this corresponds to a worst-case scenario because it yields the highest CO_2 values.

In the second scenario, the effect on the development of the CO_2 emissions of fuel efficiency improvement of gasoline and diesel ICE only in the absence of any alternative drivetrain technology is shown. The new vehicles sold in the European passenger car market continue to have higher fuel efficiency over the future decades, due to technical improvements in gasoline and diesel ICEs. On the other hand, alternative drivetrain vehicles, such as HEVs, are still very sparsely represented in the European passenger car fleets. Moreover, the trend of market penetration is small. Therefore, we will call this the 'no diffusion' scenario. Hence, the only future emission savings will arise from efficiency improvements of new vehicles entering the fleet. The potential for engine improvement is limited and is assumed to be exhausted circa 2050 (see also the input data in Appendix E). The two preliminary worst case scenarios are compared with the baseline scenario (BAU).

In the third scenario (BAU), improved ICE and new drivetrain technologies enter the market, although still without implementing any further policy incentives. A moderate diffusion of alternative drivetrains into the passenger car fleet is included. This is based on the assumption that some promoting policies will be implemented in the future, and that the perception of drivetrain technologies with reduced consumption and CO_2 emissions increases over time. The BAU is the starting point to test different policies for the promotion of alternative vehicles in subsequent sections.

Figure 5.6 shows the CO_2 emission curves for the three scenarios until 2100. The CO_2 emissions are normalised to 100% in the year 1995. We find that technological improvement and also the fleet penetration of alternative vehicles have an important effect on CO_2 emissions. In the worst-case scenario, the emissions show an increase of 156% in 2050 and of 136% in 2100 compared to reference year 1995. The increase is reduced by the ICE energy-efficiency improvement effect in the no diffusion scenario of 104% in 2050 and of 88% in 2100. In the worst-case scenario, the curve shows a monotonous increase, while the no diffusion and the baseline (BAU) scenarios have two obvious inflections around 2020 and 2060. These inflections result from the fuel efficiency increase and the technological

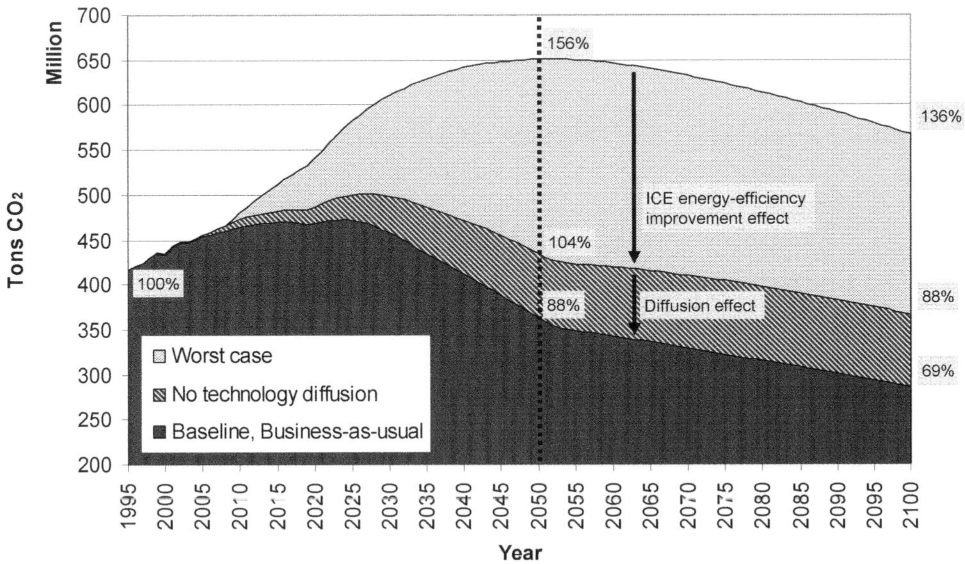

Figure 5.6 Comparison of the CO_2 emission trends in the worst case, no diffusion, and baseline business-as-usual (BAU) scenarios. The relative CO_2 emissions values are normalised to the emissions in 1995.

improvement limits, which are not effective in the worst case scenario. In the baseline scenario, the CO_2 emissions are reduced to 69% by 2100.

5.3.3 Country-specific CO_2 emission trends

Depending on the simulated fleet growth trend and the development of the diesel share, the countries selected in this chapter show different CO_2 emission perspectives over the baseline scenario. Representative modelled emission trends of Germany, France, and Poland are illustrated in Figure 5.7. Germany seems to be the only country which has already turned around its passenger car emissions in 2007. The CO_2 emission peak falls together with the gasoline ICE vehicle peak in Germany (cf. Figure 5.12). The rising diesel car share and technical improvements resulting in higher fuel efficiency for all drivetrain technologies cause the decreasing CO_2 emission trend. After mid-century, when the technical potential of ICEs is exhausted, emissions still fall. The decrease is slowed down because it is now solely due to a changing fleet composition; this is an increasing part of alternative vehicles with lower emissions (e.g. gasoline HEV, diesel HEV, and NGV as shown in Figure 5.12).

While other countries, such as France, may have a higher diesel ICE share than Germany, their fleet growth rate leads to increasing CO_2 emissions if BAU is followed. In the long run, all countries modelled (except Poland) show a

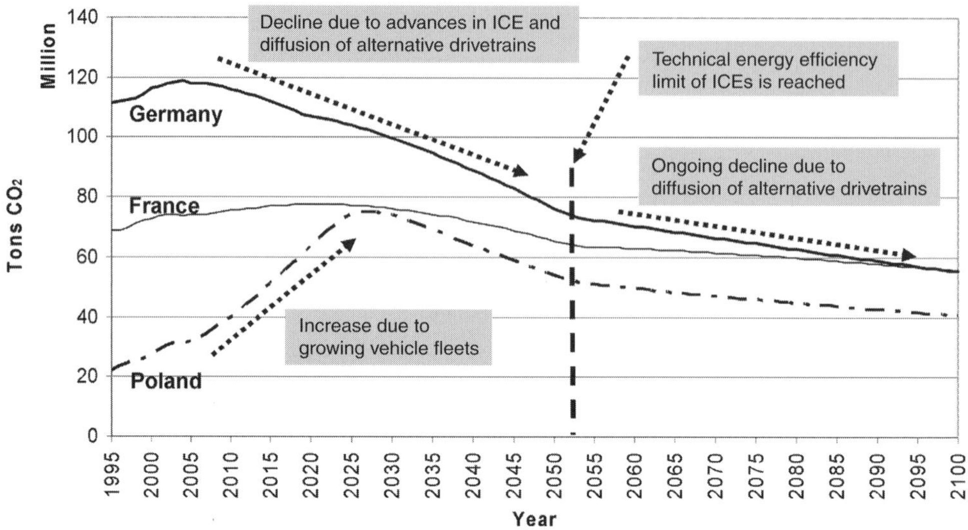

Figure 5.7 Development of CO_2 emissions for Germany, France, and Poland under the baseline scenario.

decrease in CO_2 emissions in the first half of the century, but, in the baseline scenario without incentives, only Germany, Italy, and Sweden are able to reduce their emissions substantially relative to 1995. Poland shows a strong increase since their fleet is still growing at a high rate of about 3 to 5% per annum. The fleet increase compensates all emission savings from technical advances in fuel consumption as well as from alternative vehicles entering the fleet. Toward the end of the century, Poland experiences an emission level that is almost doubled compared to 1995.

5.3.4 The role of diesel internal combustion engines in the baseline scenario

This section presents some special features of the growing fleet penetration of diesel vehicles, and the effect of a behavioural norm on the diffusion of alternative drivetrains. A more technical illustration of the behavioural norm and the critical mass is given in the detailed explanation provided below and in Figure 5.8.

Technical illustration of a behavioral norm effect This somewhat technical excurse illustrates that the critical fleet share for a self-sustaining diffusion pathway of an alternative drivetrain is higher on a direct transition path without any other competing drivetrain technologies than on a path with promotion of other alternatives. When more than one alternative with significant fleet shares is available, the behavioral norm of the dominant technology is weakened, lowering the tipping point for each alternative technology. A 3D surface can illustrate an approximated

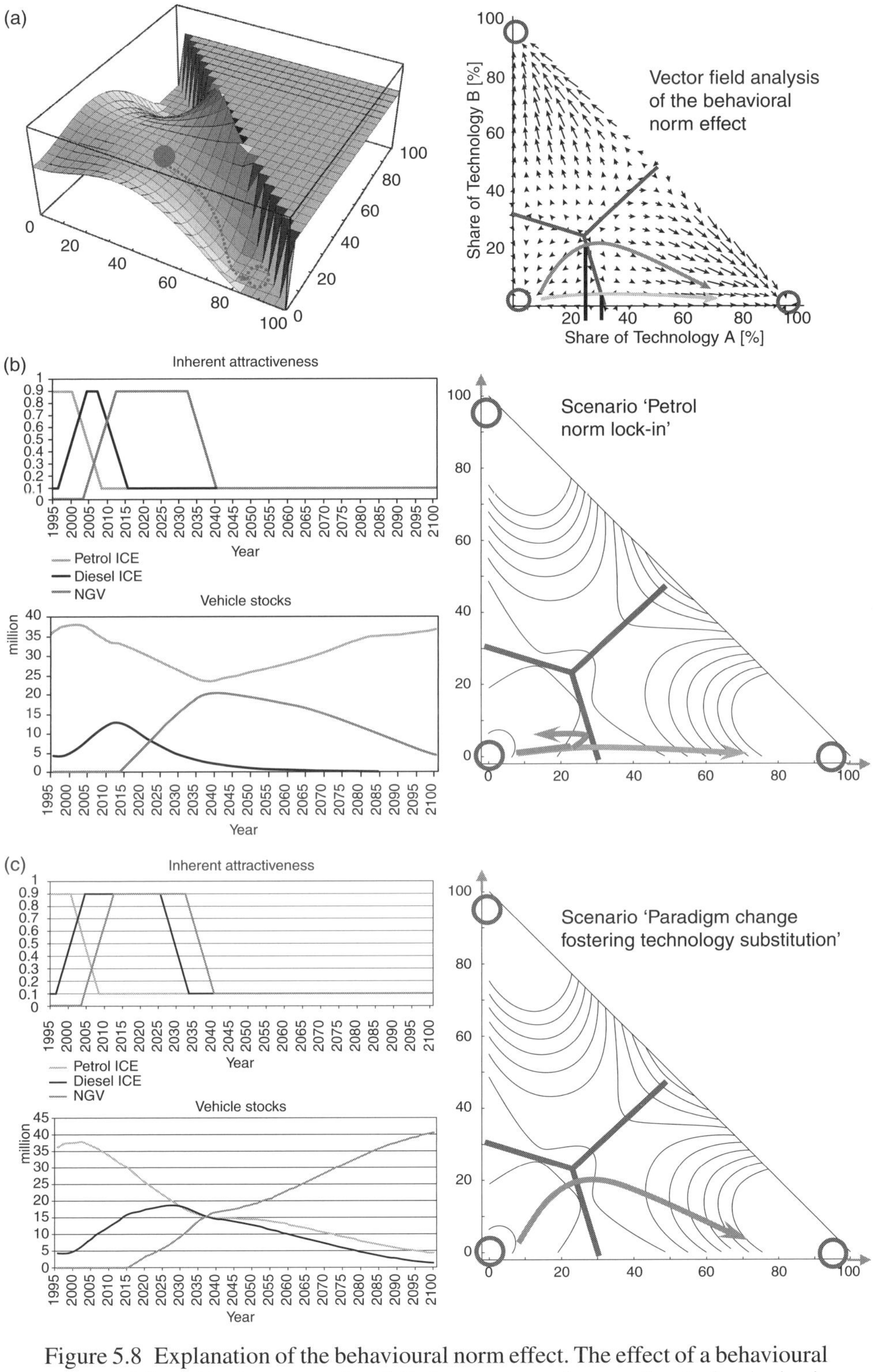

Figure 5.8 Explanation of the behavioural norm effect. The effect of a behavioural norm and the critical mass on the diffusion process of competing technologies is illustrated with the Germany case under the baseline scenario (see also colour plate 12).

auxiliary potential function for a vector field as a function of the behavioral norm for each technology (Figure 5.8a left hand side). In order to achieve self-sustaining diffusion, the ball must be pushed over the saddle point by an external force. Figure 5.8a on the right hand side shows the influence of a social behavioural norm in the process of technology adoption in the car market with two emerging alternative propulsion systems A and B. The black arrows indicate the strength and direction of the internal driving forces towards one of the attractors in the system (red circles). The red lines represent the critical shares and delimit the dominated areas. By application of an external incentive, such as subsidies, the internal forces can be overcome to achieve the critical share, which depends on the path chosen. A strong promotion of both A and B establishes a higher norm for alternatives and leads to a lower critical share for either technology (e.g. A with the green arrow), compared to the focused promotion of only one alternative (blue arrow).

This analysis framework has been applied to investigate the effect of diesel ICEs on the transition to NGV in the Figures 5.8 b & c. The graphs show values of the inherent attractiveness and stock values for petrol and diesel ICEs and NGVs in Germany. In the scenario 'Petrol norm lock-in' (Figure 5.8b), diesel ICEs disappear over time and NGVs cannot stand up to petrol ICEs in the long-term. However, in the scenario 'Paradigm change fostering technology substitution' diesel ICEs further emerge because their inherent attractiveness stays higher for a longer period, the norm of petrol ICE's is further weakened and NGVs can break through (Figure 5.8c).

Country specific diesel trends Due to the norm effect explained above, the country-specific diesel trends lead to very different outcomes in the baseline scenario with no incentives. In some countries, the diesel share continues to rise and will take the lead for a long time period despite the assumption that its inherent attractiveness will decrease (i.e. diesel attractiveness is assumed to fall to the same level as the gasoline ICE beyond 2020). In parallel, alternative drivetrain technologies end up in a niche market, although it is assumed that they become more attractive than the conventional drivetrains. A representative country for this behaviour is France in Figure 5.9. Italy, Spain, and Poland show similar behaviour.

In other countries where the diesel share is not as developed, the trend is reversed after the assumed decline of diesel's inherent attractiveness. The regime goes back to gasoline ICEs since diesel vehicles disappear, while alternative vehicles are not promoted and its expected attractiveness does not suffice (see also the explanation in Figure 5.8). A representative country is Switzerland, shown in Figure 5.10.

Despite the dominance of the gasoline ICE, alternative vehicles gain market share and prevent a stronger increase of gasoline cars after the diesel era. A big share falls upon gasoline HEVs, using the same infrastructure as gasoline ICEs. The two technologies therefore strengthen each other, while the declining diesel

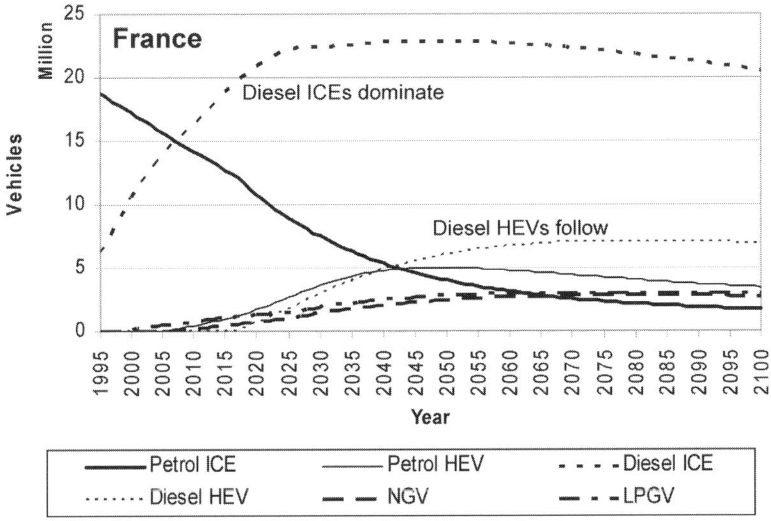

Figure 5.9 France under the baseline scenario without incentives. Alternative technologies end up in a niche market.

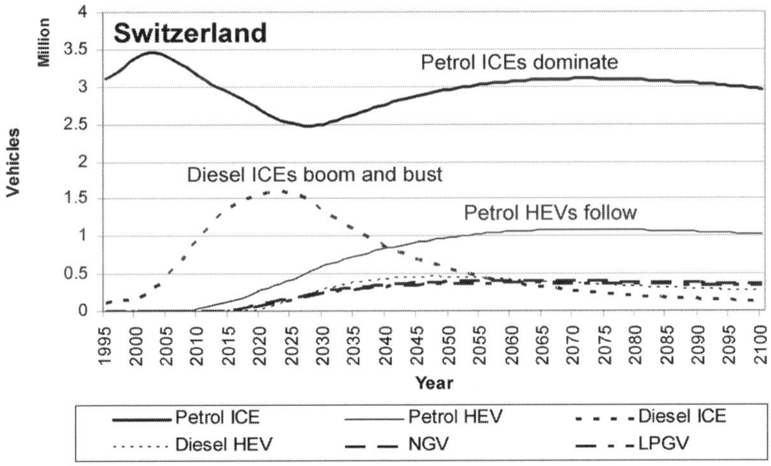

Figure 5.10 Switzerland under the baseline scenario without incentives. The gasoline internal combustion engine (ICE) recovers after diesel loses its high inherent attractiveness.

fuelling infrastructure leads to a decrease in diesel HEVs. This effect is crucial for the market introduction potential of both gasoline and diesel HEVs. Their advantage compared to gaseous-fuelled vehicles is the pre-existent fuelling infrastructure.

Yet, even a strong diesel share which seems to be very stable, as seen in the former France case, the predominance of diesel cars is not inviolable. Incentives in favour of alternative drivetrain technologies may strengthen and accelerate their market

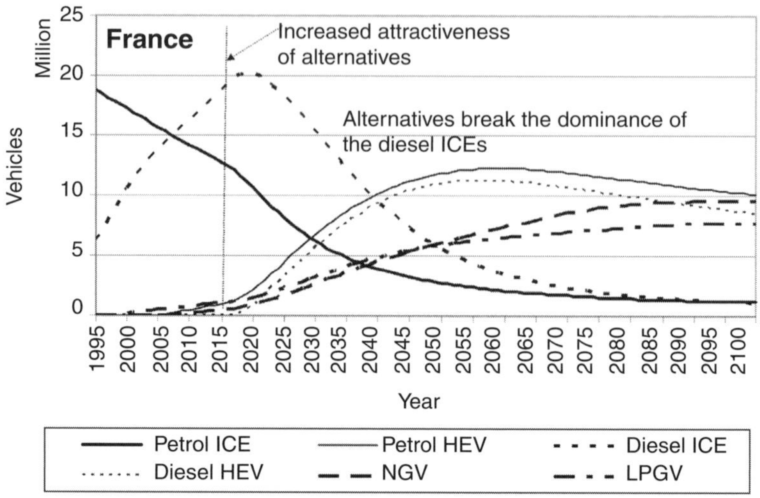

Figure 5.11 France under the baseline scenario with raised attractiveness for alternative drivetrains. The increase for alternative vehicles breaks the dominance of diesel cars.

penetration. The diesel share thus begins to fall after losing its benefit from inherent attractiveness as illustrated in Figure 5.11. We observe the behaviour discussed in the model explanation in Figure 5.8, when diesel acts as a bridge by weakening the norm of gasoline cars, paving the way for alternative vehicles. The following subsection shows in detail a systematic analysis of different policy packages.

5.3.5 Investigating different policy packages in the baseline scenario

We now focus on the diffusion results for selected, representative countries for the baseline scenario with the three different policy packages: financial support for vehicles and fuelling stations as well as fuel tax reduction (fiscal incentives), increased inherent attractiveness by means of different promoting instruments such as bans and permits (raised attractiveness), and finally the combination of the two (multi-incentives). Their effect may differ due to country-specific conditions, such as price levels and infrastructure. The results are presented with representative countries for typical behaviour or special cases revealing interesting effects. The case for Germany, which is presented for all three packages, allows a comparison of the impact of the different policy packages. We start with an overview of the results in the baseline scenario under BAU conditions, with no policy change.

Business-as-usual

The case analysis can be grouped according to country-specific characteristics concerning diesel share trends and motorisation rate. The first group of countries (mainly France and Spain) shows a diesel-dominated century with gasoline ICEs declining, as illustrated with the French fleet case in Section 5.3.4 and Figure 5.9. In this case, the diffusion of alternative vehicles is only moderate, whereas diesel HEVs perform best since they may benefit from the diesel fuelling infrastructure. The second group contains most of the countries under consideration: Germany, Great Britain, Italy, Sweden, and Switzerland. All of these countries experience a declining diesel share after its inherent attractiveness is assumed to drop to the level of gasoline ICEs. Despite the temporally increased attractiveness, the diesel norm does not gain enough strength to dominate, but is finally strong enough to break the gasoline's norm dominance. Therefore, the end of the century is characterised by a mixture of different drivetrain technologies. The simulation results show that none of the alternative drivetrains is able to rise above its competitors. The German case is shown as a representative in Figure 5.12. It illustrates the leading position of gasoline HEVs.

The third group is represented by Poland in Figure 5.13. It shows a decline of conventional ICE vehicles, but with LPGVs as the dominating alternative technology in the long run. Although diesel ICEs experience a strong rise from 1995–2025, they are not able to build up a strong enough norm to stabilise their position. Rather, they act as a bridge to LPGs, which already have a relatively high base installed by 2020 compared to other alternative vehicles. Despite the reversal of the trend in the diesel ICE's inherent attractiveness in 2020, the simulation shows that it takes another two decades until LPGVs take the lead around 2040.

In all BAU cases, we observe a remainder of gasoline and diesel ICEs toward the end of the century that is stable and not displaced by alternatives. The two technologies together still have a small market share. In Germany, for instance, they keep a sales share of some 22%, and 9% in Poland, as can be seen in Figures 5.12 and 5.13.

Fiscal incentives

The first approach to promoting alternative vehicles is through financial support for vehicles and fuelling stations. The modelled incentives have created conditions that lead to a strong rise of alternative vehicles in all countries. The differences lie in the time and order of their appearance. High fuel prices and lower income generally increase the effect of fiscal incentives and lead to a faster diffusion of alternative drivetrain technologies (provided that prices are within a range where the reductions are still significant). Great Britain, illustrated in Figure 5.14, with the highest fuel prices, reaches a market share of alternative vehicles of about 80% in 2050. Switzerland, in comparison, not displayed in this chapter, only reaches 42% market share for alternative vehicles in 2050, where 44% of these falls upon gasoline HEVs. The reason is again the benefit of hybrids from the infrastructure

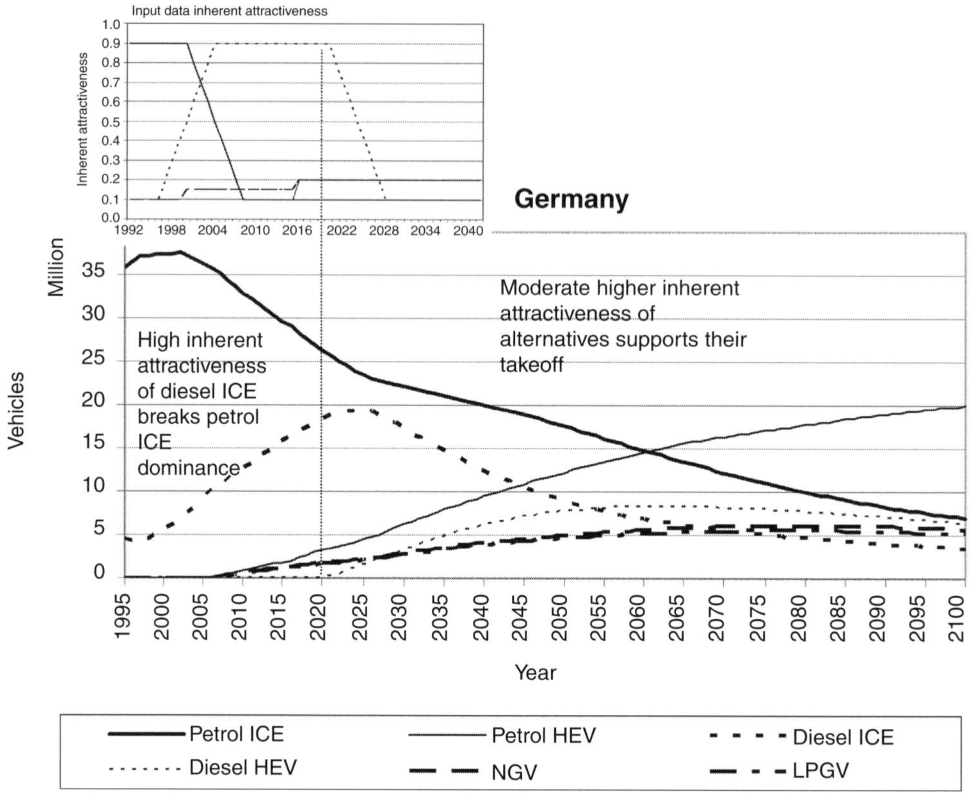

Figure 5.12 Germany under the baseline scenario without incentives. In a typical development, conventional internal combustion engines vehicles decline and a mixture of alternative vehicles characterise the market toward the end of the century.

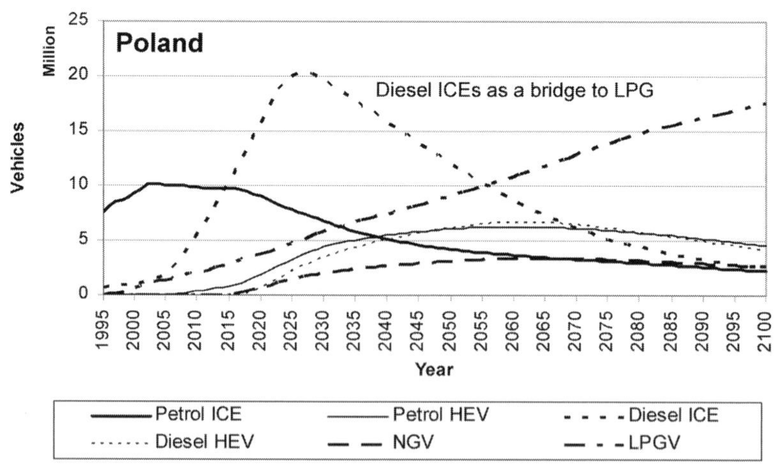

Figure 5.13 Poland under the baseline scenario without incentives. Liquid petroleum gas vehicles (LPGVs) break the strong diesel internal combustion engines (ICE) dominance.

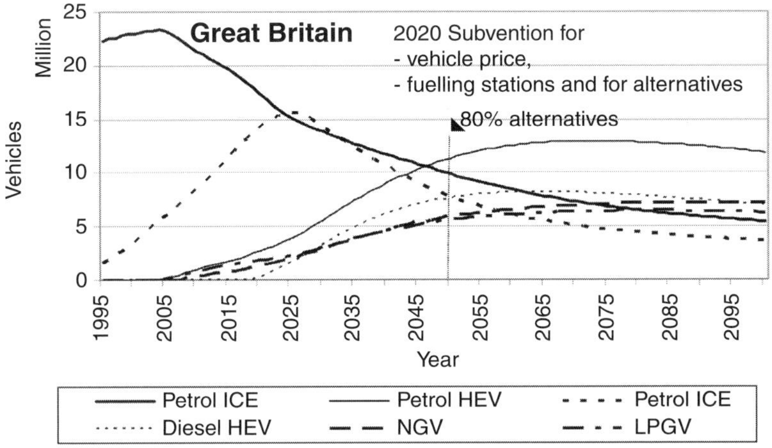

Figure 5.14 Great Britain under the baseline scenario with fiscal incentives. Comparatively higher fuel prices are assumed, as in other countries.

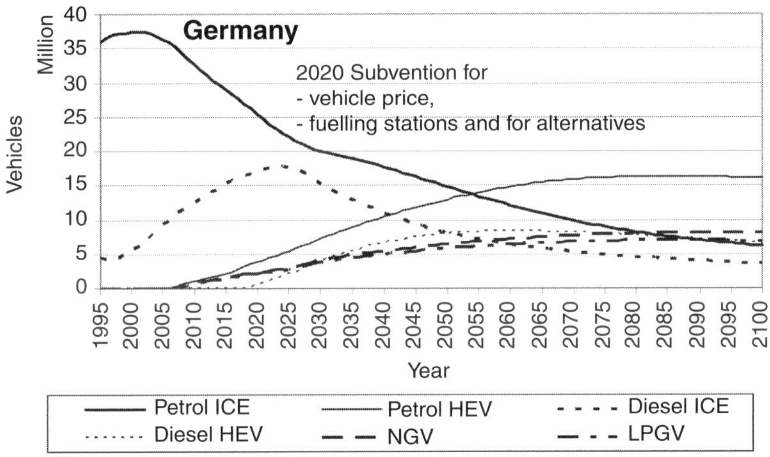

Figure 5.15 Germany under the baseline scenario with fiscal incentives. Natural gas vehicles (NGVs) and liquid petroleum gas vehicles (LPGVs) gain some market shares due to the fiscal incentives.

of the predominant gasoline ICE. The fiscal incentives are not enough to break the gasoline ICE dominance in Switzerland. Another aspect is that in countries with more balanced conditions, the differences between emerging technologies decrease, when the fiscal incentives package is applied. The subsidies and tax reductions boost other alternatives, mainly at the expense of the gasoline hybrid. The benefit from existing infrastructure becomes less important since the infrastructure of gaseous-fuelled vehicles is now promoted too. Figure 5.15 shows the German example.

Raised attractiveness

The raised attractiveness policy package influences the development of the inherent attractiveness in the model. In praxis regulatory measures, campaigns and marketing instruments (for example, fuel economy labelling) applied to advanced drivetrains and alternative fuels may have this effect since their environmental performance or comparative utility is highlighted. Some of these policies act as barriers for conventional drivetrains (i.e. driving bans and road pricing), while others are drivers for alternative vehicles (i.e. driving permits). Contrary to the fiscal incentives policy package, the raised attractiveness policy package acts on all alternative technologies equally. Gaseous-fuelled vehicles, therefore, again face an infrastructure obstacle. However, alternative vehicles in general perform better under the raised attractiveness policy package than under the fiscal incentives policy package, as Figure 5.16 shows for the German case. It is important to see in this context that all alternative drivetrain technologies together become strong enough to completely eliminate conventional ICE vehicles from the market. This introduces a fuelling infrastructure obstacle for hybrid vehicles, which show a decrease (expected) in the market share.

Figure 5.17 shows that, in Italy, NGVs and LPGVs perform even better than in Germany. LPGVs, in particular, have a significant base installed, before the turn of the millennium. To test the sensitivity of the outcome to the inherent attractiveness, we reduce it to 40% instead of 80% for all alternative vehicles. As Figure 5.18 indicates, diffusion is not affected very much. In Italy, a smaller value could bring almost the same benefit in terms of technology diffusion, meaning that the tipping point is not yet undercut by 40%. On the other hand, we observe that the lower attractiveness does not suffice to displace all conventional ICE vehicles from the market. The difference is mainly to the detriment of gaseous-fuelled vehicles.

In countries with a comparatively high income, such as Switzerland, the increased inherent attractiveness has a much higher impact than fiscal measures. In this case, the raised attractiveness assumption is strong enough to induce a real turnaround for gasoline ICEs favouring, instead, gasoline hybrids, as is shown in Figure 5.19. Again, further simulation results showed that a weaker raised attractiveness assumption would not change this outcome very much (as illustrated in Figure 5.18).

Multi-incentives

The multi-incentives policy package brings the highest benefit for the diffusion of alternative drivetrain technologies. But the additional increase of the fleet share compared to the raised attractiveness policy package is rather modest. The market penetration rates are slightly higher during the period when fiscal incentives are

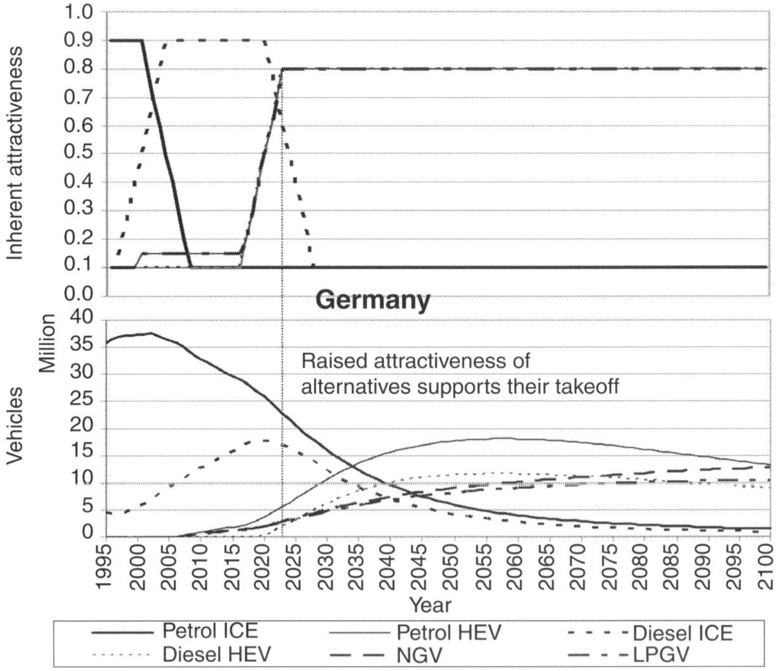

Figure 5.16 Germany under the baseline scenario with raised attractiveness. Gaseous-fuelled vehicles perform better with the raised attractiveness package than with solely fiscal incentives, and hybrids lose market share.

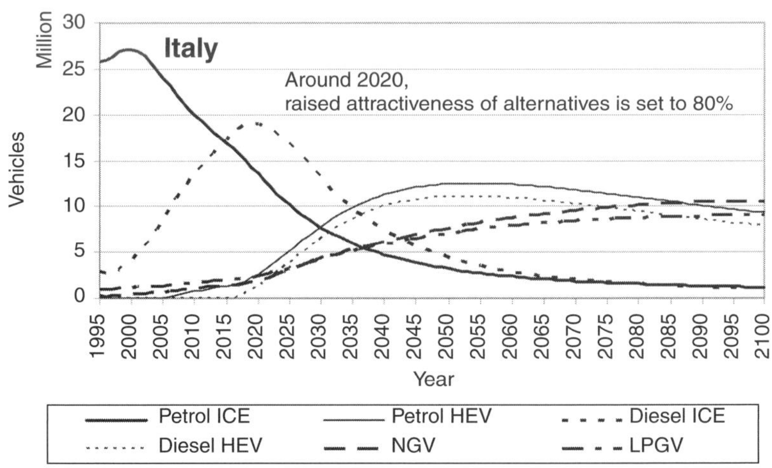

Figure 5.17 Italy under the baseline scenario with raised attractiveness. The inherent attractiveness is set to 80%.

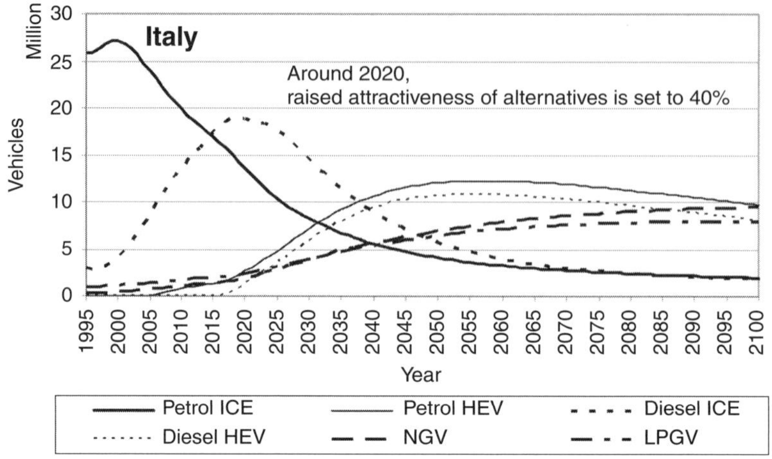

Figure 5.18 Italy under the baseline scenario with raised attractiveness at a lower level. The inherent attractiveness is set to the lower value of 40%. The reduction of the inherent attractiveness from 80% to 40% does not affect the diffusion process in Italy very much. Gaseous-fuelled vehicles react more strongly to this change than do hybrids.

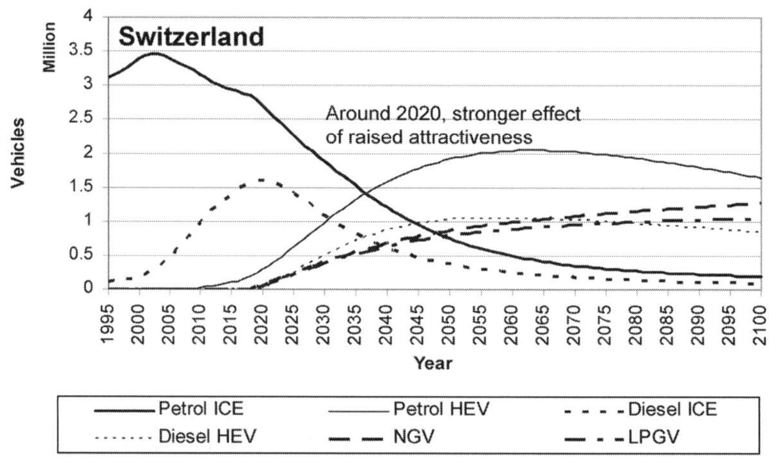

Figure 5.19 Switzerland under the baseline scenario with raised attractiveness. The case illustrates that the raised attractiveness policy package has a higher impact than with fiscal incentives, especially if high income levels are assumed. In the Swiss case the alternatives fully penetrate the market in 2100.

also granted. Toward the end of the century, diffusion rates level off for all countries, since the fleet is totally penetrated by alternative vehicles. Figures 5.16 and 5.20 provide a typical comparison. In the German case, alternative vehicles reach a fleet share of 40% and 48% in 2030 for the raised attractiveness and the multi-

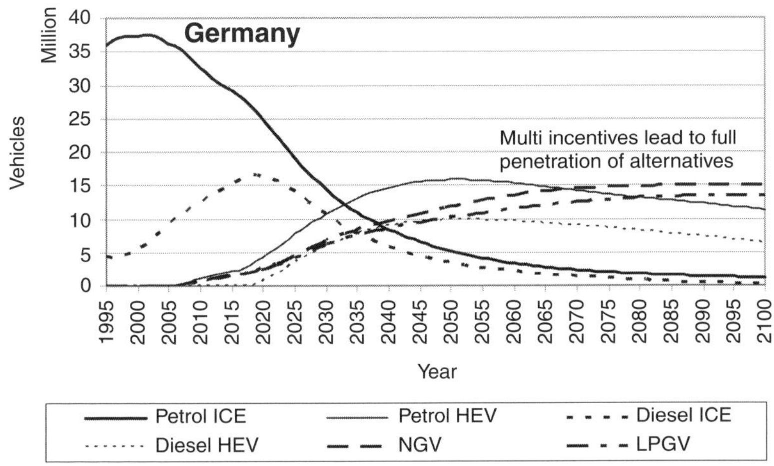

Figure 5.20 Germany under the baseline scenario with multi-incentives. The multi-incentives policy package also leads to a full penetration of alternative vehicles in Germany toward 2100. This is similar to the raised attractiveness package, although it shows lower shares of gaseous-fuelled vehicles (cf. Figure 5.19).

incentives package, respectively. In 2100, however, the alternative vehicle fleet share is 100% for both policy packages.

5.3.6 Investigating the strong promotion of a near-zero emission technology: the case for hydrogen fuel cell vehicles

Hydrogen FCVs (HFCVs) face the highest barriers for market penetration. The technology does not have an existing fleet and therefore has no behavioural norm. There is no fuelling infrastructure assumed to be available before 2010 and the technology is assumed to be only available for the mass market after 2015. Also, the investment costs for fuelling stations as well as the vehicle prices are increased. The policy package that is applied to promote HFCVs is therefore the most aggressive one. The vehicle purchase price subsidy is set to €5000, the fuel price is reduced by €0.50, and fuelling stations are granted a €100 000 subsidy. The inherent attractiveness of the vehicles amounts to 90%, corresponding to potential consumers before considering norm, cost, or satisfaction aspects. The basis for this is the assumption that HFCVs are introduced as zero-emission vehicles, fuelled with hydrogen from renewable primary energy. Although entering the market with high barriers, the promotion strategy is successful in most countries. Even in Poland, with its dominant LPGV share, the FCVs are able to take the lead after the mid-century. Figure 5.21 shows the typical development represented by Spain.

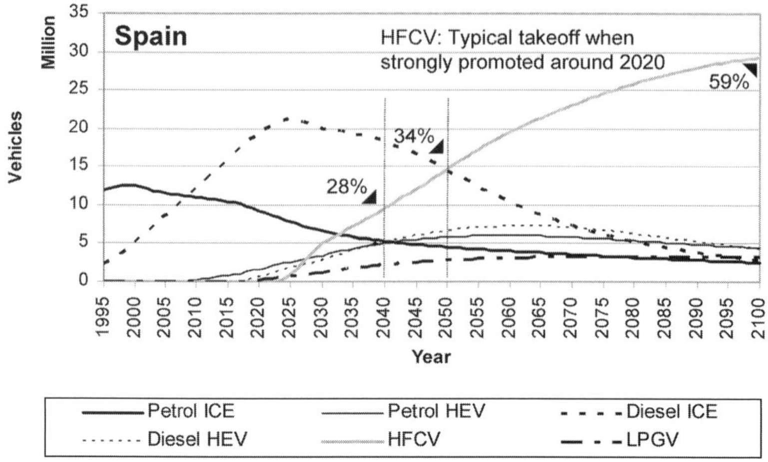

Figure 5.21 Spain under the baseline scenario with strong promotion of hydrogen fuel cell vehicles. Spain represents a typical result for most EU countries where fuel cell vehicles take off with strong promotion policies.

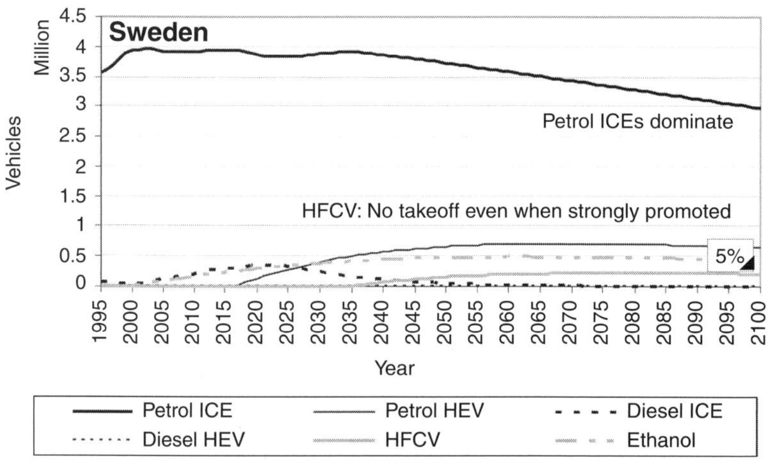

Figure 5.22 Sweden under the baseline scenario with strong promotion of hydrogen fuel cell vehicles. This is the only EU country where hydrogen does not take off. The dominance of gasoline internal combustion engines has not been broken by diesel internal combustion engine or alternative vehicles due to the restrictive promotion policy.

Figure 5.22 illustrates the case of Sweden, which is the only case where FCVs do not take off. In this case, without promotion of other alternatives, gasoline ICEs produce a strong technological lock-in situation with a high norm and an adapted, fuelling infrastructure, since high purchase prices of diesel vehicles hinder their fleet penetration (for the behavioural norm effect, cf. Figure 5.8). HFCVs alone

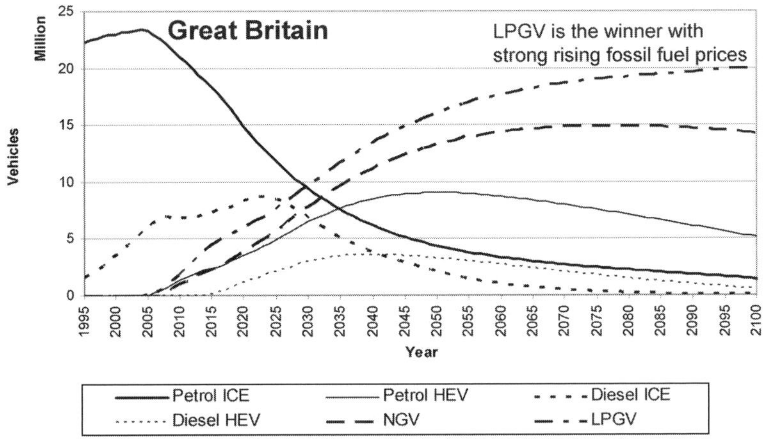

Figure 5.23 Great Britain with the application of a quadrupling of all fossil fuel prices. In the long run, we observe a strong rise in alternative drivetrain vehicles, with liquid petroleum gas vehicles as the clear winner.

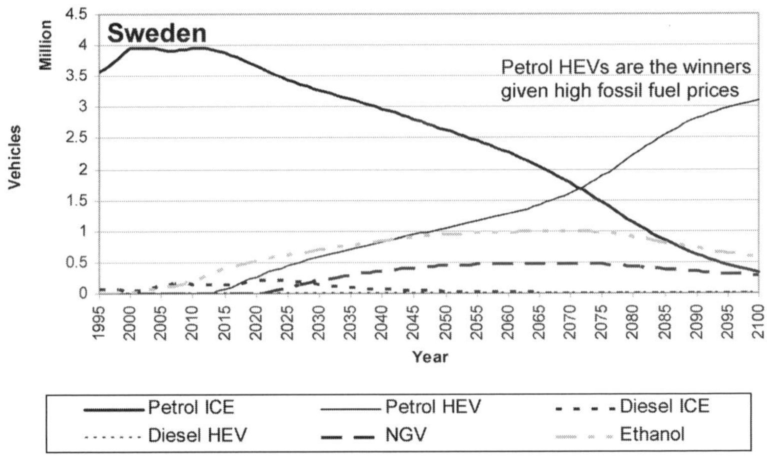

Figure 5.24 Sweden with the application of a quadrupling of all fossil fuel prices. This case is analogous to Great Britain. In the long run, we again observe a strong rise in petrol hybrid-electric vehicles (HEVs), which are able to establish an early behavioural norm.

are not able to break the dominance of gasoline hybrids, despite their strong promotion. However, strong price differences between conventional gasoline ICE and alternative fuels as well as alternative vehicles may improve the diffusion conditions for all alternatives. This will break the dominance of gasoline ICE and pave the way for the emergence of a new, dominant technology, as illustrated by the next subsections and in Figures 5.24 and 5.25.

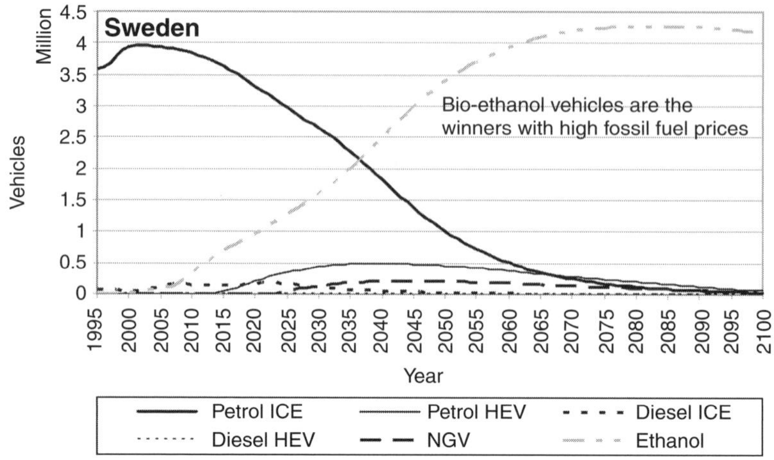

Figure 5.25 Sweden with the application of a quadrupling of all fossil fuel prices but with cheaper ethanol from biomass. The price benefit is high enough to assure an early norm, which locks out gasoline internal combustion engine, which were highly competitive in the beginning.

5.3.7 Fuel price increase

The oil price development in the first half of 2008 may outdate the fuel price assumptions in the model in the near future. In parallel to the policy packages discussed above, this section explores how the model reacts to extremely high fuel prices. We compare our model output with the results of a rather simple elasticity model for the relationship of income, fuel price, and consumption. The background for the following increased fuel price scenario is the simple elasticity model of Sterner (2003). In this case, he assumes that fuel demand G is determined by $G = Y P^{-0.8}$, with income Y and price P. It is also assumed that income will increase by 50%, while a long-term goal suggested by the IPCC is a 50% decrease in carbon emissions for the transportation sector. The question posed is what a suitable market fuel price in this case should look like to reach the emission target. If the carbon emission goal is to be achieved only by reducing the fuel demand, then the comparison of the original and the adjusted price, using the formula above, leads to:

$$\frac{P'}{P} = \left(\frac{0.5}{1.5}\right)^{-\frac{5}{4}} = 4$$

That is, an adequate fuel price would have to be four times higher than today, given the elasticities that are valid over a large price and time interval. In the absence of alternative vehicles, the emission decrease will have to be achieved by reducing the consumer's driving behaviour, i.e. notably, the average VKT. This does not seem to be very likely for the near future, even if a high fuel price is assumed. Either way, the change would take time, which cannot be considered by this simple model. With alternative drivetrains on the market, however, people can buy vehicles which consume less, such as natural gas or gasoline hybrid vehicles, to overcome the conflict between high expenses and an increasing mobility demand. This can be seen in our simulation by using the same parameter settings as in the baseline scenario, except for the increased fossil fuel prices. If we increase the prices by a factor of four as suggested by Sterner's policy experiment, the fleet share of alternative vehicles increases as people tend to buy an economical car when they purchase a new one. Figures 5.23 and 5.24 show the stock curves for the increased price scenario with quadrupled fossil fuel prices after 2007 for Great Britain and Sweden, two countries with relatively high fuel prices.

For the case of Great Britain in Figure 5.23, we observe a strong rise in alternative vehicles. Particularly, gaseous-fuelled vehicles dominate the market, where LPGVs take the lead after the mid-century, since they have already established a strong behavioural norm. The Sweden case illustrates a different development. Gasoline HEVs dominate the market early, preventing any other alternative drivetrain from significantly penetrating the market. The situation in Sweden changes completely if we assume that ethanol vehicles are fuelled with E-85 from biomass. It is assumed that this fuel would be less affected by the rising fossil fuel price. Now, ethanol has a big price benefit compared to hybrids, and they take the lead in the market and, later on, in the fleet. This interesting behaviour, as illustrated by Figure 5.25, actually lies in the fact that gasoline HEVs are not completely non-competitive; they reach a remarkable fleet share up to 2045, and develop virtually in parallel to ethanol vehicles. However, ethanol vehicles start some 5 years earlier, enabling them to establish a higher behavioural norm. The disappearance of gasoline ICEs accompanied by a decline of gasoline fuelling stations may favour this development.

The question is whether the two countries are able to reduce their CO_2 emissions by 50% compared to 1995. Figure 5.26 shows the result. The right scale applies to the two emission curves for Sweden. The CO_2 emissions are reduced to 53%, hence, the CO_2 emission reduction target is reached by Sweden in the high fuel price scenario. With the assumption of E85 bio-ethanol, emissions are reduced by a factor of more than five, to 18% compared to 1995. Great Britain, however, misses the reduction target with 70% emissions in 2100, compared to 1995.

5.3.8 *Consequences for CO_2 emissions (EU perspective)*

In this section, the consequences of the policy packages analysed under different scenarios of CO_2 emissions are presented. The first section compares different

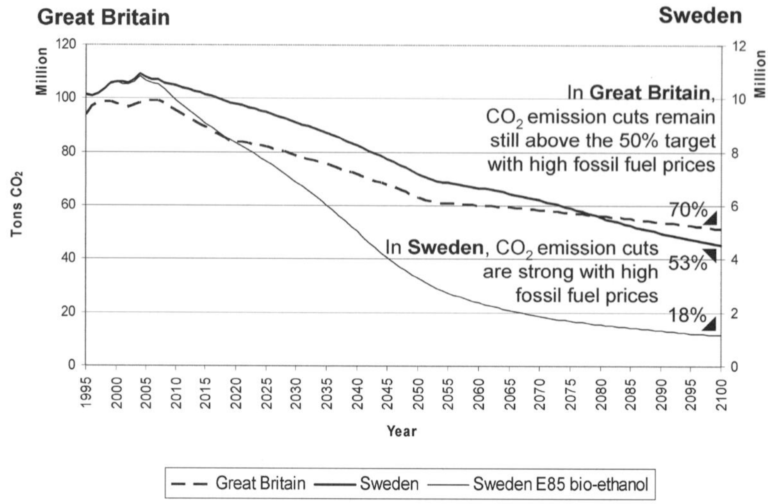

Figure 5.26 CO_2 emissions for Great Britain and Sweden in a high, fossil fuel price scenario. The right scale in this graph applies to the two curves for Sweden. With high fossil fuel prices the penetration of the alternatives leads to strong CO_2 reductions.

policy packages for the baseline scenario. Selected results are used to illustrate the effect of growing VKT or restricted fleet growth in the corresponding scenarios. The emission results presented correspond to the summarised CO_2 emissions of the six representative European countries: France, Germany, Great Britain, Italy, Poland, and Spain. All relative reduction values refer to 1995 as the reference year.

Baseline scenario

The consequences for CO_2 emissions under the policy packages analysed under the baseline scenario are illustrated in Figure 5.27. Applying the strong promotion policy package to the HFCV case reveals the highest benefit in CO_2 emission reduction, to 63% in 2050 and 28% in 2100, although it performs worse than the other policy packages before 2030, since no further technology options are promoted. The next best is the multi-incentives policy package, which shows a modest reduction surplus in the period before 2050, compared to the raised attractiveness policy package. Toward the end of the century, the emissions with the two policy packages applied reach 63%. Emission savings get smaller, if only fiscal instruments are considered, and end up at 68% in 2100, compared to 69% under BAU conditions. The earliest turnaround possible is observed with the multi-incentives assumptions in 2015.

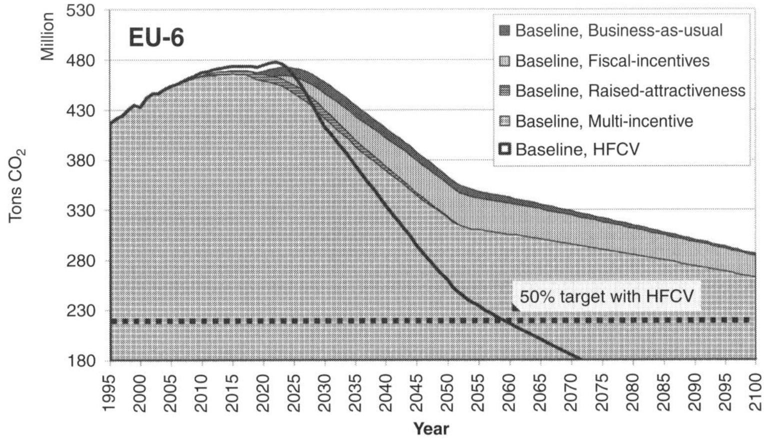

Figure 5.27 Comparison of CO_2 emissions in the baseline scenario for different policy packages applied. In the long run, the strong promotion policy of hydrogen fuel cell vehicles performs best, but since fuel cell vehicles enter the market rather late and no other alternative is promoted, emissions may be above the business-as-usual case until 2025.

Comparison to growing VKT and constant fleet scenarios

Real-world observations show that energy efficiency improvements or emissions savings are often worsened by the growth in demand for additional energy services, such as higher VKT – this is referred to as the rebound effect. In this subsection, we analyse the effect of a 10% increase in VKT until 2045, as described in the subsection on policy and scenario specification. Also, the emission effect of the constant fleet scenario, where the number of vehicles is held constant after 2008, is evaluated.

Figure 5.28 illustrates the growing VKT scenario under different policy packages. These rebound scenarios analyse the extent to which CO_2 emission benefits from efficiency improvements and technology diffusion are undermined by higher VKT. The emissions under growing VKT and BAU conditions show a 6% increase compared to the baseline BAU scenario, reaching 75% at the end of the century. Applying the growing VKT and multi-incentive assumptions would reduce emissions toward the level of the baseline BAU scenario to 69%. The best performing policy package is again the strong promotion policy package for the near zero-emission case of HFCV, reducing emissions by 69% by 2100.

Figure 5.29 sets the constant fleet scenario in relation to three different policy packages in the baseline scenario. In the baseline scenario, we assumed that the net new sales saturate over the coming decades and begin to decline toward mid-century. This is compared to the constant fleet scenario, where the number of vehicles is held constant after 2008. Although the strong promotion policy package for HFCV

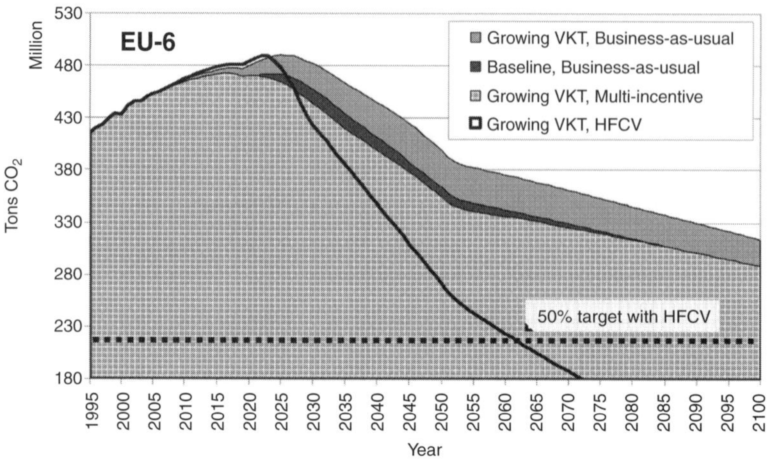

Figure 5.28 Comparison of the CO_2 emissions under the baseline and growing vehicle kilometres travelled (VKT) scenarios. The baseline scenario reaches an emission level equal to the growing VKT scenario with multi-incentives toward the end of the century.

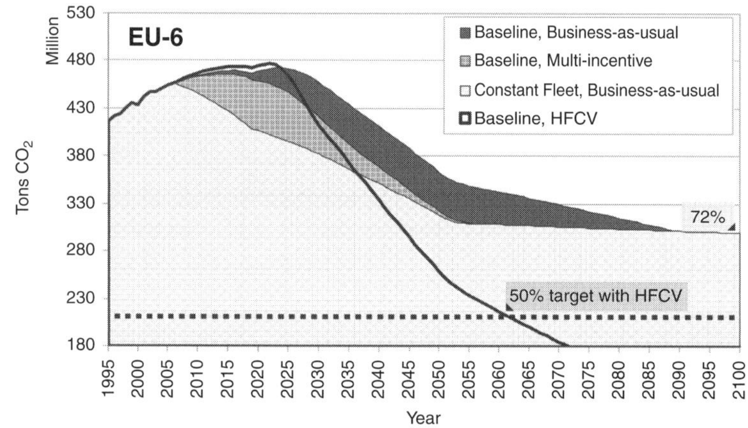

Figure 5.29 Comparison of the CO_2 emissions under the baseline and constant fleet scenario. The emission savings reached through 2035 in the baseline scenario, with strong promotion of hydrogen fuel cell vehicles (HFCVs), are also reached in the constant fleet scenario. However, the baseline scenario with the strong promotion of HFCVs yields remarkably lower emission levels.

still yields the best results in terms of emission reduction in 2100, Figure 5.29 shows that fleet size development is a crucial aspect for the development of CO_2 emissions for the future. The constant fleet scenario performs considerably better, until the vehicle size saturates and starts declining around 2050, with the baseline/multi-incentive assumption.

5.4 Discussion and conclusions

The results of policy analyses regarding diffusion and CO_2 emissions are used to derive qualitative policy implications for the promotion of alternative drivetrain technologies. In doing so, this chapter targets the trans-regional trends in EU-wide, binding policies, since the EU is aiming for harmonised standards in all areas of life. The scenarios with different parameter settings for fleet growth and the development of driven VKT, as well as the formulated policy packages, help to classify and explain the behavioural trends resulting from different incentives applied.

5.4.1 Most effective policies

Those policy packages with the strongest impact on reducing CO_2 emissions are considered most effective. The more effective a policy package is, the sooner the assumed CO_2 emission target can be met. The effect of similar incentives varies by country. Nevertheless, the same qualitative effects can be observed in all of the countries under consideration. Moreover, the introduction of incentives can be complicated and time-consuming. For example, in Switzerland, the introduction of a mandatory tax on CO_2 emissions was preceded by a voluntary phase of fuel consumption reduction agreements between the government and the automobile import sector (Schweizerische-Eidgenossenschaft, 2000). However, over the past years, it turned out that voluntary measures are not sufficient to achieve the emission goals of the CO_2 law and the vision of a 2000-Watt Society that the Federal Office of Environment has set (BAFU, 2007). Hence, political processes bring further delays which have to be kept in mind regarding the timing of the policy packages applied in the model simulations.

Fiscal instruments are a first step

The efficacy of fiscal instruments depends largely on the buying power of the consumer. The reaction of consumers to subsidies or tax reductions in countries with high average household income is less pronounced, but they still react to some extent. In Switzerland, for instance, the diesel fuel price is higher than for gasoline. The resulting savings due to the better fuel economy are much smaller than in other European countries. Nevertheless, diesel is getting more and more popular in Switzerland. This popularity may be due to an emerging appropriate norm, seeing diesel as a future technology with CO_2 emission benefits compared to gasoline vehicles. In this case, diesel will disappear from the Swiss market again when its inherent attractiveness, based on the appropriate norm, is decreased or even replaced by another one. In France, on the other hand, diesel is able to establish a social behavioural norm that is strong enough to be sustained, even after its inherent attractiveness has decreased again. Lower fuel prices and a better fuel economy give diesel a higher economical attractiveness compared to gasoline ICEs. If a technology faces high prices as an obstacle for market penetration, fiscal incentives may be granted. The impact of fiscal incentives on CO_2 emissions in the model simulations turns out to be rather small, particularly in the long run. Using only financial incentives invariably requires higher grants over a longer time period in order to achieve results that are comparable to those of multi-incentive strategies. This would bring significant additional costs that can be avoided or at least partly compensated. Policy makers must also consider that, if the incentives cannot be granted for long enough, they may cause benefits in the system that remain far below the invested capital, for example, a short-term emission reduction followed by a pickup when the system returns to its old equilibrium. Cost-optimisation models, such as MARKAL models (Gül, 2008), may provide useful additional information for the decision-making process on cost-efficient technological solutions. They are discussed in detail in Chapter 6. When comparing pure fiscal incentive policies to those raising the inherent attractiveness, it can be concluded that fiscal incentives only represent a first, necessary step, which must be followed by further measures.

High impact of inherent attractiveness

The impact of the inherent attractiveness is generally higher than that of mere fiscal incentives, as the comparison of the two policy packages, fiscal incentives and raised attractiveness, implies. Of course, it can be argued that this depends largely on the values chosen for inherent attractiveness, as well as their weighting factor. However, there are two aspects to be considered: on the one hand, the values are obtained by model calibration for each country separately – with smaller values, the model is no longer able to reproduce observed behaviour of increasing diesel shares in the various countries; on the other hand, there are other fields of policy application where non-financial mechanisms play an important role, such as environmental reporting systems. Moreover, there is some evidence that the proportion between the two types of incentives is of similar magnitude. Smoking prevention in Great Britain, for instance, was only partially successful when waged with high prices for cigarettes, until negative campaigning began and bans were introduced. However, further review of this topic is needed to draw conclusive parallels here. The comparison may not be fully adequate, because there is no substitution happening, providing an alternative to consumers. Nevertheless, this illustrates that consumer behaviour is not only triggered by fiscal incentives. Sensitivity analysis also shows that the raised attractiveness policy package might produce similar results, if attractiveness was only half as high. In the results, this is illustrated by Italy with the baseline scenario under raised attractiveness conditions.

A general problem with raising inherent attractiveness by non-financial incentives such as bans and permits is the feasibility in the long run when the number of adopters increases. We consider, for instance, a driving permit on the fast lane only for gaseous-fuelled vehicles. This would also imply a ban on driving in the fast lane for all other vehicles. Gaseous-fuelled vehicles would therefore benefit from higher speed and shortened travel time, while other drivers may be caught up in traffic jams. The inherent attractiveness of gaseous-fuelled vehicles would rise, and more and more consumers would adopt the technology. However, the rising number of gaseous-fuelled vehicles would result in a trend reversal. More and more vehicles would drive in the fast lane, and finally the advantage would be negated, having an equal distribution across all lanes. The inherent attractiveness would therefore decrease again, and the shift to gaseous-fuelled vehicles would cease. However, as this happened, a new social norm may already be guiding further adopters. Also, the inherent attractiveness is not reset to low levels in those model policy packages, assuming that a new appropriate norm supersedes the original incentive. This is also the reason why the raised attractiveness and the multi-incentives package show similar, long-term behaviour. In many cases the long-term attractiveness does not matter, since the social norm has already taken the lead. On the other hand, when no leading technology establishes a strong social norm, inherent attractiveness keeps its decisive role, in some cases leading to the coexistence of several different propulsion systems.

Two main effects of multi-incentives

Application of the multi-incentives policy package has two distinct impacts on the development of CO_2 emissions. The first effect is an improved reduction compared to the raised attractiveness package in the first half of the century. The overall diffusion of alternative drivetrain technologies is higher due to the additional fiscal incentives. This yields a small surplus in terms of emission reduction. After the fiscal incentives cease in 2030, the multi-incentives and raised attractiveness packages both maintain the same inherent attractiveness for alternative technologies. Therefore, the resulting emission reduction of the multi-incentives strategy could be expected to keep its advantage and be better at the end of century. However, the emissions of the multi-incentives package turn out to be slightly larger than those of the raised attractiveness package. The effect is rather small, but it is counterintuitive and interesting from a policy-making point of view. The main reason is that fiscal- and attractiveness-raising incentives have a different impact on gaseous-fuelled and hybrid vehicles. A high inherent attractiveness, which is applied to all alternative vehicle technologies to the same extent, favours hybrid vehicles, because they share an already installed fuelling infrastructure with conventional gasoline and diesel ICEs. Gaseous-fuelled vehicles, such as NGVs and LPGVs, face an important infrastructure barrier. Fiscal incentives, in contrast, comprise fuelling station subsidies, diminishing the disadvantage of the gaseous-fuelled vehicles. Gaseous-fuelled vehicles therefore benefit more from the multi-incentives package than hybrid vehicles, compared to the raised attractiveness package. A second reason is that the multi-incentives package is the most aggressive policy package and causes the strongest decline of conventional vehicles. This results in the reduced fuel consumption of gasoline and diesel, and a cutback of fuelling station profitability, which cannot be fully compensated by hybrid vehicles. With decreasing availability of gasoline and diesel, the position of hybrids is weakened, too.

In summary, it can be said that gaseous-fuelled vehicles are more highly promoted by the multi-incentives package, compared to hybrids, than by the raised attractiveness package. This leads to a preferred substitution of conventional vehicles by NGVs or LPGVs in the long run, and since their CO_2 emission characteristic in the model is poorer than that of hybrids, the resulting overall emissions may be increased. Although the impact on emissions is small in the simulations, it may be decisive regarding technological lock-in and the favoured long-term options. Policy makers could react to this effect by promoting different alternative drivetrain concepts and fuels.

5.4.2 Main restrictions to the successful market penetration of alternative vehicles

The model results reveal five basic obstacles to alternative drivetrain technologies' entry into the vehicle market. The first two are discussed above: both the economic aspects of a drivetrain technology and a lack of long-term prospectives in terms of an appropriate norm or adequate incentives, resulting in a low inherent attractiveness, suggest that these options are unworkable.

Rebound effect from infrastructure

An expected and rather evident, but still complex, restriction to successful market penetration is fuelling infrastructure. On the one hand, new drivetrain technology, such as hybrid vehicles, may benefit from an already installed fuelling infrastructure for conventional ICEs, resulting in faster diffusion. On the other hand, there are two possible rebounds. First, a strong penetration of hybrid technology, for example, may support fuelling infrastructure and fuel availability, slowing down the disappearance of conventional ICE vehicles. Second, a quick phase-out of the incumbent ICE technology due to an emerging variety of other alternatives may lead to a decline of the infrastructure. If not yet strong enough, hybrids would suffer from an infrastructure barrier.

The simulation results for Germany and France, for example, illustrate the coupling of HEVs to the incumbent gasoline and diesel ICEs. In the Baseline scenario under BAU conditions, the diesel ICE loses its dominating position in Germany after a decrease in its inherent attractiveness. Gasoline ICEs fill the emerging gap and become the leading technology. The decrease of gasoline ICEs is slowed down compared to France, where diesel ICEs become the leading drivetrain technology. In France, diesel HEVs benefit from the shared diesel fuelling infrastructure. In order to motivate the consumer to switch to hybrid vehicles and avoid a rebound effect, additional incentives must be granted. Moreover, from the strategic point of view for gaseous-fuelled vehicles which need a separate fuelling infrastructure, replacing existing gasoline- and diesel-filling facilities instead of simply adding the gas infrastructure would speed up the diffusion.

Rebound effect from improved fuel economy

Improving fuel economy, which is one among several important measures to reduce consumption and GHG emissions applied today, can cause another rebound effect. If the emphasis on reducing fuel consumption is increased concerning the power–fuel–economy trade-off, it should lead to an increase in the inherent attractiveness of conventional ICEs. The savings that could be realised by the technical achievements may in turn tempt people into driving longer distances, thus compensating or, in the worst case, even over-compensating, for any emission reductions. Additionally, consumers may feel that technology substitution is obsolete. This could slow down the technology transition process and lead to effectively higher overall CO_2 emission. Hybrid technology has the advantage of benefiting from the same technological improvements as conventional ICEs. It could therefore still penetrate the market if accurately promoted. This effect is hypothetical since there is not yet any empirical evidence (de Haan *et al.*, 2007).

Early technological lock-in

Finally, as was illustrated in previous sections, an early technological lock-in is a very strong hindrance to market penetration. If a single bridge technology can build up a new leading social norm early, the introduction of other alternatives will be constrained. An extreme example is the dominance of gasoline hybrid vehicles in Sweden, which leads to an early lock-in, anticipating the emergence of hydrogen fuel cell vehicles. The HFCV policy package in the baseline scenario for Sweden shows a marginal penetration for HFCVs, while in all other countries considered in this study, HFCVs were able to become the leading technology, at least in the second half of the century. The reason for this may be partly explained by vehicle prices in Sweden. The data obtained for Sweden contradicts the purchase price order of all other countries. Gasoline HEVs are much cheaper compared to Germany, for example. It enables gasoline HEVs to fully benefit from its inherent attractiveness. They subsequently establish an early, dominating social norm, which cannot be broken by the late entrance of HFCVs. Also important, the high fuel prices in Sweden contribute to the economic attractiveness of gasoline HEVs, despite a rather high mean household income. In Switzerland, by contrast, none of the promoted alternative drivetrain technologies can benefit from high fuel prices and establish an early norm. HFCVs face a market situation with gasoline ICEs as a clear leader at the time of their introduction. Their advantages carry more weight compared to gasoline ICEs than gasoline HEVs, which facilitates their breakthrough. The challenge for policy makers is to pave the way for the most promising drivetrain technologies, while anticipating an early lock-in of drivetrain systems that were introduced as bridge technologies to break the dominating norm.

The consequence of rising fuel prices

This section highlights the effect of rising fuel prices, although the future price of oil and fuel is very difficult to estimate. The strong rise during the first half of 2008 nearly exceeded the assumptions underlying the VSM. It is therefore useful to explore the effect that an extreme rise in fuel prices would have in the model simulation, as was done for Sweden and Great Britain. The sudden rise in fuel prices by a factor of four causes the increasing diesel trend to slow down. In parallel, a short-term increase in gasoline vehicles is observed immediately after the price increase. This is due to the new level of prices which reduces the relative advantage of diesel compared to gasoline vehicles. With the rise of alternative propulsion systems, ICEs lose their dominance on the car market. Although the diesel share continues to rise at first, it eventually starts to decrease as soon as the alternative vehicles become strong enough.

For Sweden, we again observe the gasoline HEV lock-in. This lock-in can be broken by ethanol vehicles with the assumption that they are fuelled with E85 based on biomass, and are thus less affected by the high fossil fuel prices than HEVs. In Great Britain, gaseous-fuelled vehicles, particularly LPGVs, take the lead since they represent the cheapest option, purchase price included. However, in all cases, the pressure of high fuel prices leads to quite an early technological lock-in of bridge technologies. Policy makers and strategists should be aware of this price effect. Both Sweden and Great Britain achieve high emission reductions. The increased use of cleaner vehicles leads to reduced fuel consumption and lower CO_2 emissions, hence the CO_2 emission reduction is achieved by changing the consumers' purchase preferences without any further constraint on their personal mobility. In Sweden, the emission reduction effect of the dominant ethanol vehicles is strongly increased by their high share of CO_2-neutral fuel additives.

5.4.3 Impact on CO_2 emissions

The impacts on CO_2 emissions, so far not covered by preceding sections, are rigorously discussed here. First, we describe the qualitative development of CO_2 emissions that can be expected for the coming decades under different assumptions of technical improvement and moderate technology substitution. A special emphasis lies on the role of zero-emission vehicles in the context of reduction targets and their market introduction strategy.

Qualitative development

As the emission trends show, technical improvement is important to limit annual CO_2 emissions. To achieve decreasing emissions with an ongoing trend toward the end of the century, continuous technology diffusion is necessary. Even in the baseline BAU scenario with only moderate technology diffusion, the contribution of alternative drivetrain technologies is crucial for long-term emission reductions. Since improvements in fuel economy also affect alternative technologies, a surplus is yielded by the combination of the two effects. The resulting CO_2 emission curves generally show a characteristic pattern, as illustrated in Figure 5.30.

Although diffusion is a smooth process in the simulation runs, there are two characteristic bends observed. The first one occurs at position A in Figure 5.30, around 2020. It can be explained with the development of VKT. The VKT is assumed to decrease slightly in all countries, leading to reduced overall fuel consumption and an effective emission reduction, levelling off around 2020. The following phase between A and C comprises a rise and subsequent decline of emissions, separated by an overall emission turnaround. First, emissions rise since the fleet continues to grow, which is no longer compensated by a decreasing VKT. However, technical

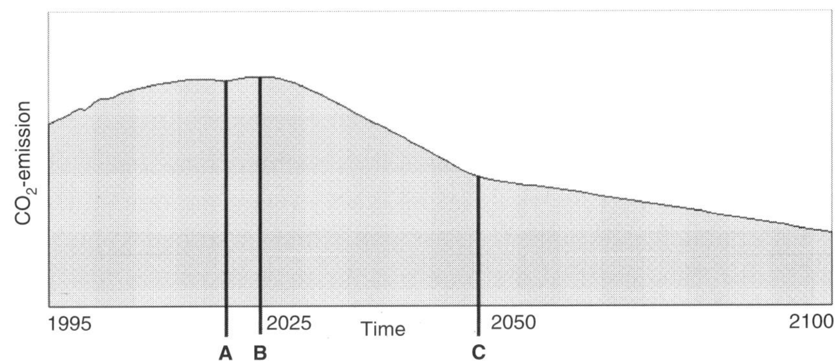

Figure 5.30 The typical, qualitative development of CO_2 emissions from the vehicle substitution model.

improvements, as well as diffusion of alternative drivetrains, introducing more low-consumption vehicles into the fleet, lead to a reversal of the emission trend at B. After tapping the full technological improvement potential by C (around 2050), technology diffusion is left as the only mechanism causing further emission reduction. The development is slowed down significantly in this last phase, since most markets saturate in the second half of the century and alternative vehicles have fully penetrated the market.

Need for near zero-emission vehicles

Near zero-emission vehicles, particularly HFCVs, have a tremendous emission reduction potential, assuming that hydrogen is produced from renewable energy resources. Their advantage disappears if hydrogen is produced by electrolysis using electricity generated from coal, for example. This would only shift the production of CO_2 and degrade the vehicle's zero-emission characteristic. Although the question of hydrogen supply is not completely answered yet, the results emphasise that near zero-emission vehicles are needed to meet emission reduction targets. But there are challenges to their market introduction, since HFCVs are more expensive and lack a fuelling infrastructure. This delays their fleet penetration and leads to a very low contribution to emission reduction during the early phase. Although introduced in 2015 and promoted with a high incentive policy package, the market penetration of HFCVs in the simulation runs remains marginal until 2025. In the short run, this leads to a worse performance of this strategy in terms of CO_2 emissions compared to policy packages promoting several alternatives such as hybrids or gaseous-fuelled vehicles from the beginning. However, the CO_2 emission reduction achieved with HFCVs in the long run are greater than those of less disruptive technologies, reaching almost 50% of 1995 emissions toward the end of the century. This is an important insight for decision makers and political strategists. If goals such as the EU 50% reduction target are to be reached by 2050, an early introduction and consequent promotion is indispensable. This particularly involves a stringent promotion for a parallel build-up of fuelling infrastructure.

EU perspective

The EU projection of CO_2 emissions emphasises two important aspects. First, constant vehicle fleet and growing VKT are important levers in the system. The first can maximise all emission reduction effects early on, while the second is able to increase emissions in the same order of magnitude as the multi-incentives policy package reduces emissions. Growing VKT can easily override all benefits from technology substitution. Second, in the baseline scenario under BAU conditions, the EU emission reduction targets are clearly missed. Countries with a high growth potential, such as Poland, increase their emissions from road transportation in our model runs, such that the overall reduction potential remains small. To meet a 50% CO_2 emission reduction target by 2050 therefore still needs further effort, such as a change in the modal split together with the promotion of zero-emission vehicles. Although the EU fleet scenario results cannot be directly compared with the finding of the MIT fleet model exercises (Bandivadekar *et al.*, 2008b), the policy implications concerning the role of advanced ICE and bridging technologies for meeting short-term administered targets are comparable. However, our study provides a clearer picture on reaching critical climate change CO_2-emission reduction targets and highlights the imperative of pressing and committed efforts toward radical, advanced, near zero-emission technologies (i.e. clean battery electric and FCVs), undertaken today.

5.4.4 Conclusions

The VSM described in this chapter is able to map theoretical concepts of technology diffusion. It has been used to test different policy instruments to promote alternative drivetrain technologies for their effectiveness and their impact on CO_2 emissions.

Coming back to the questions raised in this chapter, we can draw the following conclusions.

1. When do stable states and self-sustaining diffusion processes emerge in the system?

In the long term, stable states and self-sustaining diffusion processes emerge by the social norm and are separated by critical fleet shares. To break the norm of the incumbent drivetrain system and change the technological paradigm, the dominant norm has to be broken. This is helped by competition among a variety of alternative technologies. Nevertheless, policy makers must decide the most favourable technology path, preventing a suboptimal technology lock-in.

2. What are the important levers in the system?

Technology substitution is needed in addition to technological drivetrain improvement, showing a high potential regarding long-term fuel consumption and CO_2 emission reduction. Fuelling infrastructure and vehicle type spectrum play decisive roles in market penetration. The emission benefits from technology substitution may be compensated by a growing vehicle fleet or the amount of VKT.

3. What policies help to reach emission reduction targets; and on what time scale?

Fiscal incentives remain an important category of instruments. However, policy packages raising the inherent attractiveness have a higher, long-term impact than mere fiscal incentives. With the emergence of zero-emission vehicles, CO_2

emission reduction targets become feasible within decades. The time of their introduction and the aggressiveness of promotion are decisive in reaching reduction targets as scheduled.

Raising inherent attractiveness is able to compensate for lower consumer satisfaction due to the higher cost of certain highly efficient technologies. By building up an appropriate social norm, people become willing to pay more and adopt a technology which they would not have chosen otherwise. The emergence of tipping points found by Janssen *et al.* (2006) for the market penetration of NGVs in Switzerland could be explained and generalised with the concept of the social behavioural norm. The simulation runs generally support the argument that policies should always comprise a package of different instruments to get the highest added value in the short- and midterm and to decrease the cost per tonne of CO_2 that is saved (European Commission, 1995; Sterner, 2003). A single fiscal or other incentive is, in most cases, not enough to reach the critical mass for a break out of a lock-in situation and the subsequent establishment of a new norm toward self-sustaining development. Managing undesirable effects is necessary and must be part of any revised policy package.

6

Long-term scenarios of the global energy and transport system

TIMUR GÜL AND HAL TURTON

6.1 Introduction

The earlier chapters considered a number of important factors for the transition to alternative fuels and vehicles, including issues of market introduction, environmental and resource impacts, possible vehicle configurations, and comparative pollutant emissions. These are highly relevant and practical considerations for assessing the options and processes to facilitate a shift to alternative fuels. In this chapter, we expand on these earlier chapters to consider a possible transition to alternative fuels and transportation technologies in a broader, global, and longer term context. Specifically, we consider energy system interactions and technological change, major driving forces of global economic and demographic change, resource availability, and climate change policy. This complements and extends the approach in Chapter 5 which analysed factors driving deployment of new vehicle technologies from consumer and regulatory perspectives.

Let us briefly introduce each of the important global and long-term issues mentioned above, considering first energy system interactions. To assess the suitability of different technological options for the transport sector over the longer term, it is also necessary to consider technological options in the energy sector (for supplying fuels to transportation), and trends in other end-use sectors (which compete for the same energy carriers used by transportation). In the energy sector, the pace of technological change, bottlenecks to the deployment of new technologies (such as a hydrogen refuelling network), and technological lock-ins (for example, the current integrated system of oil production, distribution, and refining) are further factors likely to influence the suitability of potential alternative fuels and technologies in transportation.

Transition to Hydrogen: Pathways Toward Clean Transportation, ed. E. Wilhelm and A. Wokaun. Published by Cambridge University Press.

In addition to these energy system interactions, a number of international factors and trends are important, such as economic development and increasing demands for mobility and energy in today's developing countries, and developments in major markets (which may determine the availability of transport technologies). Finally, adopting a global perspective requires us to also consider physical limitations on the availability of energy resources (particularly oil and gas) and global environmental policy. We incorporate all of these factors in the analysis of transitions to alternative transport technologies in this chapter, but focus particularly on the role of climate policy, and technological development and change.

6.1.1 Background

Global trends in energy demand are already posing major challenges to long-term, economic and environmental sustainability. For instance, between 1971 and 2005, global primary energy consumption increased from 229 EJ up to almost 480 EJ (IEA, 2007), requiring greater resource extraction and resulting in increasing pollution emissions. According to the IEA as well as the US Energy Information Administration (EIA, 2007a), a continuation of current trends is likely to increase global primary energy demand to approximately 740 EJ by the year 2030, driven by strong growth in China and India, and despite the impact of a number of policy measures expected to be taken in OECD countries.[1] One of the main challenges arising from this rapid growth in energy demand is that, today, around 81% of primary energy is supplied by fossil fuels – namely coal, oil, and natural gas – emissions from which contribute to global climate change.

The transport sector, and, in particular, personal transport, plays a pivotal role in these energy consumption trends, mainly due to two key factors. First and foremost, it relies almost entirely on one fossil resource, i.e. petroleum and its products. Petroleum supplies 95% of the total energy used by world transport (IPCC, 2007a), and this high reliance on petroleum fuels translates into high CO_2 emissions as a result of the combustion process. The transport sector produced about 6.3 Gt CO_2 emissions in 2005, or 23% of global energy-related CO_2 emissions (IEA, 2008a), and the total amount of transport-related CO_2 emissions has more than doubled over the past 40 years (2.8 Gt CO_2 in 1971).

The problem of CO_2 emissions from transport is likely to increase in the future as a result of the second key problem in transportation: the global demand for transport is projected to increase strongly in the decades to come as a result of

[1] OECD is the Organisation for Economic Co-operation and Development, a Paris-based international organisation with currently 30 mainly industrialised member countries.

economic development and growth (WBCSD, 2004; EIA, 2007a). Much of the growth is anticipated to take place in the developing world, notably in growing economies such as China, India, and Brazil. Goldman Sachs (2004) expects China and India to emerge as the world's leading car markets, overtaking the United States in 20 (China) to 30 (India) years; whereas the IEA projects that China will overtake the United States as the largest car market in the world even sooner, by as early as 2015 (IEA, 2007).[2]

6.1.2 Approach and objectives

These developments in transport and energy demand are very likely to exacerbate the impacts on global climate change, as suggested by the latest Intergovernmental Panel on Climate Change report (IPCC, 2007b). As a consequence, responding to the challenge of satisfying increased demand for energy and mobility, while at the same time combating climate change and maintaining and improving current levels of energy security, is high on the agenda of policy makers. As a means for tackling this challenge, alternative fuels, such as hydrogen and biofuels, have been advocated as potential substitutes for petroleum fuels in personal transport, but there is still a considerable need for an understanding of drivers and bottlenecks for their deployment. Accordingly, we focus further in this chapter to consider two key questions:

- What is the impact of pursuing different climate policy targets on the prospects of alternative fuels for reducing global CO_2 emissions from personal transport?
- What are key drivers and bottlenecks for the deployment of alternative fuels, in particular, for hydrogen?

The analysis is pursued using the GMM model, which provides a technology-rich depiction of the global energy system. This framework enables us to consider the range of factors described earlier, including energy system interactions and resource limitations. We describe GMM in more detail in the next section. We use this model to develop, quantify, and analyse a number of long-term scenarios of future global energy system and transport technology deployment. Scenarios allow us to ask 'what if' questions about the future, and identify possible technological trends associated with certain assumptions about future economic development and transport demand. Here we consider scenarios of climate policy and their implications for alternative fuels and technologies in transport (and energy system

[2] It should be noted that, during the writing of this book, China overtook the United States as the largest car market (at least temporarily) (Tian, 2010).

development), and focus on deployment of hydrogen. We return to these scenarios in more detail in subsequent subsections of this chapter.

6.2 Introduction to the Global Multi-regional MARKAL model

The GMM model was originally developed by Barreto (2001) and has since been enhanced and applied for numerous energy policy analyses (Barreto and Kypreos, 2004; Rafaj, 2005; Rafaj and Kypreos, 2005; Rafaj *et al.*, 2005). GMM belongs to the MARKAL (MARKet ALlocation) family of models, which is a group of bottom-up, perfect foresight, cost-optimisation models that identify least-cost solutions for the energy system under given sets of assumptions and constraints and for a given time horizon (Fishbone and Abilock, 1981; Fishbone *et al.*, 1983; Loulou *et al.*, 2004). For the analysis presented here, GMM was further enhanced starting from the version described in Rafaj and Kypreos (2005) by increasing the geographical detail (see Figure 6.1), extending the representation of alternative fuel production pathways (hydrogen and biofuels), the personal transport sector, and incorporating a representation of technological learning to mimic elements of long-term technological change.

The reference energy system (RES) of GMM is depicted in Figure 6.2 and covers the entire energy system from the extraction of resources to conversion of primary energy carriers to the use of final energy carriers in end-use technologies across eight different demand sectors. Among these demand sectors, personal transport is

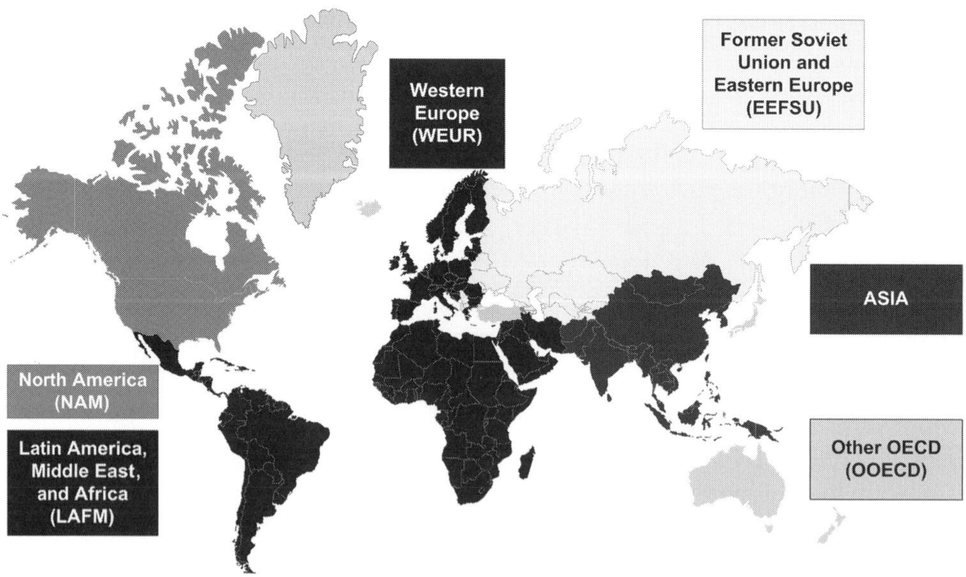

Figure 6.1 The six world regions represented in the global multi-regional MARKAL model.

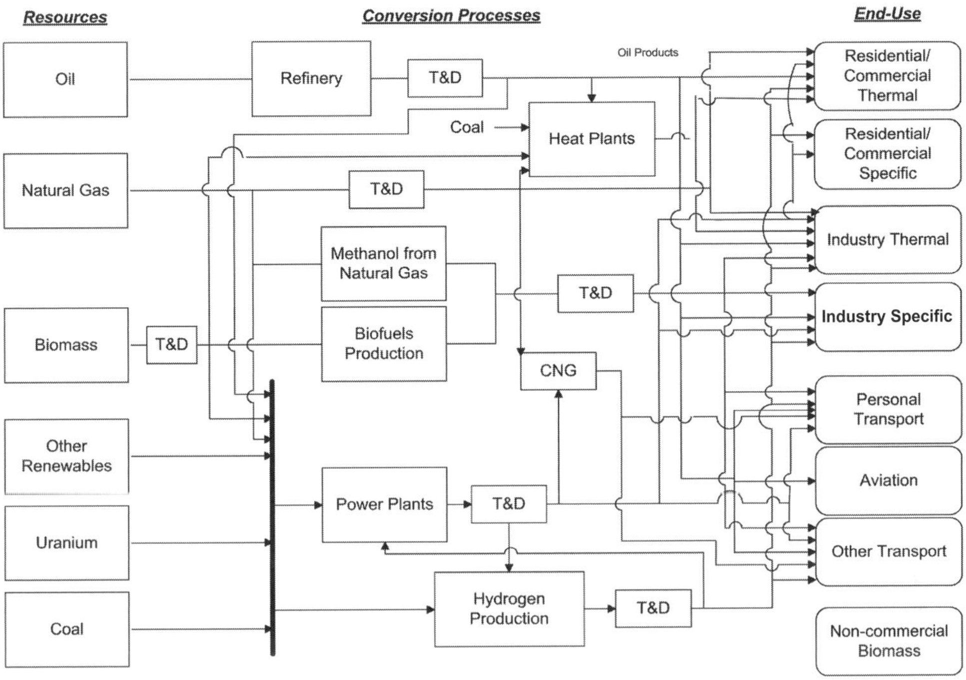

Figure 6.2 Simplified reference energy system as applied in global multi-regional MARKAL model.

modelled in detail. For the other demand sectors, we define a set of generic standard and advanced end-use devices (e.g. Rafaj, 2005).

As previously mentioned, GMM incorporates a high level of detail in the depiction of the supply sector and energy conversion processes. Technology representation in electricity generation includes conventional and alternative options based on fossil fuels, nuclear and renewables, and is described in more detail in Rafaj (2005). For analysis of a possible transition to alternative transportation technologies presented here, GMM was extended to consider hydrogen and biofuels fuel chains as well as personal transport in much more detail (see description of the hydrogen and biofuels modules below, and transport technology characteristics in Figure 6.4). Another feature of GMM to note is the model's very long time horizon (100 years), allowing us to explore longer term technological trends, and the long-term implications of economic development and limited resource (principally oil and gas) availability.

The representation of different geographical regions in GMM (Figure 6.1) allows simulation of bilateral or global trade of energy or environmental commodities. Energy commodities traded in GMM include hard coal, natural gas, liquefied natural gas, oil, diesel, and gasoline, as well as biofuels.[3]

[3] In GMM, we also incorporate estimates of trade costs based on Manne *et al.* (1995).

GMM is calibrated to statistics for the year 2000 from the IEA (2002a, b) on both the production and demand sides. Future demands for energy in GMM are based on the B2 scenario in the IPCC's SRES, and this scenario is characterised as a 'dynamics-as-usual' development of the world, with median population and economic growth, moderate economic catch up by developing countries, and a continuation of historical trends in energy intensity (IIASA, 2007). Future demands for personal transport are based on IEA/WBCSD (2004) until the year 2050.[4] Thereafter, demand developments are based on Turton (2006a). For other transport sectors, future demand is assumed to grow at the same pace as GDP in the SRES-B2 scenario.

We now briefly discuss three additional important features of GMM which are particularly relevant for transitions in the transportation sector: resource availability, climate change, and technological change.

6.2.1 Resource availability

Over the long term, depletion of fossil fuel resources is likely to have a strong influence on the development of energy and transport systems. Fossil resource estimates in GMM are based on Rogner (1997) and distinguish different categories of coal, oil and natural gas. These categories reflect different levels of certainty and accessibility, with extraction costs increasing for more speculative or less accessible resource categories (Rogner, 1997).[5] Other resources considered in GMM include uranium and biomass. While uranium is assumed to be abundant, the global biomass potential has been limited to 195 EJ per year, distinguished according to six different biomass types.[6] Importantly, there is a large degree of uncertainty about this biomass potential, and a possible range of (technical) primary biomass potentials of 33 to 1135 EJ per year for the year 2050 has been suggested by Hoogwijk *et al.* (2003). The potential assumed in our analysis is therefore rather at the conservative end of this range, accounting for the controversy about potential land use conflicts with food production. Biomass feedstock costs were derived from the FAO as weighted averages for the years 2000–2003 (FAO, 2006), using producer prices including biomass production, harvesting, pre-treatment, transport and storage, as well as farmers' margins. Prices vary by feedstock and world

[4] Note, the population growth and car transport demand from IIASA (2007) and IEA/WBCSD (2004) assumed in this chapter are lower than assumed for Western Europe in the analysis in Chapter 5. This reflects some of the uncertainty about many of the factors affecting future transport demand. Such a larger population and car travel demand would further increase the challenges posed by transport to long-term economic and environmental sustainability, likely necessitating an even more rapid switch to alternative fuels and drivetrain technologies than described in this chapter.

[5] For example, oil extraction costs range from around US$20 to US$60 per barrel, depending on the resource category and world region.

[6] Wood residues, rapeseed/soybeans, corn grains, sugar cane/sugar beet, stover, waste.

region and are in the range between 1.5 US$/GJ (sugar beet in OOECD) to 11.2 US$/GJ (soybeans in ASIA).[7] For further details, see Gül (2008).

For some other renewable technologies, explicit assumptions have been applied to reflect deployment potentials, which limit the market penetration of renewable energy technologies such as solar, wind, and hydro power in GMM (see Gül, 2008, for more details). Furthermore, intermittent electricity generation technologies such as solar PV and wind power are assumed to be limited to a maximum penetration level of 25% of electricity generation by region. This is a rough estimate of the maximum level of intermittent generation that can be integrated while maintaining grid stability and avoiding the need to dump substantial amounts of generation, although there is some uncertainty (Hoogwijk *et al.*, 2007).

6.2.2 Climate change modelling

The energy sector, and transportation in particular, currently relies to a large extent on the combustion of fossil fuels, which produces CO_2 and other GHG emissions. To account for the impact of these CO_2 emissions on the climate, GMM has been linked to the MAGICC model (Wigley and Raper, 1997; Wigley, 2003). This also allows the assessment of the impact of climate policy on the energy system, including on the introduction of alternative fuels in transport.

6.2.3 Technological learning and cost assumptions

One further feature of GMM that makes it well suited to the analysis of the future role of alternative transportation technology and fuel options is the representation of some elements of technological change. Specifically, GMM seeks to represent the process of technological learning, whereby experience gained in the production and use of a new technology leads to improvements in the performance and reductions in the cost of the technology (Messner, 1997).

In GMM, this technology learning has been applied by creating technology clusters, i.e. families of technologies evolving and diffusing together (Barreto and Kypreos, 2006). Similar technologies from the same technology family may benefit from experience gained through the deployment of other technologies in the family, leading to improvements in cost or performance for all the technologies in the same technology cluster. Thus, the clusters approach has been applied in this analysis to a number of technologies by identifying several 'key components',

[7] Throughout, US$ refers to year 2000 United States dollars.

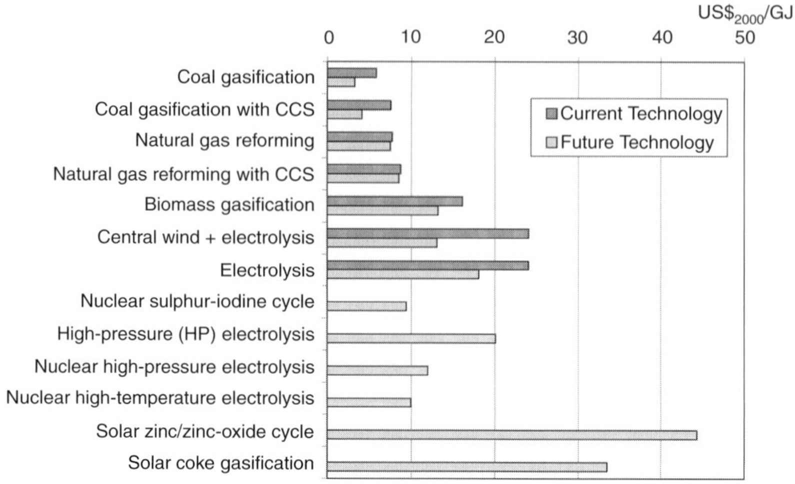

Figure 6.3 Hydrogen production costs from different technology options in Global Multi-regional MARKAL model. Adapted from H2A (2006a) and Felder (2007).

i.e. components that are common to several technologies (e.g. a battery system, which is used in a variety of vehicle types).

Technology learning is implemented in GMM for a range of technology options, including electricity generation, hydrogen production, biofuel synthesis, and vehicle technologies. It is important to note that the learning rates for these technologies are highly uncertain, with the estimates applied here falling within the literature ranges (e.g. McDonald and Schrattenholzer, 2001). Specifically, we used estimates in Barreto and Kypreos (2006) for electricity, Ragettli (2007) for biofuels, and an analysis of H2A (2006a) for hydrogen. Figure 6.3 illustrates the resulting cost estimates for different current and future hydrogen production pathways.

Personal transportation in GMM is treated using 'generic' vehicles, i.e. no distinction is made between different vehicle sizes, and vehicle technologies are specified so as to ensure that different drivetrain options provide a similar level of service in terms of power output and vehicle range. All cost data for learning components are based on Kasseris and Heywood (2007), Kromer and Heywood (2007), and Turton (2006a), and are presented in Table 6.1 distinguishing the initial and floor cost[8] assumptions along with the learning rates used in the personal transport sector in this analysis. Floor cost assumptions in personal transport are based on the same literature sources and represent estimates for what is achievable within a few decades with reasonable technology development progress. In our modelling analysis, whether and when these floor costs are achieved is determined

[8] A floor cost represents the assumed lowest, ultimately achievable cost of a technology.

Table 6.1. *Investment cost clusters in personal transport.*[a]

Technology	Engine size	Learning rate	Initial cost	Floor cost	
Fuel cell	40 kW[b]	15%	250	50	US$/kW
Reformer	40 kW	15%	90	25	US$/kW
Hybrid battery system	28 kW	15%	2500	800	US$/vehicle
Fuel cell battery system	42 kW[b]	15%	3250	1200	US$/vehicle
Battery electric	48 kWh	15%	16250	12000	US$/vehicle
Plug-in hybrid	8.2 kWh	15%	6500	2800	US$/vehicle

[a] The learning rate is the rate of cost decline per doubling of cumulative capacity installed.
[b] The assumed specifications of FCVs (with a 40-kW fuel cell system and a 42-kW lithium-ion battery) differ from some of the prototypes under development and are discussed in Chapter 3, which envisages a relatively larger fuel cell system. The specification described in the table above has been adopted on the basis that the fuel cell represents one of the most expensive components of the FCV, and installing a larger battery enables a smaller fuel cell to be used, significantly reducing the cost of the vehicle.
Source: Kasseris and Heywood (2007), Kromer and Heywood (2007), and Turton (2006a).

endogenously, depending on whether investing in these new, initially expensive options proves to be cost effective. Thus, we assume a range of possible future costs for different drivetrain technologies, as illustrated in Figure 6.4, which shows the additional cost of alternative drivetrains compared to the advanced gasoline ICEV anticipated for 2030. Figure 6.4 also presents assumed efficiencies in terms of gasoline-equivalent litres of fuel per 100 km of travel.

6.2.4 Hydrogen and biofuels modules

In the course of developing the analysis presented in this chapter, GMM was extended by two modules for representing the production and delivery of hydrogen and biofuels – both potentially important long-term options for transportation. Technology specifications in these modules are based on an extensive review of available literature and discussion with technology experts. For biofuels, the module considers biodiesel, Fischer–Tropsch–diesel, ethanol, methanol, dimethyl ether, and SNG derived from biomass (bio-SNG), based on a review by Ragettli (2007) of 50 biofuel production processes. In GMM, the costs of biofuel production depend on the type of fuel produced as well as the cost of the biomass feedstock, i.e. it differs by world region, and ranges from 2.3 US$/GJ for methanol derived from wood residues in Latin America, Africa, and the Middle East (LAFM) (refer to Figure 6.1) to 20.5 US$/GJ for ethanol from corn grains in Western Europe

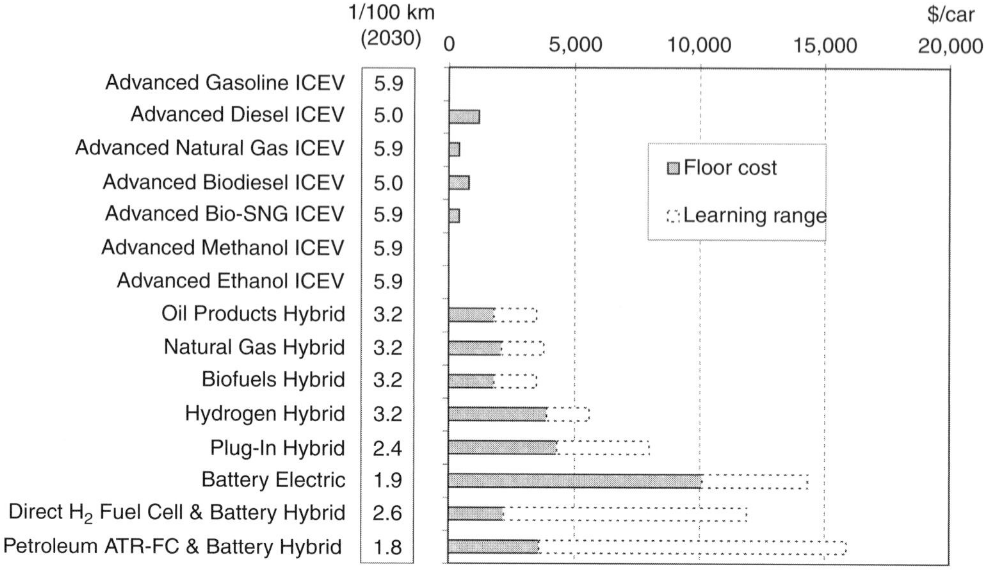

Figure 6.4 Efficiency and technology costs of various drivetrain options in the Global Multi-regional MARKAL model, additional cost compared to advanced gasoline internal combustion energy vehicles. Note: ATR-FC refers to autothermal reformer-fuel cell. Efficiencies are given in terms of gasoline-equivalent litres per 100 km, based on real-world driving conditions. Based on Kasseris and Heywood (2007), Kromer and Heywood (2007), and Turton (2006a).

(WEUR). On top of these costs, additional costs for delivery are in the range of 3.49 US$/GJ for biodiesel to 8.4 US$/GJ for bio-SNG based on IEA (1999).

For hydrogen production, a wide range of central hydrogen production technologies have been identified and considered in the analysis in this chapter, covering coal gasification both with and without CCS, natural gas reforming with and without CCS, alkaline water electrolysis, wind turbines dedicated to producing hydrogen from electrolysis, three different advanced nuclear hydrogen production technologies, biomass gasification, and two solar hydrogen production technologies. Coal gasification is the cheapest technology option with expected long-term future costs of 3.25 US$/GJ; solar hydrogen production is the most expensive at almost 70 US$/GJ (see Figure 6.3, and refer also to Chapter 2). A detailed discussion of these costs is beyond the scope of this chapter, but can be found in Gül (2008).

For the delivery of hydrogen, an extensive network of technology options was considered in the present analysis. Figure 6.5 depicts the applied network comprising pipeline delivery, truck delivery (gaseous/liquid), and combined systems, where hydrogen is delivered by pipeline to a terminal at the gates of a load centre (e.g. a city) and then further distributed by truck (liquid/gaseous).

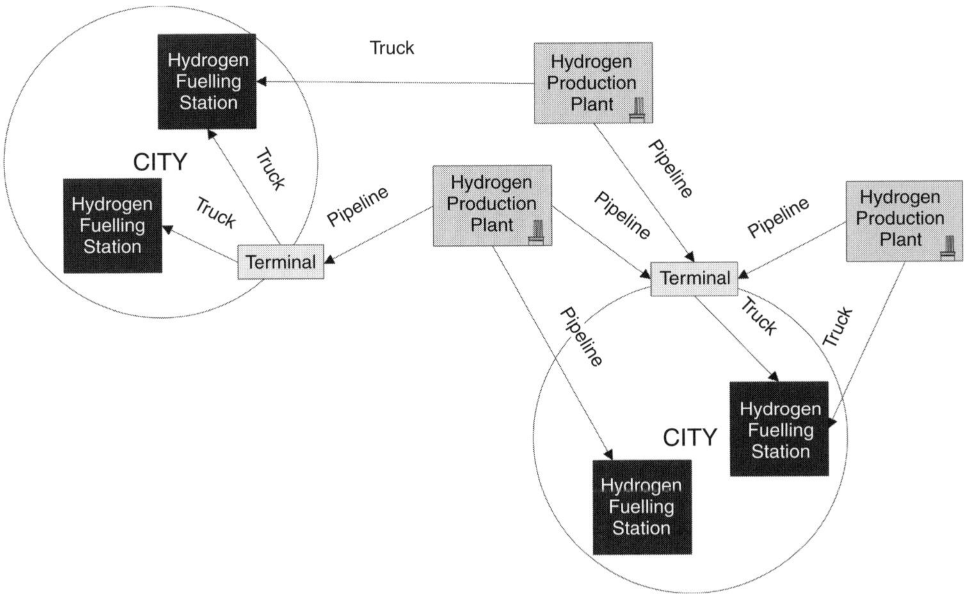

Figure 6.5 Hydrogen delivery infrastructure.

The hydrogen delivery infrastructure was modelled using H2A (2006b) for a demand centre size of 250 tonnes of hydrogen per day[9] and an average delivery distance of 80 km. Costs of the different hydrogen delivery options vary between 9.9 US$/GJ (pipeline delivery) to 17.5 US$/GJ (gaseous truck delivery) and are additive to the cost of hydrogen production.

Finally, the hydrogen supply chain is complemented by decentralised options for hydrogen production, comprising reforming of natural gas, gasoline, and methanol as well as alkaline water electrolysis based on H2A (2006a) and Simbeck and Chang (2002). Among these, natural gas reforming is the cheapest option, with expected future costs of 16.3 US$/GJ of hydrogen; the most expensive option is gasoline reforming at a cost of 26.1 US$/GJ. Again, a detailed discussion of costs is beyond the scope of this chapter, but may be found in Gül (2008).

The preceding discussion provides an overview of many of the relevant features of the GMM model, which we use to analyse scenarios of long-term and global transitions to alternative transportation options. As mentioned earlier, it is important to note that scenarios provide a means to explore 'what if' questions about the future, and some of the significant uncertainty that exists over the development of new and emerging technologies, climate policies, and transport demand. Here we pay particular attention to uncertainty about future climate

[9] This could correspond, for example, to a city size of about 800 000 inhabitants, if all hydrogen demand was from fuel cell vehicles.

policy, and the implications for the deployment of alternative fuels and transportation technologies. We start by presenting a baseline scenario, describing a world in which we assume no climate policy. This provides a point of reference to understand subsequent scenarios incorporating a climate policy and determine how this may affect the deployment of alternative fuels and technologies in transportation.

6.3 Baseline scenario

6.3.1 General structural changes in the energy system

The baseline scenario of GMM considers a future world in which there is no effective climate policy (and thus current policy efforts such as Kyoto or post-Kyoto measures for reducing CO_2 emissions are ignored). Consequently, CO_2 emissions in the baseline continue increasing along current trends and reach 21 Gt of carbon per year by 2100, resulting in an atmospheric CO_2 concentration of around 770 ppmv by the end of the century. Figure 6.6 displays global CO_2 emissions by region, showing how anticipated economic growth in developing regions such as ASIA and LAFM results in them becoming the main contributors to global CO_2 emissions.

Under the baseline scenario assumptions, annual primary energy consumption is projected to undergo significant growth, reaching almost 1600 EJ by 2100. In the

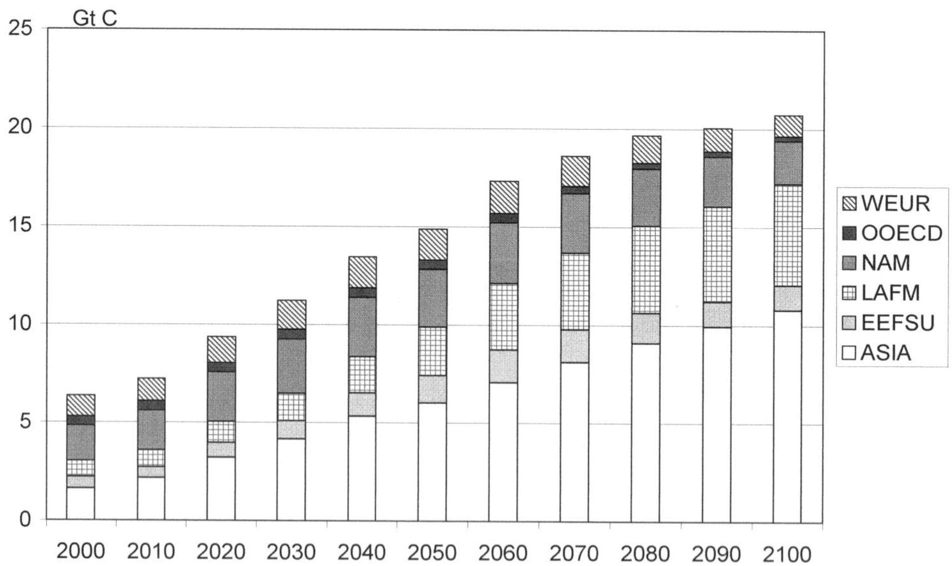

Figure 6.6 Global CO_2 emissions in the baseline scenario (in Gt C).

absence of carbon mitigation policies or policies that internalise external costs, fossil fuels remain the predominant source of energy, and coal makes up 41% of total primary energy consumption in 2100 as a result of heavy utilisation in electricity production. (These results are not presented here, but it is worth noting that global electricity generation increases by a factor of more than seven in the baseline scenario.) Even though the relative shares of oil and natural gas in total primary energy consumption are modelled to decline in this scenario, the absolute levels of consumption increase until the year 2050; thereafter, consumption declines, driven by resource depletion and correspondingly higher oil and gas prices. Oil and natural gas together account for 19% of primary energy consumption by 2100, of which 158 EJ is from oil, and 139 EJ from natural gas. By the end of the century, oil is consumed almost entirely in aviation, which experiences rapid growth in this scenario.

Most of the growth in primary energy consumption is projected to take place in ASIA as well as in LAFM, as shown in Figure 6.7, in line with the CO_2 emission developments described earlier.

6.3.2 Personal transport and alternative fuels

Personal transport undergoes considerable changes in the baseline scenario. With vigorous growth in demand for mobility in developing countries, oil availability and price become an important issue for the personal transport sector, leading to a number of developments. These developments are illustrated in Figure 6.8, which shows how conventional ICEVs are gradually replaced by biofuels ICEVs and by hybrid vehicles, in particular, oil product, and synfuel hybrids.[10] In the long run, biofuel hybrids become an important technology in personal transport. The attractiveness of hybrid vehicles is driven partly by improvements arising from technology learning, making hybrids cheaper, and partly by increasing fossil fuel prices, and fuel scarcity. It is worth noting in this context that early utilisation of biofuel hybrids is predominately in North America (NAM) where assumed longer driving distances per vehicle, compared to other world regions, help to overcome the higher upfront cost of these vehicles earlier. However, under this scenario, NAM is overtaken by the developing regions LAFM and ASIA toward the middle of the century, i.e. by regions with high availability of low-cost biomass.

As noted, alternative fuels such as biofuels and synfuels play an important role in transportation in this scenario, and global production of these fuels increases significantly. In the baseline scenario, the main alternative fuels produced are liquid fuels derived from coal, constituting 80% of alternative fuel production in

[10] Throughout this chapter, the term 'synfuels' refers to liquid fuels derived from coal.

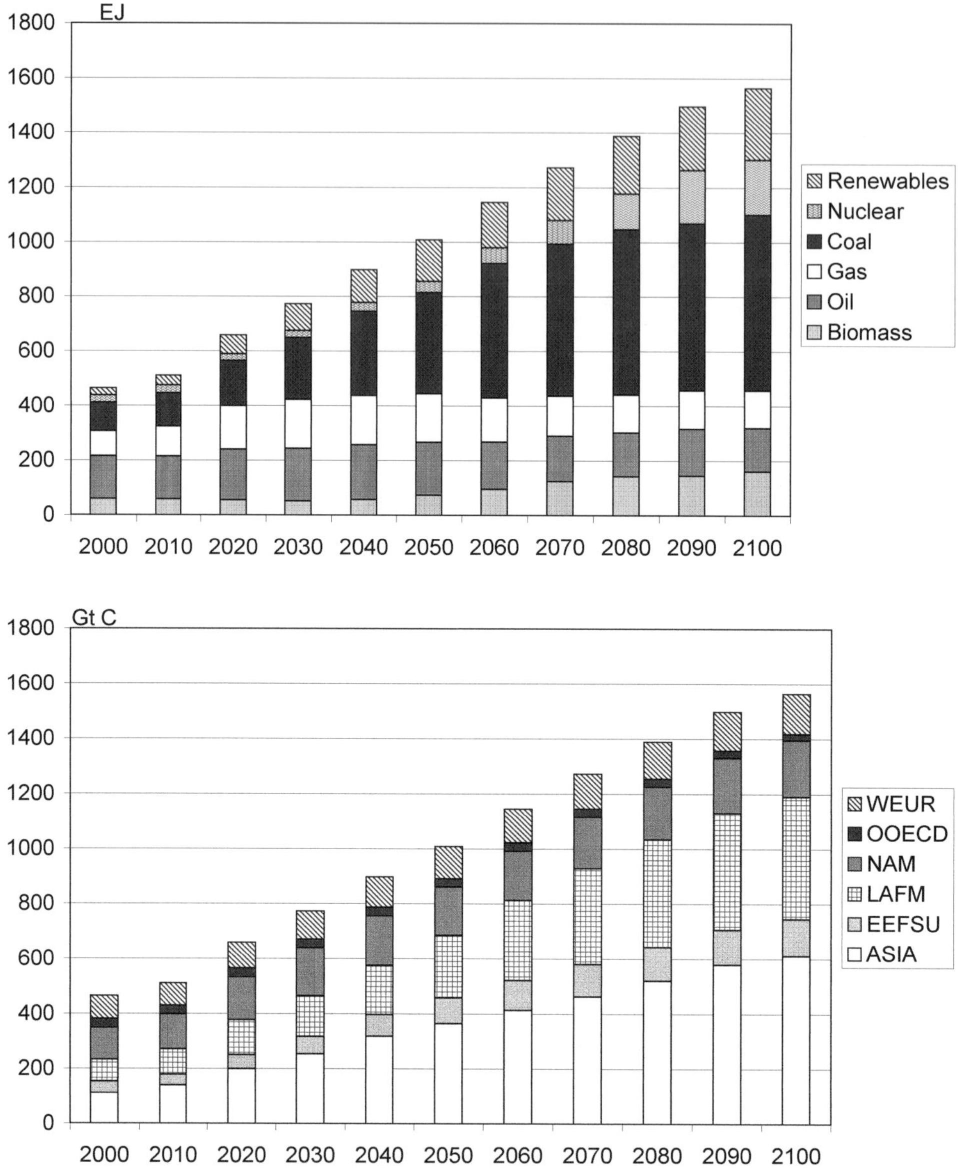

Figure 6.7 Primary energy consumption in the baseline scenario.

the year 2100 with 115 EJ produced. Total global biofuels production reaches 29 EJ in the year 2100, occurring mainly in LAFM as a result of high availability of low-cost biomass resources, in particular, wood residues, sugar cane, and stover. In the absence of policy support, hydrogen production remains marginal at about 0.6 EJ in the year 2100 in the baseline scenario and is used in other transport niche markets in ASIA only.

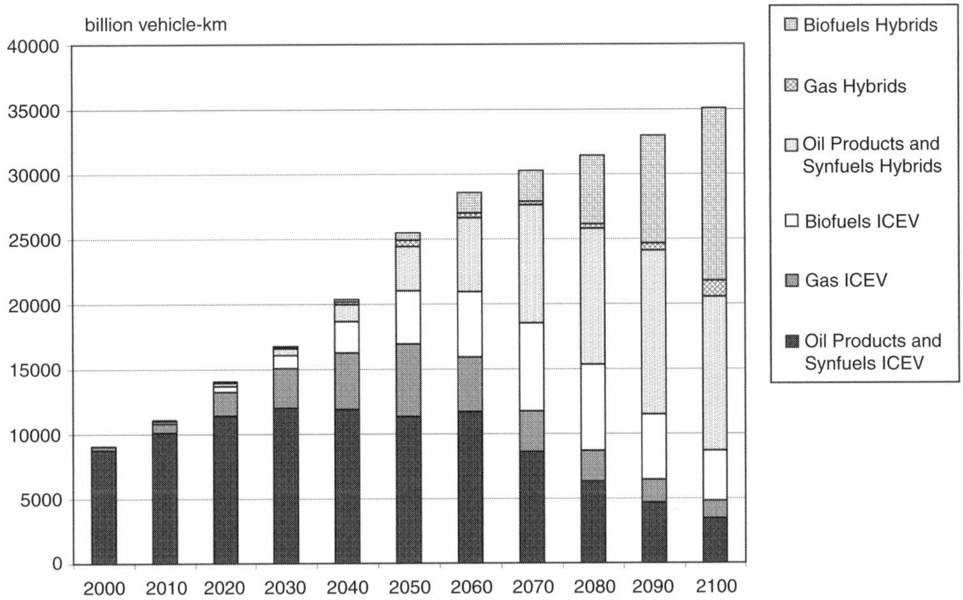

Figure 6.8 Personal transport development in the baseline scenario.

The baseline scenario, thus, suggests that biofuels are a competitive option in personal transport in the long run, when petroleum becomes a limited and costly resource. Even though this scenario only considers the biomass potential deemed to be available for energy purposes, it is still important to remember that biomass costs are assumed to stay at current levels, i.e. constant in real terms throughout the century. With developing biomass markets, however, the situation may alter and market prices of biomass may increase. This in turn may reduce the attractiveness of biofuels.

The following section now aims to explore the impact of policies addressing climate change on the market penetration of alternative fuels. This analysis makes use of one important feature of GMM, which is the possibility of assessing the impact of climate change policies with the climate model MAGICC. The results shed light on the competitiveness of alternative fuels to respond to this energy system challenge.

6.4 Climate change scenarios

The analyses conducted in this section focus on three increasingly stringent climate stabilisation scenarios:

- 650 ppmv CO_2 concentration in the atmosphere,
- 550 ppmv CO_2 concentration in the atmosphere, and
- 450 ppmv CO_2 concentration in the atmosphere.

Emission limits in GMM were implemented as cumulative emission constraints. The analysis therefore considers 'when' and 'where' flexibility for CO_2 mitigation. Flexibility mechanisms such as 'where' flexibility are defined in the Kyoto Protocol (UNFCCC, 1999) in the form of GHG emission trading methods designed to allow mitigation targets to be achieved at lower cost. Applied to the modelling analyses with GMM, 'when' and 'where' flexibility means that the scenario results presented here are cost optimal with regard to timing and location of CO_2 mitigation measures. In this light, the results presented here can be regarded as least-cost solutions, not only from a perspective of optimal technology choices, but also from optimal timing of CO_2 abatement.

6.4.1 General structural changes in the energy system

As a result of pursuing increasingly stringent climate policy targets, the global energy system undergoes substantial transformation in the climate mitigation scenarios modelled with GMM. Primary energy consumption is likely to be directed toward increasing shares of carbon-free energy carriers, in particular, newer renewable energies (solar, wind, etc.), biomass, and nuclear energy. The use of coal is gradually reduced as we move toward more stringent climate policy.

These developments coincide with an increasing level of electrification, with electricity becoming the main energy carrier toward the end of the century throughout all scenarios investigated. Hydrogen plays a progressively more important role in the second half of the century, especially when stringent climate policy targets are pursued. Biofuels are an important energy carrier in all scenarios; however, under the most stringent climate policy target, a reduction of biofuels consumption is observed, as available biomass is consumed for direct uses instead.

6.4.2 Personal transport and alternative fuels

Personal transport contributes to the climate policy targets analysed in these scenarios through the deployment of an increasing share of HFCVs, as shown in Figure 6.9. However, we can see the level of deployment varies with the stringency of the climate policy target: FCVs do not enter the market in the 650 ppmv climate policy target scenario, but reach a market share of 78% by the year 2100 under a 450 ppmv policy regime.

The assessment of the impact of climate policy on personal transport shows that, with a mild climate policy, biofuel hybrids as well as biofuel ICEVs increase market share at the expense of oil product and synfuel hybrids (compared to the baseline scenario; see Figure 6.8). Under more stringent climate policy targets, however, biofuel vehicles gradually lose market share, while HFCVs become

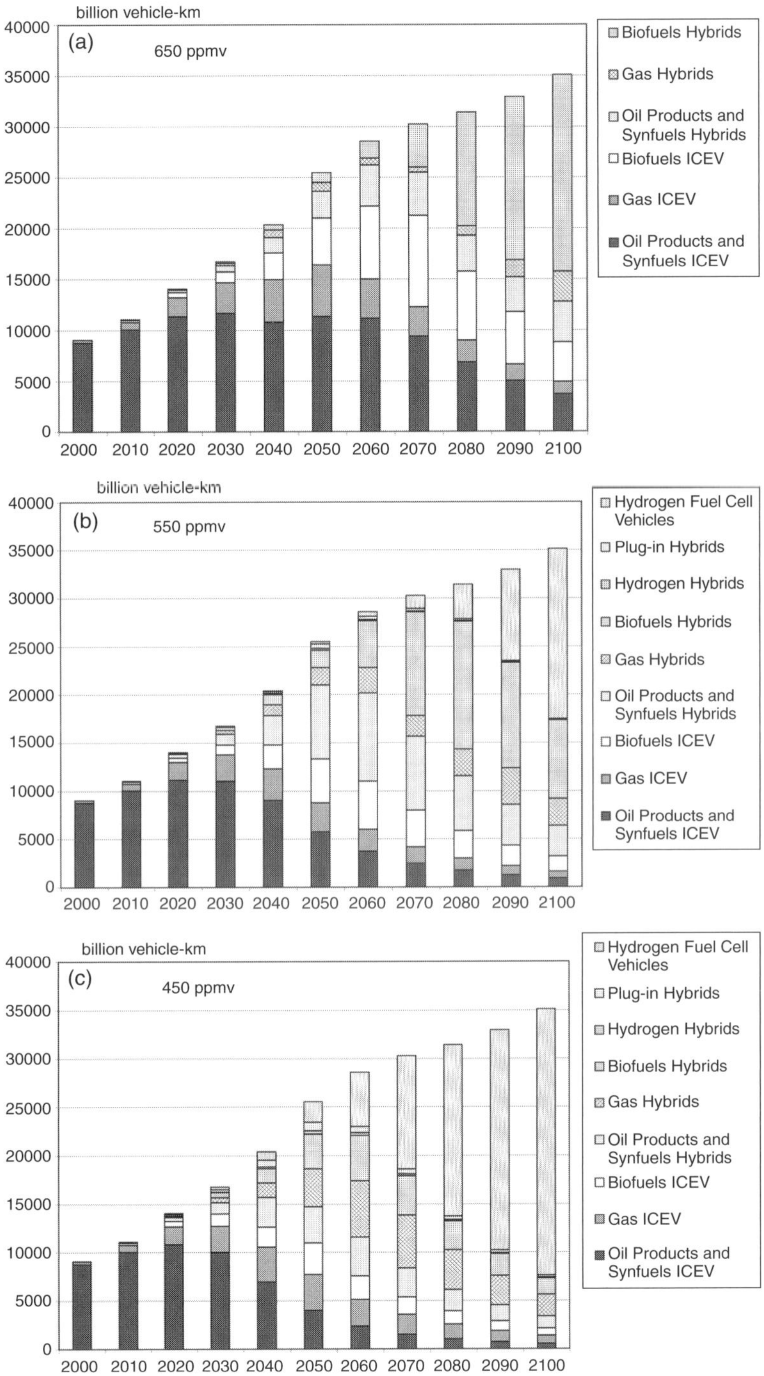

Figure 6.9 Personal transport across scenarios (a) 650 ppmv, (b) 550 ppmv, (c) 450 ppmv.

increasingly important. In these scenarios, biofuels represent a transition fuel and share the market with natural gas, oil products, and synfuels (mostly hybrid vehicles), until hydrogen fuel cells reach maturity and take over the market. These patterns of deployment are similar to those presented in Chapter 5.

Figure 6.10 presents hydrogen and biofuels production under the scenarios described above. It shows that hydrogen production increases in line with the

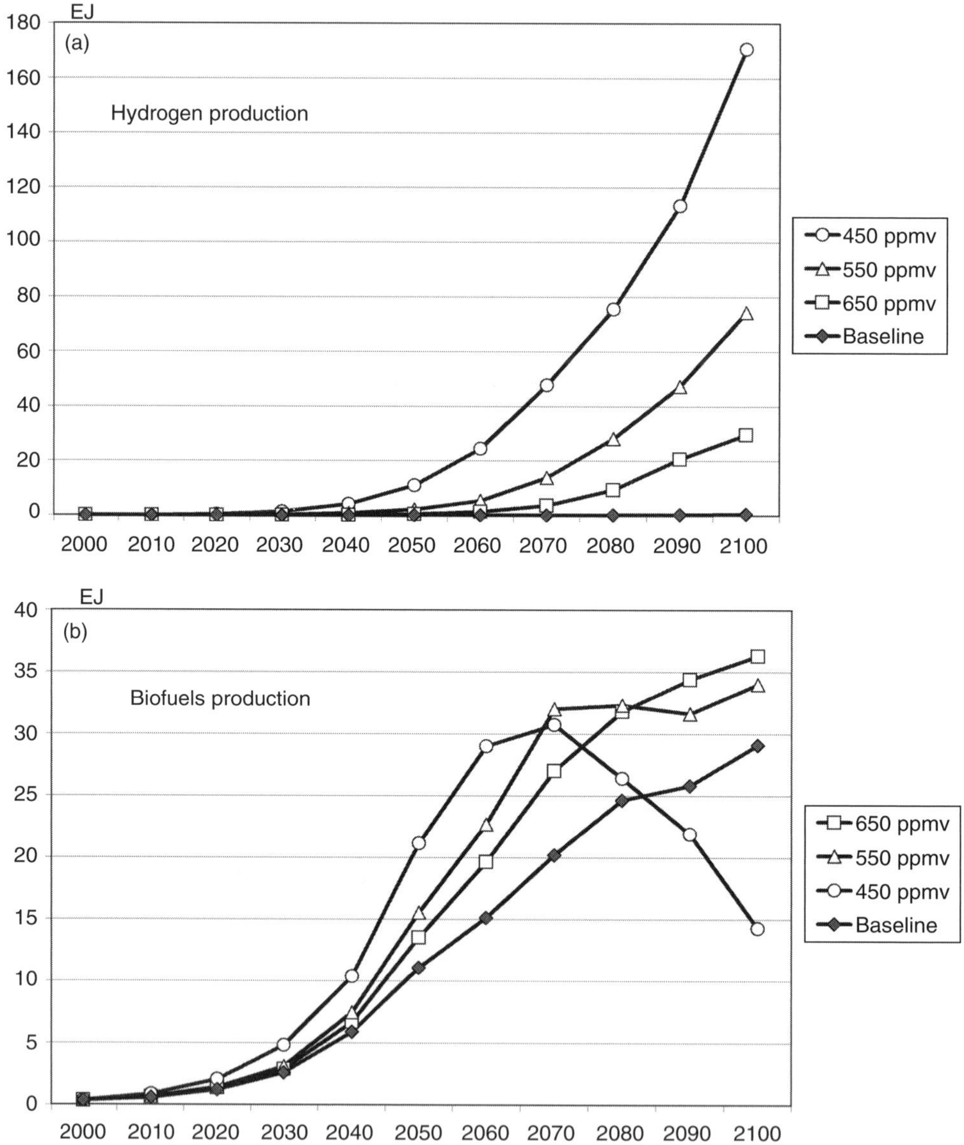

Figure 6.10 Total hydrogen (a) and biofuels (b) production for energy use under different climate policies.

stringency of the climate policy targets, reaching up to 171 EJ of hydrogen production by the year 2100 under the strongest climate policy target.

The trends for biofuels are less obvious. All of the climate policy targets modelled here support the earlier and stronger deployment of biofuels, although much of this deployment is in other subsectors of transport, rather than in passenger cars. However, for the strongest climate policy target (and to a lesser extent also for the 550 ppmv climate policy scenario), the deployment of biofuels decreases in the second half of the century, which is when hydrogen becomes available on a larger scale.

It is also worth mentioning that the scale of production by the end of the century is significantly different for hydrogen and biofuels. Biofuel production levels remain at the same order of magnitude as in the baseline, while hydrogen production is boosted substantially in order to achieve the climate policy targets. The key limiting factor for a further deployment of biofuels is the availability of biomass: in a carbon-constrained society, biomass is a valuable energy carrier, and there appear to be other cost effective options for using this fuel in electricity and heat production.[11]

The deployment of biofuels is largely concentrated in two world regions: ASIA and LAFM. This result occurs for two main reasons: first, both regions are likely to experience the most significant growth in demand for energy services across the century; and second, LAFM is the region with the most abundant availability of low cost biomass.

A major uncertainty affecting the realisation of the high levels of utilisation of HFCVs observed here is whether costs can be brought down sufficiently. Given the potential importance of the fuel cell cost, we conducted a sensitivity analysis in which more pessimistic assumptions were applied for the cost that could ultimately be achieved for FCVs. This showed that, with a stringent climate policy, i.e. a 450 ppmv CO_2 concentration target, hydrogen ICE-HEVs then would be the technology of choice. The reason that hydrogen is still attractive even without cost-competitive fuel cells can be attributed to the ability to produce H_2 from low-emissions sources; while other alternative drivetrain and fuel options are less attractive. For instance, biofuels are limited through the available biomass potential, and battery vehicles are too expensive under the assumptions in this analysis.[12] This makes hydrogen an attractive energy carrier for stringent climate change mitigation targets, *independent* of whether it is used in fuel cells or ICEs. Under a

[11] Note that GMM considers biomass IGCC plants with carbon capture, which is a highly competitive technology under strong climate change mitigation policy. The results were confirmed by a sensitivity analysis that excluded this technology.

[12] As discussed in Section 6.2.3, vehicle characteristics are specified such that the different vehicle types are all able to provide a similar service, particularly in terms of power output and vehicle range. This ensures that different vehicle technologies represent substitutes. For battery vehicles, we have thus assumed a battery size of 48 kWh to provide sufficient driving range. This large battery makes the technology fairly costly (see Figure 6.4).

550 ppmv concentration target, however, biofuels and natural gas hybrids were found to dominate future transport, with the same assumption about fuel cell costs. This shows that the utilisation of hydrogen as a fuel in personal transport is, from the perspective of cost-optimisation planning, not only dependent on the future costs of fuel cells, but also on the degree of the climate policy target.

As a final remark, it is important to note that the future of HFCVs is also dependent on the future costs of plug-in hybrids as well as battery cars. Sensitivity analysis conducted for the 450 ppmv concentration target showed that these vehicles become competitive at battery floor costs of about 100 US$/kWh. These costs, however, seem very optimistic from today's perspective, for example, the IEA sees future battery costs rather in the order of 300 US$/kWh (IEA, 2008b).

6.4.3 Transitioning to alternative fuels

The discussion of the implications of climate policy targets suggests that the energy system needs to undergo significant structural changes to shift away from fossil fuels as part of efforts to mitigate climate change. To get a better understanding of the scale of the challenge to transition to alternative fuels, a useful indicator is the level of investment required to achieve climate policy targets. Figure 6.11 compares the investment needs for hydrogen and biofuels production and delivery technologies to investment needs for other 'conventional' fuels, including fossil fuels and electricity, obtained from the analysis with GMM. It further compares the baseline scenario investments with those of the 450 ppmv and 550 ppmv climate policy scenarios. Figure 6.11a represents cumulative undiscounted investment needs until the year 2050; Figure 6.11b cumulative undiscounted investment needs until 2100.

The analysis shows that, under baseline conditions, fuel supply will require substantial investments on the order of 39 trillion US$ (to 2050) to meet the increasing demand for energy services, of which, roughly half is in the power sector and the other half in fossil fuel supply; investment needs for hydrogen and biofuels under baseline conditions are marginal as a result of little incentive for their utilisation. Note that the observations for the power sector are similar to projections of the IEA, which suggested that power generation, transmission, and distribution are likely to account for 10 trillion US$ until the year 2030 (IEA, 2003).

Mitigating climate change and shifting to alternative fuels entails significant need for additional investments, especially if stringent climate policy targets are pursued. Until 2050, investments in the power sector, in particular, for clean coal technologies, renewables, and nuclear energy, increase by more than 50% (or around 9 trillion US$) for the most stringent climate policy target compared to

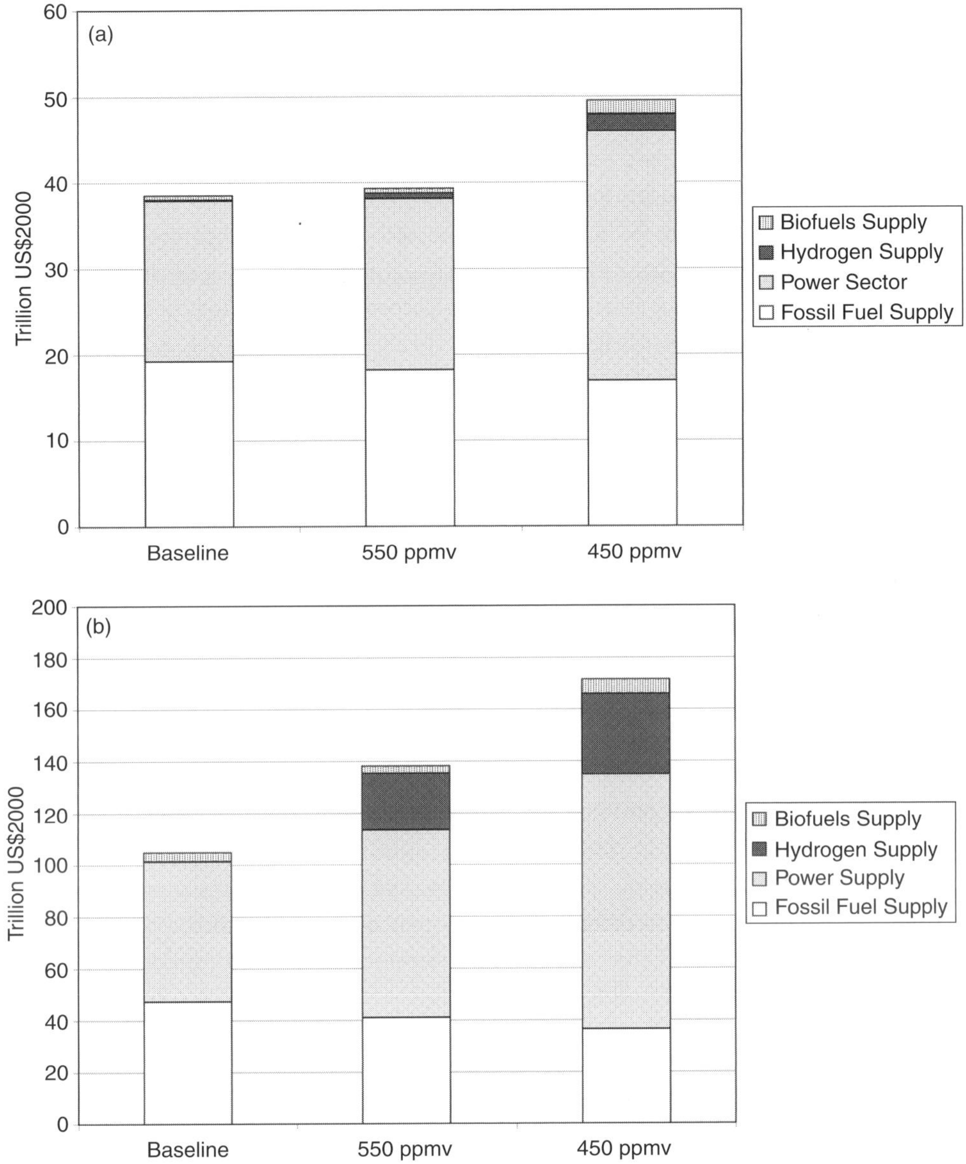

Figure 6.11 Investment needs for conventional and alternative fuel supply under different scenarios: (a) until 2050; (b) until 2100.

the baseline. Investment needs for alternative fuels under stringent climate policy until 2050 are significantly lower (i.e. below 2 trillion US$ for hydrogen as well as for biofuels). These additional investments for electricity, hydrogen, and biofuels supply are partly offset by a decreased need for fossil fuel supply, although total investment in energy supply technologies until 2050 for achieving a 450 ppmv climate policy target is about 29% above the level in the baseline. Under the more moderate climate policy target (550 ppmv), total additional energy supply investment requirements until 2050 are on the order of 2% compared to the baseline case.

Investment requirements over the entire time horizon until 2100 show a somewhat different picture. Additional investments in the power sector amount to 83% for the 450 ppmv climate policy target, and 34% for the 550 ppmv target compared to the baseline.[13] Requirements for investment in hydrogen infrastructure are boosted under climate mitigation policy scenarios.[14] For biofuels, additional investments remain comparatively low even for the 450 ppmv case, again, because of a more cost-effective utilisation of biomass in other sectors. As above, total additional investment needs are partly reduced by a decreased demand for fossil fuels.

Such results may sound daunting at first glance, but need to be seen in the appropriate perspective. The following key observations can be made:

- In the first half of the century, investment needs in the power sector outweigh those for hydrogen and biofuels by an order of magnitude. This observation is a result of the vigorously increasing demand for electricity worldwide, evidenced in the results above.
- In the long run, developing hydrogen as a fuel entails substantial investments. However, the bulk of investment under optimal CO_2 policy design and timing is likely to occur in the second half of the century, when technologies for hydrogen production and utilisation have reached a significant degree of maturity (itself requiring early investment), accelerated by learning-by-doing mechanisms. In this light, keeping the 'hydrogen option' open does not require a major investment in the first half of the century, and supports efforts to pursue a stringent climate policy in the second half of the century.
- More generally, pursuing climate policy targets and shifting from fossil to alternative fuels is costly, even under an optimal CO_2 allocation scheme as applied in this analysis here ('when' and 'where' flexibility for reducing CO_2 emissions). However, investment needs are relatively modest compared to the underlying level of economic activity and growth assumptions (with cumulative costs amounting to 0.6% of cumulative GDP for the 450 ppmv scenario). Nonetheless, mobilising this investment is an important policy challenge.

[13] It is worth noting that discounted total energy system costs are only 4.8% higher in the 450 ppmv scenario, and 1.2% in the 550 ppmv scenario, compared to the baseline scenario.

[14] For the case of hydrogen, it is interesting to note that more than 60% of investment required until 2050 is dedicated to developing hydrogen production technologies, in particular, to reducing costs of current, predominantly low carbon technologies. In the long run, the bulk of the investment is required for developing an appropriate hydrogen delivery infrastructure network.

6.4.4 Where will hydrogen come from?

The results presented in this chapter suggest that hydrogen can play an important role in future energy systems as a result of climate policy, and the more stringent the climate policy target, the more significant the role of hydrogen. Inevitably, this poses the question, 'Where will hydrogen come from?'

A considerable number of studies were reviewed for the definition and implementation of hydrogen production and delivery technologies in GMM; and the results presented here are based on this techno-economic assessment. Figure 6.12 shows hydrogen production modelled under the most stringent policy regime, i.e. a 450-ppmv CO_2 concentration target.

With a stringent climate policy, hydrogen is produced on a large scale and in central production facilities. The favoured production technology is coal gasification with CCS due to its low cost and the ability to co-produce electricity and store CO_2. Other low emissions production technologies are also attractive, including production from nuclear and dedicated wind power production with electrolysis.[15] It should be noted, however, that there are some uncertainties regarding the future acceptability of some of these low emissions options, including nuclear and CCS, indicating that other more expensive options (see Figure 6.3) may also be attractive under certain conditions, particularly if there are technology breakthroughs. Figure 6.13 compares the shares of the different hydrogen production technologies across the different climate policy scenarios, highlighting once more that coal gasification is the most competitive technology, providing, for example, for almost all hydrogen production under a mild climate mitigation policy.

Centrally produced hydrogen needs to be delivered to the consumer, and the various hydrogen delivery routes considered in this analysis were mentioned in Section 6.2.4. Figure 6.14 depicts the shares of different hydrogen infrastructure options utilised under the stringent climate policy case. It shows how early hydrogen deployment takes place through various delivery routes, making use of pipelines and combined systems with pipeline delivery to terminals and gaseous truck delivery to fuelling stations in pilot regions on the one hand, and direct liquid truck delivery on the other.

The results of this analysis suggest that a shift to hydrogen in personal transport is supported by stringent climate policy, and that hydrogen can be an important energy carrier in a future, more sustainable energy system. From the perspective of a cost-optimising social planner, hydrogen is most competitive when produced at central production facilities. The analysis of hydrogen delivery routes suggests that

15 Hydrogen from biomass does not play a role in hydrogen production in this scenario, as biomass is utilised more cost effectively in other sectors. A sensitivity analysis that excludes biomass IGCC with CC from the electricity sector confirmed this result, i.e. hydrogen from biomass gasification was not deployed.

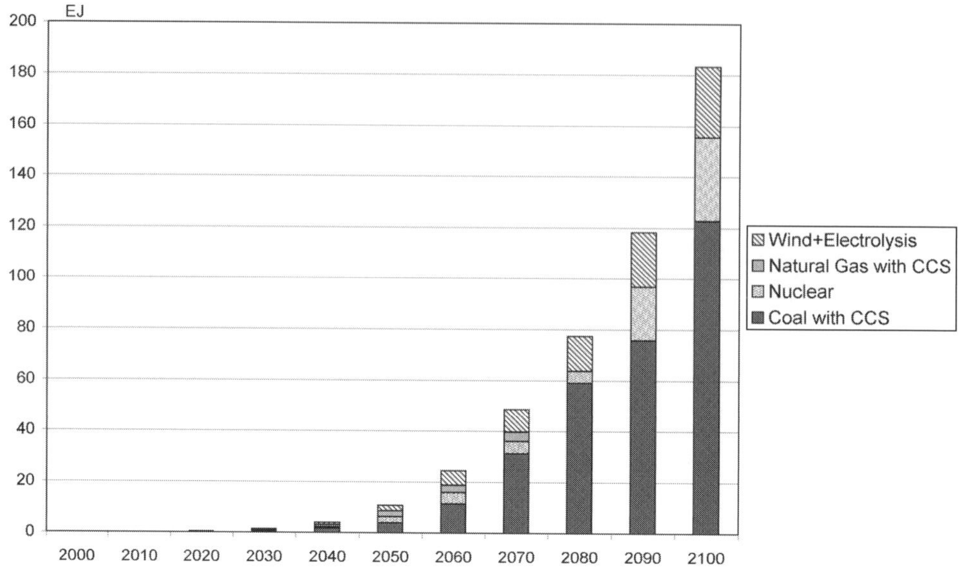

Figure 6.12 Hydrogen production under a 450-ppmv CO_2 policy regime.

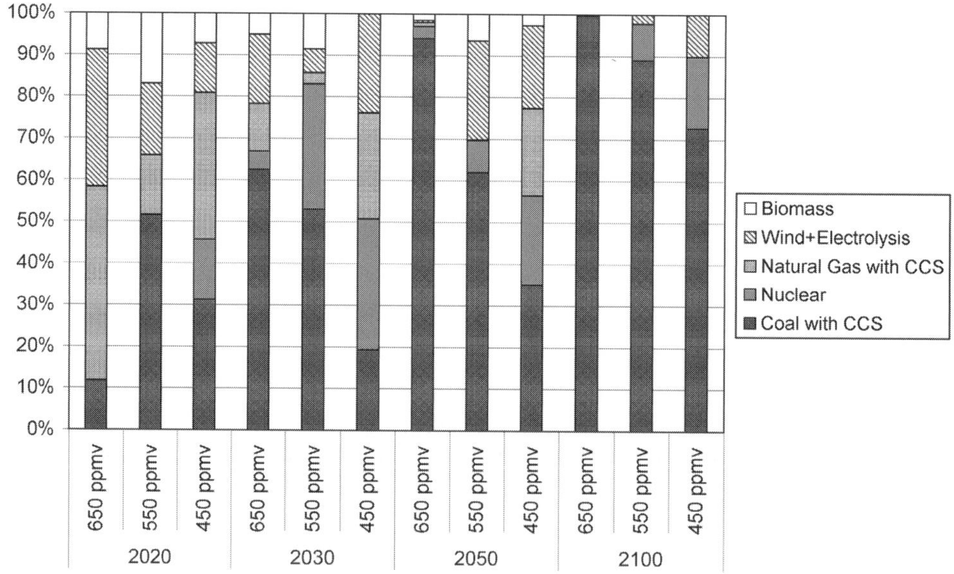

Figure 6.13 Hydrogen production across climate policy scenarios.

large-scale hydrogen deployment is initiated mostly by pilot projects, supplying single load centres by pipeline systems or combined pipeline and truck delivery systems. Additionally, flexible systems such as trucks carrying liquefied hydrogen deliver approximately 19% of the hydrogen to additional load centres in 2030, and

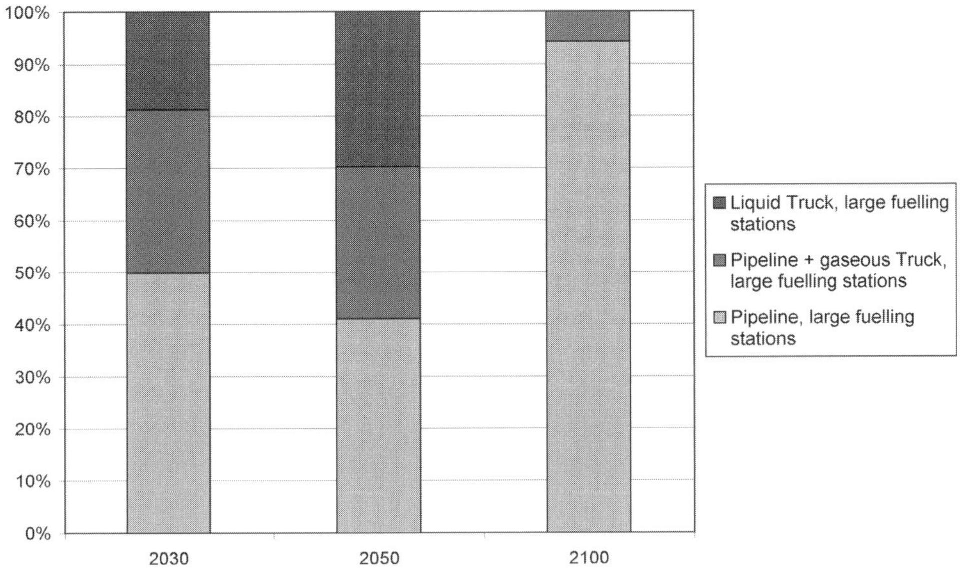

Figure 6.14 Hydrogen delivery routes in the 450-ppmv climate policy case.

this share is gradually expanded toward the middle of the century with increasing demand for hydrogen. In the long run, however, an extensive pipeline network is constructed as the least-cost solution for hydrogen delivery.

It is important to note in this context that this analysis does not necessarily suggest that this is the best potential route to develop large-scale hydrogen infrastructure toward what is often called a 'hydrogen economy';[16] rather, it is found that the cost-optimal solution to develop hydrogen is from pilot regions, i.e. local applications, to more large-scale global hydrogen utilisation. However, large-scale implementation of a hydrogen infrastructure, in particular, pipelines, requires significant efforts, and the chicken-or-egg problem as to whether hydrogen demand needs to develop first in order to make a hydrogen supply infrastructure develop, or whether hydrogen infrastructures need to be built first to spur demand for hydrogen, is an ongoing debate. Most authors suggest that hydrogen needs to develop out of niche markets in order to become an option for large-scale deployment in transport. A study for the Netherlands (Smit *et al.*, 2007), for example, suggests that hydrogen fuel cell car market penetration could follow the deployment of hydrogen in the residential and commercial sector in CHP systems, and would initially be fuelled

[16] The term 'hydrogen economy' in this context is somewhat misleading, as it suggests that hydrogen could become the dominating energy carrier in the future. The results of the analysis here, however, suggest that a hydrogen economy would probably rather evolve toward a 'hydrogen + electricity economy', as both energy carriers were found to play an important role in mitigating climate change. They should, thus, be seen as complements rather than competitors.

by hydrogen from forecourt natural gas reforming. Moreover, a review of analogies between hydrogen and early gasoline refuelling methods in Melaina (2007) notes that, much as early gasoline demand was covered from latent gasoline production capacity in the kerosene industry, early hydrogen demand could be covered in a similar way (i.e. from by-products from the chemical or refining industry), at least during a transition phase.[17]

In the analyses presented here, an increase in hydrogen production is the cost-optimal solution to satisfy climate policy targets, suggesting that hydrogen can be a viable option to decrease global GHG emissions. However, the large-scale deployment of central hydrogen, delivered in the long run mostly through pipeline networks, raises the question as to whether there are barriers to the implementation of such extensive delivery networks. In fact, the scale of deployment observed in the modelling of the stringent climate policy case is equivalent to as much as 1.23 million km of pipeline in place on a global level in 2100, compared to 16 000 km of hydrogen pipelines currently in place around the world (Simbeck, 2004). If pipeline expansion takes place in a linear way and starts with 16 000 km in the year 2000, this would translate into about 121 000 km of pipeline that would need to be constructed per decade, or around 1500 pipelines of 80 km lengths as they were used in the analysis, in order to reach the suggested levels in the year 2100.

Realising this level of pipeline deployment becomes less daunting when we consider the long-term time horizon of the analysis. For comparison, in the United States alone, more than 486 000 km of natural gas pipelines are currently in place (EIA, 2007b). That is, one country alone has, over roughly the last century, deployed around 40% of the pipeline length projected to be needed globally for hydrogen distribution over the current century. Moreover, there is reason to believe that existing natural gas pipeline networks could be used for delivering hydrogen in blends with natural gas in early phases of hydrogen deployment for use in stationary applications, as described in, for example, Haeseldonckx and D'haeseleer (2007). This would reduce the need for initial pipeline infrastructure expansion, and underlines that, in order to facilitate the use of hydrogen, creative solutions will be required.

However, such observations still seem somewhat daunting and imply the need for further analysis of the viability of this effort. In order to understand whether the expansion of the delivery network is a bottleneck for large-scale hydrogen deployment, the following section analyses the practicability and scale of the effort required to facilitate a transition to hydrogen.

[17] As noted in Chapter 2, these sources may represent a significant early niche supply of H_2, potentially sufficient in Germany alone to fuel 300 000 cars (Roads2HyCom, 2009).

6.4.5 Hydrogen infrastructure deployment

In this section, the practicability of deploying hydrogen on a large scale, such as observed for the 450-ppmv climate policy target, is analysed in order to assess the role of infrastructure as a possible bottleneck.

An important feature of the deployment of hydrogen infrastructure that we have considered here is the lumpy nature of such investments. By 'lumpy', we mean that infrastructure units are built in certain minimum sizes, and thus a minimum amount of investment is needed irrespective of the quantity of hydrogen delivered – a good example is for hydrogen delivery infrastructure such as pipelines, where a minimum length and size of pipeline is required. For the purpose of this analysis, a minimum, lumpy capacity level corresponding to 250 tonnes of hydrogen per day was applied for all hydrogen infrastructure options, in line with the underlying cost assumptions used for the delivery network. Note that this implies a minimum investment level for truck delivery as well, which was deemed necessary because the truck delivery routes make use of terminals for loading/unloading of trucks, which have been designed at this scale and would benefit unevenly from scale-economies if excluded from the minimum investment requirements.

In addition to defining minimum investment levels, three cases were analysed distinguishing allowed maximum installations of individual delivery options, to represent limits to the available manufacturing and installation capacity. Specifically, maximum 'blocks' of 10, 50, or 100 units of delivery infrastructure capacity were allowed per decade and region. This means, for example, that a maximum of 10 (or 50, or 100) units of pipelines with a capacity of 250 tonnes of hydrogen per day can be implemented per decade in each world region.

Figure 6.15 depicts the obtained scale of hydrogen production for the 450-ppmv and 550-ppmv climate policy targets, comparing the hydrogen deployment levels of the previous analyses ('no blocks') with those using different maximum 'blocks' of investment per decade.

The results indicate that the total hydrogen production level is reduced if a minimum investment is considered for hydrogen infrastructure, i.e. hydrogen deployment takes place on a smaller scale. This is particularly the case in the 550-ppmv climate policy scenario, where hydrogen production is reduced by almost 50%.

In the more stringent climate policy case, hydrogen production levels are reduced as well, but to a lesser extent. Moreover, total hydrogen production is not affected by the maximum level of investments allowed per delivery infrastructure option under a 450-ppmv climate policy target, i.e. the cases distinguishing 10, 50, and 100 'blocks' result in nearly identical total hydrogen production. The reason for this observation is illustrated in Figure 6.16, which shows how deployment constraints on infrastructure for hydrogen distribution from centralised sites motivate the

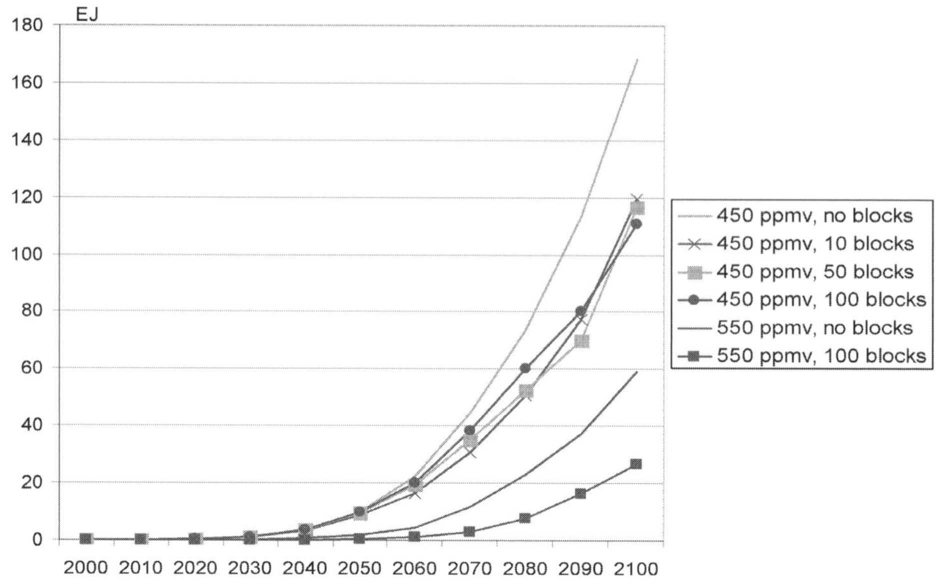

Figure 6.15 Hydrogen production with and without ('no blocks') lumpy investment in infrastructure, and with different maximum deployment rates. Lumpy investment assumes the minimum construction of 250 t H_2/day, while the maximum deployment rates are specified in terms of the number of blocks of 10, 50, or 100 units of infrastructure with a 250 t H_2/day capacity that can be deployed per decade.

deployment of decentralised hydrogen production facilities, which do not require hydrogen distribution infrastructure since the hydrogen is produced on-site.

The analysis of minimum investment and maximum deployment levels suggests two main things: first, the development of an appropriate hydrogen infrastructure *is* an issue for large-scale deployment of hydrogen. Throughout all scenarios it was found that the total hydrogen production level decreased through invoking a minimum lumpy investment and maximum deployment level on infrastructure deployment. The degree to which hydrogen production is reduced is linked to the climate policy target: the more stringent the target, the less hydrogen production is reduced. Second, it was found that imposing a maximum investment per decade and region leads to a shift from central hydrogen production to decentralised options. In this case, decentralised hydrogen is initially provided from natural gas reforming, until a shift to electrolysis takes place toward the end of the century.

It is implicit that constraining the deployment of hydrogen infrastructure comes at a cost, and the lowest limit of 10 pipeline blocks adds about 1.5% to total discounted energy system costs over the twenty-first century, which is in the order of 154 trillion US$ for the 450-ppmv case without investment constraints on hydrogen.

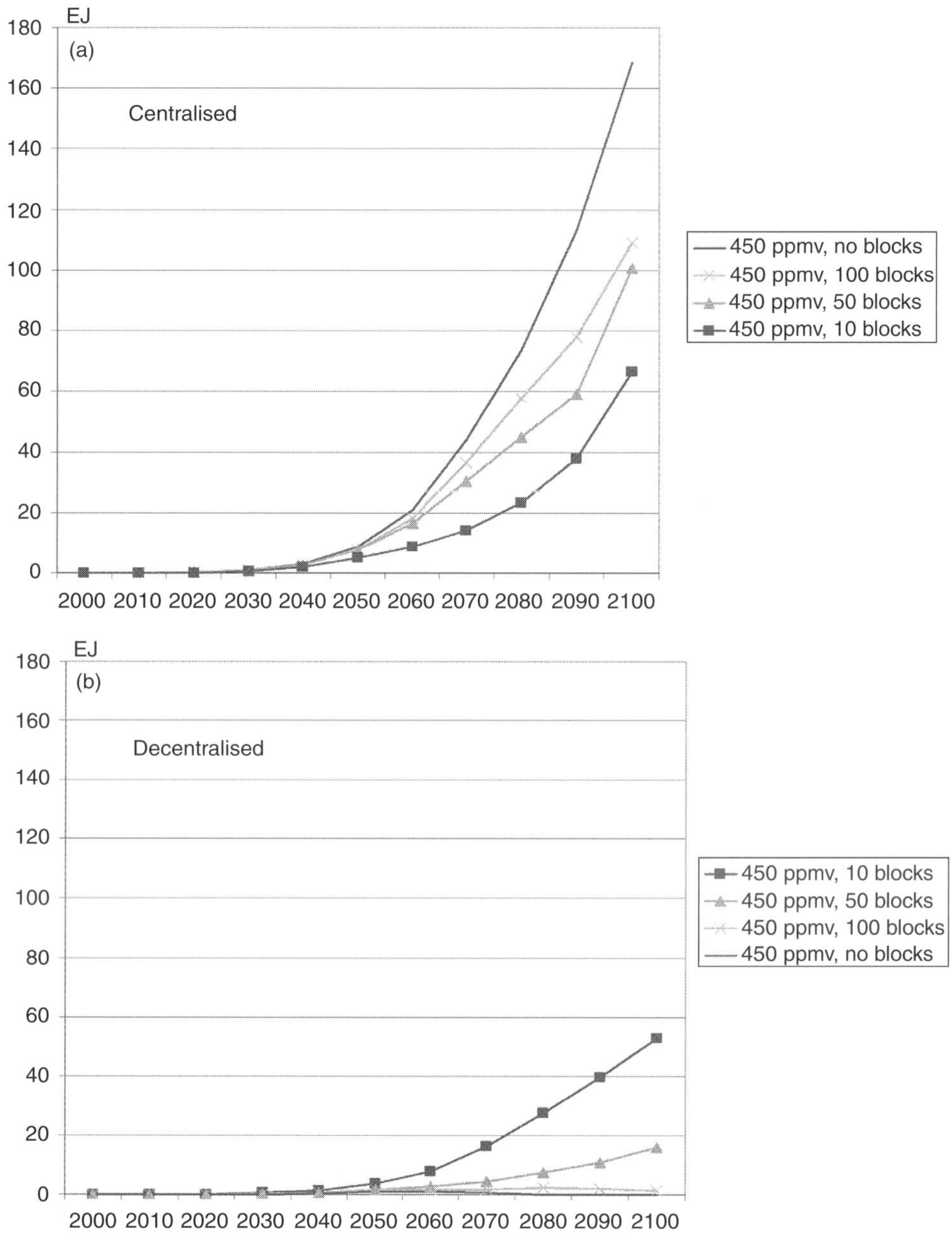

Figure 6.16 Hydrogen from centralised (a) and decentralised (b) production facilities with lumpy investment and different maximum deployment rates. Lumpy investment assumes the minimum construction of 250 t H_2/day, while the maximum deployment rates are specified in terms of the number of blocks of 10, 50 or 100 units of infrastructure with a 250 t H_2/day capacity that can be deployed per decade.

Limiting the scale of investment in hydrogen infrastructures not only limits the level of hydrogen fuel production, but consequently also the degree to which hydrogen is utilised. While we saw above that hydrogen is an attractive option in personal transport under a 550-ppmv as well as a 450-ppmv climate policy target, an important change is observed when we account for the possible impact of minimum investment levels. Figure 6.17 shows how the contribution of hydrogen fuel cells to the 450-ppmv target is decreased only modestly compared to the results in Figure 6.9, even with a maximum of only 10 blocks of investment in infrastructure (i.e. a total capacity of 2500 t/day for each delivery option). In comparison, under a 550-ppmv climate policy scenario, hydrogen fuel cells do not play a significant role in personal transport when hydrogen infrastructure deployment is limited. In this case, hydrogen is still used in other sectors, such as freight transportation.

6.5 Summary and discussion

The analyses conducted in this chapter have explored technology and fuel choice in transportation under different climate policy targets. In addition, bottlenecks to the implementation of different transport options have been investigated, with a special focus on hydrogen.

We have seen that climate policy targets are one of the key factors supporting the deployment of alternative fuels such as hydrogen and biofuels in transportation. Biofuels are a particularly cost-effective option for meeting *moderate climate policy* targets, (e.g. 650 down to 550 ppmv), consistent with the findings of other authors. such as Turton (2006b). The results obtained here, however, indicate that there is no single optimal way of utilising biomass. In a future sustainable energy system, biomass is likely to become an important energy carrier, and its importance is increasing with increasing climate policy targets. Biofuels can play a role in personal transport; but the degree of their utilisation depends on the availability of sustainable produced low-cost biomass, an issue which has been widely debated recently in the context of competition with agricultural land uses for food production and increasing food prices, potentially induced by an increasing utilisation of biofuels (Swissinfo, 2007; Gucciardi Garcez and de Souza Vianna, 2009). For some world regions which lack abundant and cheap biomass resources, an increased utilisation may also result in an increased need for biofuel imports. Notably, the extent to which biofuels can become a competitive option for transport is linked to whether and when hydrogen can be developed as a transportation fuel. The present analysis suggests that the scarce biomass resources may be better utilised to reduce CO_2 emissions in electricity and heat production, similar to observations of other authors (Azar *et al.*, 2003; Grahn *et al.*, 2007). In such scenarios, biofuels should be considered as a bridging option until hydrogen fuel cells become competitive in personal transport.

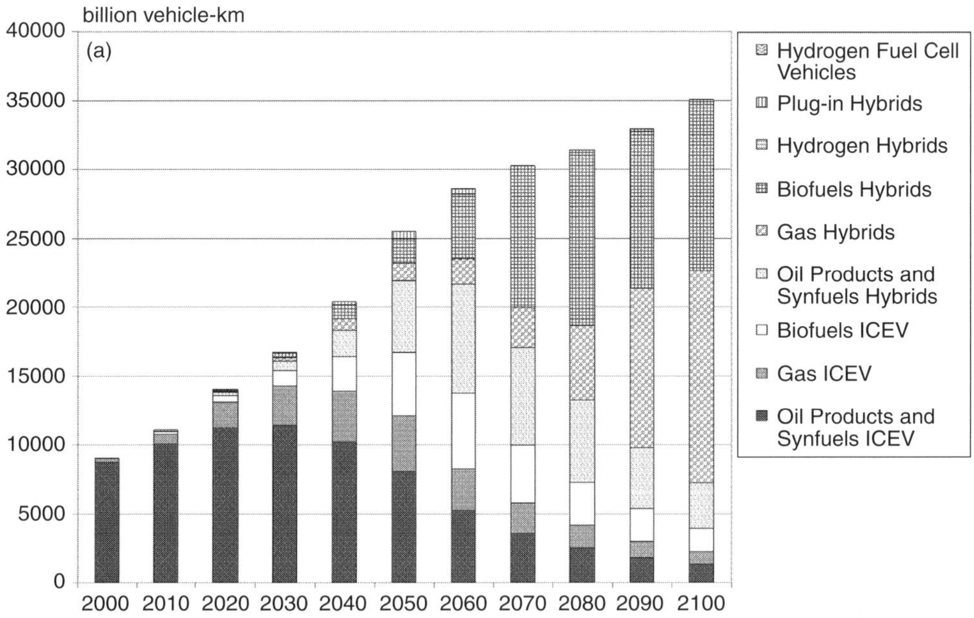

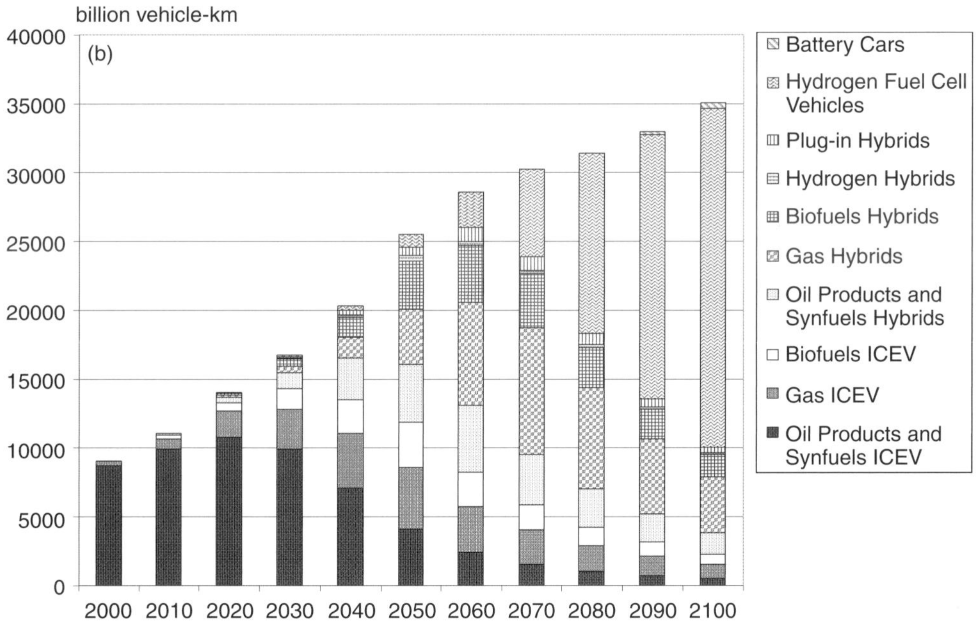

Figure 6.17 Share of hydrogen fuel cells in personal transport with lumpy infrastructure investment and deployment maximums. (a) The 550-ppmv scenario with a maximum of 100 units of investment. (b) The 450-ppmv scenario with a 10-block maximum deployment rate. Compare to Figure 6.9 for the equivalent results without lumpy investment or infrastructure deployment limits.

A transition to hydrogen was seen to depend strongly on future climate policy targets, with hydrogen favoured by more stringent targets (from 550 down to 450 ppmv). The detailed analysis in this chapter of hydrogen production technologies and delivery infrastructure tells us that one key to the deployment of hydrogen is the availability of low- or zero-emission technologies for hydrogen production. Here, we identified coal gasification with CCS (and electricity co-production) as a highly promising technology in this regard, along with nuclear hydrogen production and wind generation dedicated to hydrogen production via electrolysis (although, with technological breakthroughs, other options, such as solar thermal production, could also play a role).

The other key to a successful deployment of hydrogen is the build-up of a network for hydrogen delivery. An analysis accounting for the need to build a minimum distribution network, and likely limitations on the speed at which such a network can be rolled out, showed that the possibility to quickly expand large scale hydrogen distribution networks is a prerequisite for hydrogen deployment. It was found that limitations to the rate at which a hydrogen distribution network can be constructed may be a bottleneck for hydrogen deployment and prevent market penetration of hydrogen fuel cells. Very stringent climate policy (450 ppmv), however, was found to promote alternative hydrogen production and delivery modes, in particular, hydrogen production at refuelling stations, thereby facilitating the deployment of fuel cells in personal transport. Such an option, however, is a more costly means to deploy hydrogen.

The analyses pursued in this chapter have identified technology and fuel options that are competitive for meeting a range of climate policy scenarios. Clearly, however, there is a considerable degree of uncertainty involved in such an assessment, and technology breakthroughs such as rapid improvements in the cost and performance of batteries, or adoption of vehicle-to-grid systems, could alter the picture in favour of other fuels, such as electricity. Furthermore, emergence of new business models and niche markets could substantially change the competitiveness of alternative fuel options. For example, the emergence of a significant group of consumers willing to accept vehicles with only a short driving range could provide additional support for alternative vehicles, particularly pure electric and plug-in hybrid vehicles. Similar niche markets for low-emissions vehicles may be created by government initiatives, such as those aimed at reducing urban air pollution. Finally, strong support from car manufacturers for a particular drivetrain technology, coupled with interests of suppliers of fuels, including electricity, are also likely to play a role in determining the future drivetrain technology pathway.

A similar level of uncertainty exists regarding barriers to the deployment of alternative fuels. Although we have investigated some of these barriers, there are many more obstacles to overcome (including additional factors related to consumer and manufacturer acceptance of alternative technologies and fuels, and encouraging sufficient R&D and deployment while technologies are still immature) for which additional policy support will be needed. In this context it is important to appreciate that, historically, even strong policy support has not necessarily been sufficient to promote the desired levels of alternative fuels or advanced vehicles. By way of an example, McNutt and Rodgers (2004) reviewed experiences with US legislation during the 1990s and discussed why the Energy Policy Act of 1992, which established a goal for alternative fuel use of 10% by the year 2000, and 30% by 2010, did not achieve the desired outcome. The developments of alternative fuel use have by far fallen short of these targets, despite significant financial and policy investments. The authors identify several reasons for this development, among others a chicken-or-egg problem with the development of a refuelling infrastructure. More importantly, however, the industry with a substantial stake in petroleum fuels and vehicle technologies responded by implementing notable improvements for conventional fuels and vehicles, thus delivering significant emission reductions and weakening the policy argument. Moreover, alternative fuels failed to develop out of niche market applications into the mainstream, despite heavy government support.

Such experiences show the extent of the challenge facing biofuels and hydrogen. It is not only technology development, such as achieving low-cost fuel cells with 50 US$/kW that are needed to make new technologies economically viable. It is also an appropriate policy environment that is required to facilitate a switch to alternative fuels, together with industry involvement, e.g. public–private partnerships, particularly given the large investment needs discussed in this chapter. One key lesson from past experiences is that promoting the adoption of alternative fuels in transport is an effort that requires the involvement of all stakeholders; otherwise, existing infrastructure lock-ins are hard to overcome. A good example for such a collaborative effort is the development of a hydrogen-fuelled economy in Iceland, where several industrial players are collaborating with academia and policy makers to gain experiences with the application of hydrogen fuel cells in buses and with the development of a hydrogen infrastructure. Such projects represent precisely the type of niche applications that are needed to develop hydrogen further.

The findings of this chapter support the notion that policy makers need to work on two fronts simultaneously to address energy system challenges associated with climate change. First, they need to provide the required RD&D programmes to support the development of biofuels as a transition fuel in personal transport. Additionally, an appropriate regulatory framework could help to increase the share of biofuels in transport, e.g. by invoking minimum shares of biofuels in transport as recently done in the European Union for the year 2020 (European Commission, 2008), or by setting a minimum standard for biofuels blends into conventional fuels.[18] Such policy action could help to overcome existing infrastructure lock-ins in transport fuel supply and gain experiences with the deployment of alternative fuels.

In addition, policy makers need to support further research on hydrogen and fuel cells, particularly through the implementation of demonstration projects. Technologies that require particular attention in this light appear to be hydrogen fuel cells, both for mobile and stationary applications; and hydrogen production technologies, particularly involving gasification, but also electrolytic hydrogen production.[19] Ideally, such research projects are realised as efforts involving all the various stakeholders involved, ranging from industry to academia and politics, but also consumers for gaining insights into the implications of hydrogen use in daily applications and the satisfaction of consumers with the 'commodity, hydrogen'. Early involvement of all stakeholders allows minimisation of the risk of failure when it comes to the actual market introduction of hydrogen in the transport sector.

A combination of such policy measures addressing both biofuels and hydrogen simultaneously can help spur the use of biofuels as a transition fuel until the feasibility of reducing hydrogen fuel cell costs is better understood, and first experiences with hydrogen on a pilot project level have been gained.

In any case, hydrogen appears likely to be a long-term option for transport under the assumptions in our analysis, as it represents a radical departure from today's transport fuel chains, and there is still a considerable need for RD&D and an analysis of the potential of individual technologies, in particular, fuel cells. However, facilitating hydrogen requires efforts today: the analysis presented here showed that, to realise stringent climate policy targets, investment needs for gaining experiences with hydrogen production technologies and setting up pilot projects with an appropriate hydrogen supply infrastructure are high. Nevertheless, the scale of the effort over the next 50 years is small compared to the additional investment needs required in the power sector, and would allow us to 'keep the hydrogen option open' for the second half of the century, thus potentially allowing future

[18] Which policy is applied is obviously a question of the target that is pursued. If, for example, reducing CO_2 emissions from transport and decreasing the use of oil products in transport is the target, then a performance-based target that mandates, for example, a maximum cap on CO_2 emissions per kilometre driven may be a more appropriate choice. A minimum share of biofuels in personal transport, as suggested here, additionally induces the development of biofuel production technologies, but does not necessarily reduce CO_2 emissions to the extent a performance-based target does.

[19] This could also lead to spillover benefits supporting the more widespread use of renewable energies in the electricity sector by providing storage capacity to intermittent resources. An important technology here could be proton exchange membrane electrolysis, i.e. electrolysers that work like a reversed fuel cell.

generations to reduce CO_2 emissions drastically and satisfy transport demand in a more sustainable manner than today.

6.6 Acknowledgements

The authors are grateful to Socrates Kypreos and Leonardo Barreto for preliminary discussions of concepts and early results. We would also like to thank the Swiss National Centre of Competence in Research on Climate (NCCR Climate) for their financial support.

7

Integrated assessment of hydrogen in transportation

ALEXANDER WOKAUN

The analyses presented in this book have resulted in several overarching conclusions, which may be briefly summarised as follows:

- Increasing transportation energy security and reducing the environmental impacts of personal vehicles are important high-level goals.
- To achieve these goals, there is no single solution or 'silver bullet' pathway.
- Hydrogen will only represent an attractive transportation fuel to meet stringent CO_2 reduction targets if it is produced from primary energy resources with zero or low CO_2 emissions in the future energy system.
- The paradigm shift that is needed to make the transition from the current, problematic global transportation system will be led by the example of visionary communities, groups, and countries, using innovations that will necessarily be created by the developed world.
- Both short- and long-term policy measures are needed to support the development of future mobility technology, in order to internalise costs and provide a pathway to clean transportation.

In addition, the following themes are central to the role that hydrogen can play as a transportation fuel and have been explored throughout this book:

- The life cycle impacts of the full hydrogen fuel cycle were compared with conventional fuels.
- Changes in the hydrogen emissions due to transport were analysed, along with their effects on atmospheric chemistry.
- The 'optimum' design of a vehicle for any single buyer depends strongly on both the desired mix of performance and utility and on the available fuels. A

Transition to Hydrogen: Pathways Toward Clean Transportation, ed. E. Wilhelm and A. Wokaun. Published by Cambridge University Press.

multi-criteria methodology for an individualised assessment of vehicle design options has been developed.

- Introducing hydrogen will require profound changes in the transportation system. The dynamics of the way innovations diffuse into transportation markets was therefore studied, considering barriers, opportunities, and feedbacks.
- Energy–economic models have been used to generate scenarios maximising inter-generational welfare. Resource scarcity and climate protection goals were identified as important drivers promoting changes in the transportation system.

7.1 Review of results

The key contributions that the individual studies reported in this book have made to these overall conclusions are outlined below.

7.1.1 Comparison of alternative and conventional fuels by LCA

The potential life cycle impacts of hydrogen as an energy carrier were compared to conventional fuels, using a well-to-tank analysis. Two primary energy forms are considered to contribute toward the cumulative primary energy demand: fossil and nuclear (Figure 2.14). The ratio of these *non-renewable* primary energy forms to the final energy content of hydrogen delivered is highest (between 3 and 6) for hydrogen produced from fossil electricity. Production from two renewable pathways (gasification and reforming of biomass and STD of zinc oxide followed by hydrolysis) show non-renewable CED ratios of between 2 and 3, due to the auxiliary energy demand of the processes, and are therefore quite comparable to steam reforming of methane or coal gasification. It is only the electrolysis pathways starting from renewable and nuclear electricity that excel by producing CED ratios below 0.5. Nuclear has a fossil CED as low as hydroelectricity.

This finding has direct consequences on GHG emissions (Figure 2.15). While petrol and diesel burned in ICE engines produce emissions of 11 kg CO_2 equivalent per energy equivalent to 1 kg of hydrogen, the corresponding values are 30–70 kg CO_2 eq./kg H_2 for hydrogen produced by fossil electricity, 26–27 kg CO_2 eq./kg H_2 for coal gasification, and 13–20 kg CO_2 eq./kg H_2 for production from methane, biomass, and STD of water. These emissions may be strongly reduced if CCS is applied, and are lowest for the electrolysis pathway using renewable electricity.

Among the non-GHG pollutants, total suspended particulates (TSP) and SO_2 emissions are largest for pathways using electricity from coal, while the differences among the other options are less pronounced (Figures 2.16 and 2.25). The direct production of hydrogen from natural gas or coal, even when central flue gas cleaning technologies are applied, produces slightly higher NO_x emissions than the renewable pathways.

7.1.2 Multi-criteria assessment of alternative drivetrains

The introduction of alternative concepts such as FCVs provides an opportunity to redesign the vehicle in its entirety, rather than making incremental improvements starting from existing models. This is the context of the heuristic vehicle design performed in this study. Groups of technology options (e.g. size class of vehicle, fuel type, and chemical energy converter), exogenous variables (such as fuel prices), and endogenous technology options (e.g. materials used, tank size, rated power of the chemical energy converter, hybrid architecture, and battery capacity) were combined, using heuristic rules respecting existing relationships between the design choices and basic physical principles. Criteria of interest to drivers, manufacturers, and regulators such as fuel consumption, GHG and pollutant emissions, total cost of ownership, range, acceleration capability, and safety are generated as outputs from the powertrain simulation and life cycle models.

Detailed technical modelling of the powertrain options has been performed for standard driving cycles. To ensure a fair comparison, it proved essential to employ an optimum control strategy for each hybrid architecture design. Looking first at fuel and hybridisation criteria, it was confirmed that hybrids, in particular, series hybrids, achieve lower life cycle CO_2 emissions (Figure 3.23). Electric vehicles perform well due to their high efficiency, even if the current US electricity mix is assumed. Future (2035) emissions from biodiesel except for CO_2 are only slightly lower than those of gasoline vehicles while hydrogen fuel produced using renewably generated electricity could result in significant emissions reductions (Figure 3.24). A significant improvement in life cycle NO_x emissions is expected between 2010 and 2035 for both conventional and alternative fuels, and for hydrogen in particular (Figures 3.25 and 3.26). This is in large part due to tightening emissions standards around the world, but also because of the assumption of renewable and nuclear primary energy for hydrogen and electricity.

Results are then shown in the form of trade-off curves between pairs of different criteria. As an example chosen from a large set of possible comparisons, Figure 3.17 illustrates that the choice of powertrain technology has a pronounced influence on vehicle acceleration and fuel consumption. Reducing the 0–100 km/h acceleration time by increasing the engine displacement results in a heavy penalty in fuel consumption. On the other hand, by changing the hybridisation ratio (i.e. increasing battery size *at a given engine displacement*) it is possible to achieve better performance with similar or only slightly increased CO_2 emissions.

This example also illustrates why it is important to consider all the heuristically reasonable combinations of a very broad range of individual vehicle options. By plotting CO_2 emissions against acceleration times for all of the heuristically combined vehicles in the design set (Figure 3.19), it can be seen that, within a given powertrain architecture, it is possible to both lower the acceleration time *and* simultaneously decrease the emissions. The Pareto frontier allows the designer to identify, for a desired emission level, the option delivering the best performance, or to find the vehicle with lowest emissions for a given acceleration.

Preliminary stakeholder profiles have been used to generate MCDA results, allowing for different weightings of economy, ecology, and performance criteria according to personal preference. Emphasising a single criterion results in polarised design choices, which are critically dependent on the assumptions and input data (e.g. biofuel subsidies, vehicle and battery costs). On the other hand, varying the relative weights of two criteria (e.g. performance and environment preference) changes the preferred choice in a logical order (Figure 3.30). In the case where a particular solution has a clear advantage (e.g. inexpensive biodiesel with moderately better CO_2 emissions), some fairly robust solutions emerge (Figure 3.31). These results should be considered as illustrative, as the full utility of the multi-criteria analysis approach is only obtained when larger stakeholder groups are polled and allowed to iterate on their preference selections.

In summary, the results show that the 'optimum' vehicle strongly depends on stakeholder priorities for some of the technological criteria (e.g. type of the chemical energy converter, choice of fuel), while others are preferred by most stakeholders (e.g. lightweighting technology). While petrol and diesel hybrids continue to prevail for high payload and range requirements, FCVs emerge as a superior option for typical passenger car applications, and purely electric vehicles find their optimum application where low range and performance are tolerated.

7.1.3 Hydrogen in the atmosphere

H_2 may act as an indirect GHG, by decreasing the concentration of the important OH radical in the atmosphere, and thereby increasing the residence time of the GHG methane. In order to address this concern in the context of a future hydrogen economy, the present inventory of H_2 emissions was compared with scenarios for the future.

The approach taken to assess the impact of introducing hydrogen as a transportation fuel was to compare the emission factors of ICE and FCVs, and then estimate their shares in future transportation system scenarios. H_2 emissions over a standard driving cycle were found to be much smaller for diesel than gasoline cars, while motorcycles and scooters have significantly higher emissions (Figure 4.8). The start-up phase under cold-start conditions often makes a dominant contribution to overall H_2 emissions, as do fuel-rich conditions.

An examination of the available test results shows that future road emissions will be dominated by two-wheelers and FCVs, and efforts must be made to control these emissions (Figure 4.9). These direct H_2 emissions would, however, be lower than the losses associated with the production, transportation, distribution, and storage of H_2 considering the high end of the projected loss rates.

Estimated loss rates from the literature on various production, distribution, storage, and end-use systems have been critically reviewed. The lower range of loss rates quoted by industrial sources appears to be more realistic. Combining these reasonable assumptions with the potential share of hydrogen in a global energy system at the end of the twenty-first century, it is expected that H_2 emissions from transport should remain within the realm of current estimates of total anthropogenic H_2 release into the atmosphere. This emerges from the sensitivity analysis presented in Table 4.4, where the transport and overall anthropogenic emissions are compared for reasonable assumptions of the hydrogen loss rate during production and distribution, and for the FCV emission factors.

7.1.4 System dynamics analysis of market penetration

A system dynamics model for vehicle substitution in the European fleet was developed to map the concepts of technology diffusion and substitution. This captured the non-linear characteristics of a behavioural norm developing within a society, which influences the inherent and perceived attractiveness of alternative propulsion modes.

The expected technical development, including technology diffusion, is taken as the baseline. Which policy packages, levels of stimulation, and inherent attractiveness are necessary for the European car stock to reach its emission reduction targets were investigated. Figure 5.27 illustrates that fiscal incentives alone are not very effective unless the inherent attractiveness of alternative propulsion technologies (such as hybrids or natural gas cars) is increased. But it is only with the introduction of a 'near-zero emission' vehicle (here represented by a FCV running on renewable hydrogen) that the goal of halving passenger transport emissions after 2050 could be reached. In addition, the huge impact of the vehicle fleet size and the 2035 VKT becomes evident (Figures 5.28 and 5.29).

Given the generic nature of the system dynamics model, some general conclusions can be drawn concerning the potential acceptance of hydrogen-based transportation. As with other technologies, there is a network of interconnected feedback loops involving vehicle manufacturers, fuel distributors, and vehicle users. These include variables such as the spectrum of vehicles offered, the availability of fuelling infrastructure, and the utility derived by the consumer. These feedback loops may be self-reinforcing, and become stronger as the share of the new technology increases. However, in order to get the market penetration started, some prerequisites must be met. Inherent vehicle attractiveness is most important, as promoted by the social norm of a climate-conscious society. This is closely followed by the right price signals, based first on promotion and/or regulation and, in the long term, by an attractive total cost of ownership.

Supposing that the emission challenges cannot be met by incumbent technologies, it is interesting to analyse which type of transformation would be expected to occur within the car manufacturing industry. Although the possibility of a disruptive transformation was considered, where a new market entrant conquers the market, it was judged more likely that the existing car industry would realise the need to achieve stringent emission targets and the incumbent players would carry out the necessary, fairly radical innovations in an endogenous transformation.

7.1.5 Long-term scenarios of the global energy and transportation system

Scenarios for the evolution of the world energy system have been developed to analyse the conditions under which alternative fuels and vehicle drivetrains may be competitive, while accounting for resource and climate constraints. As a key result, it was found that the use of both hydrogen and biofuels in transportation can contribute to achieving climate stabilisation targets on a global level. Biofuels, and hybrid cars employing these fuels, are found to be competitive under moderate climate policy targets stabilising atmospheric CO_2 concentration at 650 ppmv (Figure 6.9). One key factor for the deployment of biofuels is the global availability of biomass and its cost. The contribution of biofuels to transportation is reduced with more stringent climate policy targets, with biomass utilised in these cases for the decarbonisation of other sectors of the energy system, such as heat or electricity production. Hydrogen is competitive under the most stringent climate policy targets, such as 450 ppmv CO_2 (see again Figure 6.9). Subsequent analysis of conditions under which battery electric vehicles could play a larger role indicates that substantial reductions in battery costs are likely to be needed. In agreement with the global studies, the enforcement of ambitious climate targets in Europe will necessitate profound changes in the vehicle fleet.

7.1.6 Increasing transportation energy security

Although increasing transportation energy security is one of the main motivations of this book, this issue has been indirectly addressed within the context of each chapter, rather than the direct focus of a specific analysis. The insecurity of supply associated with our present, oil-based transportation system is based on two major factors: the geographic unevenness of the global petroleum resource distribution and the shift from dwindling known reserves with low extraction costs to newer reserves with ever rising exploration and extraction costs. As this book has shown, we expect that the shift from a petroleum-based transportation system to a sustainable system using hydrogen and/or electricity as energy carriers will be gradual, and that many different primary energy resources and carriers will co-exist for some time in a mix that will vary over time and is based on regional differences. In general, however, we can expect that this transition from oil toward gas, coal-based synfuels, and nuclear on the one hand, and biomass as well as other renewables on the other hand, will be in the direction of primary resources that have large reserves and availability as well as broader geographic distributions. This shift will not necessarily be direct. In fact, decreased security of supply and higher prices in the near to intermediate future will likely provide important signals that will help to drive the transition. The issue of geographic distribution is also not entirely clear cut. There are obvious differences in the distribution of both remaining fossil resources and renewable resources, but the broader primary energy mix and the ability to transport and stockpile some resources (in particular, nuclear fuels) means that there should be less overall dependence on specific regions and longer term price stability. The shift from refined oil as an energy carrier toward synthetic and/or biofuels, hydrogen, and electricity will also mean that production and distribution of energy carriers could become more regional, which should also tend to increase security of supply.

7.2 Barriers to hydrogen as a transportation fuel

Three main classes of barriers to hydrogen as a transportation fuel have been identified, which may be summarised as follows:

- The creation of a hydrogen distribution infrastructure.
- The higher cost of hydrogen vehicles.
- The necessity of winning consumer acceptance, which is related to both cost and fuel availability, but can be considered as a separate issue due to the inherent difference between fuel cell technology and ICE-based vehicles.

Both in Europe and in the United States, clear visions have been formulated for the way in which a build-up of the hydrogen distribution infrastructure could take place. The associated scenarios and concepts are described in Section 6.4.5. For a large-scale deployment, central production facilities, pipelines to major distribution ports outside city centres, and a truck-based distribution network serving individual fuelling stations are envisioned. This is supplemented by decentralised production, which becomes more important if fewer resources for pipeline infrastructure projects are available (Chapter 6, Figure 6.16).

Barriers in this context are definitely the required additional investment costs, as exhibited in Figure 6.11. These costs are large, in the order of 25 trillion $US\$_{2000}$ in total by the end of the century. However, they are not completely out of scope, considering that a contribution of 100 billion $US\$_{2000}$ per year from industrialised to developing countries has been discussed for the period of 2020 onward to support the adaptation to climate change.

One might ask whether, in addition, the construction of hydrogen pipelines would represent a societal acceptance challenge. Given the fact that the natural gas grid, which is more extensive, would likely transport mixtures of methane and hydrogen during the second half of the century, this appears to be less of an issue.

On the consumers' side, a critical factor is the price increase of an FCV, as compared to an advanced vehicle with an ICE. From a purely rational point of view, total cost of ownership is the decision criterion. The cost of operating a vehicle fuelled by gasoline or diesel would depend on the climate targets, as reflected in imposed carbon taxes. Energy–economic optimisation modelling shows that, without a climate policy, there is no incentive to change to HFCVs. For the adoption of FCVs by 2050, a switch could take place if a fuel cell system costing below 50 US$/kW is reached by 2020 (Figure 7.1). The more stringent the climate target, the higher the acceptable cost of the alternative drivetrain. For a 60% CO_2 reduction target, FCVs would be attractive even for a final fuel cell investment price as high as 70 US$/kW.

As discussed in detail in Chapter 5, car purchases are often not based on rational decisions. In economic terms, this corresponds to amortising the investment not over the useful life of the car but over a very short time on the order of 3 years. As an orientation for the acceptable price premium of an alternative technology, one might use the surcharge the consumers are prepared to pay today for a hybrid vehicle, which in turn depends on its size. From this guideline, one arrives at a crude estimate that an acceptable price increase might be on the order of 3000 $US\$_{2000}$ per car, which would translate into 100 US$/kW for a 30 kW fuel cell system, and 50 US$/kW for a 60 kW fuel cell system.

7.3 Opportunities offered by hydrogen

7.3.1 Regional

In contrast to early visions of a 'world hydrogen economy', the present consensus shows that there will not be a single transportation solution, but that continents and regions will develop according to their specific boundary conditions.

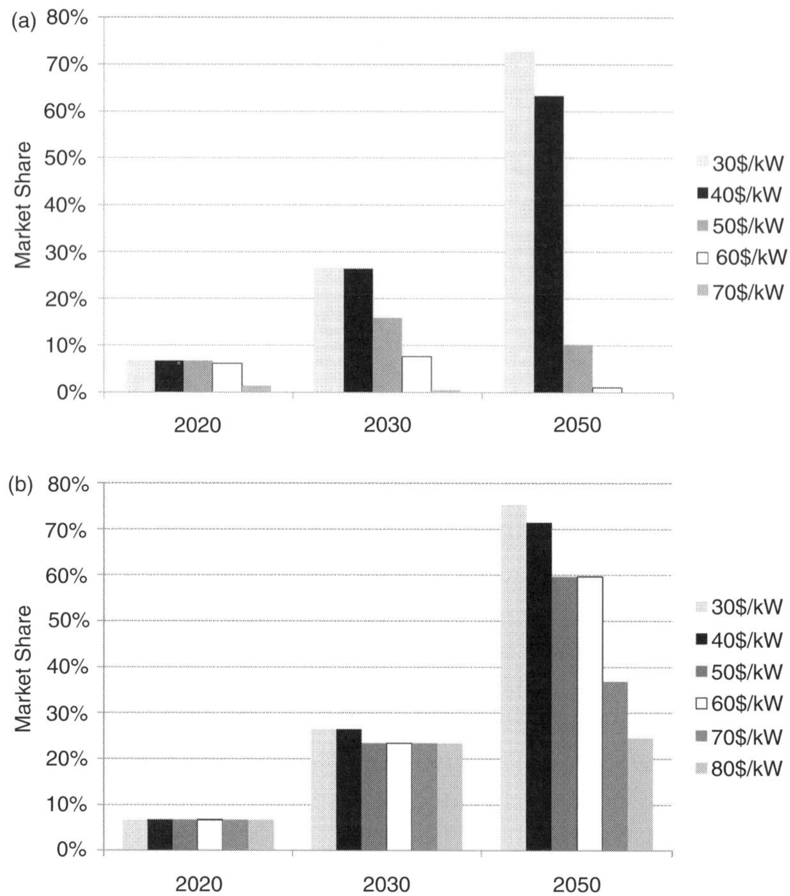

Figure 7.1 Time-dependent market share of fuel cell vehicles for various assumptions of the floor cost of fuel cell stacks, which are assumed to be reached in 2020. (a) 50% CO_2 reduction target of the European Union (EU-29) until 2050. (b) 60% CO_2 reduction target.

HFCVs are particularly attractive for regions that

- do not have fossil fuel reserves of their own, or do not have sufficient available biomass resources,
- follow stringent national climate policies,
- have sufficient economic power to invest in the build-up of a hydrogen infrastructure, and
- have travel distances that cannot be easily met by BEVs.

Based on these criteria, Europe and Japan appear as potential early markets. For North America, the last two criteria certainly apply, and a strong climate policy is emerging. The dependence on imported oil is seen as a big threat, whereas large coal reserves are available. In this situation, a large market for FCVs could develop with major quantities of hydrogen being produced from coal.

There are three types of counter examples where HFCVs appear less attractive:

- densely populated mega cities or metropolitan areas where commuting distances can be covered by battery electric vehicles,
- countries that continue to have their own residual oil reserves, and
- regions presently under development with large biomass resources, which may be used for the production of biofuels.

7.3.2 Global

Summarising the global scenarios presented above, we recognise significant contributions from three options. Biofuels will contribute to achieving climate protection goals, provided that the biomass resource is available for transportation use at low cost in sufficient quantities. Hydrogen is competitive under the most stringent climate policy targets. Battery electric vehicles will play a major role, in particular, for city transport, provided that costs of high-energy batteries can be reduced and lifetime expectations can be met.

Combining local and global considerations, it appears likely that hybrid vehicles propelled by fossil fuels (in world regions where they are still available), hybrids running on biofuels, and HFCVs will all coexist in different regions of the world for several decades. They may be supplemented by purely electric transportation in metropolitan regions. This emphasises and supports the often-portrayed insight that no single 'silver bullet' solution exists.

7.4 The role of hydrogen in transportation

7.4.1 Risks and trade-offs

In agreement with the global studies, the enforcement of ambitious climate targets by regions or nations will necessitate profound changes in the vehicle fleet. It is safe to say that, after the age of 'cheap oil', alternative fuels based on renewable energies, nuclear electricity, or coal gasification will necessarily play an important role. In this context, it is interesting to consider the option of binding hydrogen (which was produced by a low CO_2 route) to either CO or CO_2, for the production of synthetic liquid fuels. While this approach would avoid the necessity of installing a hydrogen distribution infrastructure, it is clearly inferior to the direct use of hydrogen due to its lower efficiency: The Fischer–Tropsch type synthesis of long chain hydrocarbons has a chemical efficiency of ~70%, and the well-to-tank efficiency of diesel vehicles in the driving cycle takes typical values of 30%. The direct use of hydrogen in an FCV with tank-to-wheel cycle efficiencies between 40 and 60% (Chapter 4) offers definite advantages.

7.4.2 Synergies with hydrogen use in other sectors of the economy

Hydrogen is presently used in large quantities in the chemical industry, mainly for ammonia synthesis by the Haber–Bosch process, and in oil refining. Present industrial production amounts to ~50 Tg (50 Mt) of hydrogen per year, overwhelmingly used for non-energetic applications. This quantity corresponds to a heating value of 6 EJ, or ~1% of present global energy demand. In the baseline scenario without climate policy, a five-fold increase in the production of hydrogen by the end of the twenty-first century would yield an energetic equivalent of ~30 EJ (Table 4.3). In striking contrast, the 450 ppm climate policy scenario would imply a 30-fold increase in the production relative to 2100, equivalent to ~170 EJ, or in the order of 10% of the expected global energy demand for environmentally conscious development scenarios. This appears as a reasonable range for the fractional use of hydrogen as an energy carrier; increasing industrial production by a factor of 30 within 90 years would be certainly feasible.

The use of hydrogen is expected to grow in importance for the production of (biomass- or CO_2-derived) feedstocks for the chemical industry, and for the reduction of oxides without the use of carbon. Adding hydrogen, in the order of 20% by volume, into natural gas offers attractive advantages with respect to cleaner combustion in boilers and gas turbines (Griebel, 2007; Daniele *et al.*, 2009).

The chemical industry is reported to currently produce significant excess quantities of hydrogen. Estimates given in the reports of the EU project Roads2HyCom (Roads2HyCom, 2009) suggest that the surplus of hydrogen produced in Europe presently amounts to 0.44 billion m_N^3 per year. With a heating value of approximately 3 kWh/m_N^3, this corresponds to an energy content of 1.3×10^9 kWh. As the consumption of an advanced FCV (assuming a driving distance of 15 000 km/year and a consumption of 30 kWh/100 km) amounts to 4500 kWh/year, this surplus alone would suffice to fuel 300 000 FCVs. With full use of the existing production capacities, and assuming a surplus share between 10% and 20% of total production, the total surplus production potential was estimated between 2 and 10 billion m_N^3 per year, i.e. a factor of 5 to 25 times more than the present value.

In view of these trends, it may be expected that synergies can be developed between the needs for hydrogen in chemistry, transportation, and the energy industry. Sharing the cost of building hydrogen distribution infrastructure would clearly improve the competitiveness of hydrogen-based transportation.

Appendix A

Summary of fundamental assumptions

The results presented in this book were generated using a very large set of modelling assumptions. The purpose of this appendix is to highlight some of the more common, and some of the most important assumptions made in the various chapters. Where discrepancies exist, explanations have been provided.

Chapter	Section	Value	Units	Assumption/Notes
1	1.1	–	–	Study is restricted to LDVs (cars and light trucks)
2	2.1.1	1	kg	Purified gaseous hydrogen at 45 MPa is assumed as the basis for all well-to-tank calculations
2	2.1.1	–	–	Euro 3 ICE vehicle used to benchmark the conventional fuel emissions
2	2.1.2	–	–	Modelling is performed using assumptions for European boundary conditions
2	2.2.4	100	years	Time frame for evaluating the impact of emissions
2	2.3.3	–	–	Generation 2 nuclear for 2005, Generation 3 nuclear in 2035
2	2.3.4	95	%	Capture rate of CCS technology combined with SMR
2	2.3.7	3.4	l/100 km	Fuel efficiency of FCVs
2	2.3.7	100–1000	kg/day	Range of demand for H_2 at filling stations
2	2.3.9	200	km	Basis for the transport of hydrogen
3	3.2	40	%	Manufacturer margin assumed on production to sale for LDVs
3	3.4.1	–	–	Simulation was performed on the NEDC
3	3.4.2	–	–	The state of charge for all hybrid vehicles before/after the simulation is unchanged
3	3.5.1	5–50	%	Discount rate assumed (5 – rational consumer, 50 – real consumer)
3	3.3.2	–	–	All costs are normalised to the 2010 US$

Chapter	Section	Value	Units	Assumption/Notes
3	3.3.4	–	–	All fuels are analysed based on their lower heating values (LHV)
3	3.5.3	16848	veh-km/year	2035 VKT
3	3.5.3	9.6	years	Vehicle replacement interval
4	4.3	50	%	Emissions from biofuels vehicles relative to conventional gasoline vehicles
4	4.3	80	%	Emissions from biofuels hybrids relative to conventional gasoline vehicles
4	4.3	87.3	%	Emissions from gasoline hybrids relative to conventional gasoline vehicles (from Chapter 3)
4	4.3	80	%	Emissions of natural gas hybrids relative to conventional gasoline vehicles
4	4.3	56	%	Emissions of plug-in hybrids relative to conventional gasoline vehicles (from Chapter 3)
4	4.3	–	km	Global LDV kilometres travelled by world region according to Fulton and Eads (2004) in 2010 and 2020
4	4.3	–	mg/km	Same H_2 emission factors for similar vehicle technologies for 2050 and 2100 scenarios
4	4.3	–	km	Global LDV kilometres travelled in 2050 and 2100 according to GMM
4	4.3	–	–	% share of each LDV technology for 2010, 2020, 2050, and 2100 based on GMM results
4	4.3	–	–	Technologies other than conventional gasoline and diesel incorporated only into developed world fleets in 2010 and 2020
4	4.3	10	year	Vehicle replacement interval
4	4.3	–	mg/km	E_{H2} for vehicles other than conventional gasoline vehicles and 2-wheelers constant with time
4	4.3	–	mg/km	E_{H2} for all diesel vehicles constant with time
4	4.4.3	–	Tg/PJ	H_2 production for petroleum refining based on global oil supply figures (Gül, 2008)
4	4.4.3	–	%	Economic growth (IPCC, 2000) used as guide for industrial H_2 production (excluding refining) after 2020
4	4.4.3	47	million Tns	Global hydrogen production in 2010 (Bond *et al.*, 2011a)
5	5.2.1	12–14	year	Vehicle replacement interval
5	5.2.1	–	–	Vehicle efficiency is 30% better in 2045 than 2007, characteristics come from VW Golf IV
5	5.2.1	–	–	Fleet growth based on countries trends
6	6.2.1	195	EJ/year	Global biomass availability, with assumptions for oil, gas, and coal. Based on Rogner (1997)
6	6.2.3	–	–	Assumptions regarding advanced powertrain cost characteristics in Table 6.1
6	6.2.4	3.25–70	$/GJ	Assumptions regarding hydrogen production cost characteristics in Figures 6.3 and 6.4

(*cont.*)

Chapter	Section	Value	Units	Assumption/Notes
6	various	–	–	Fuel costs are determined endogenously based on supply and demand. Supply is affected by resource availability and extraction costs (from Rogner, 1997); whereas demand for fuel is determined endogenously by the choice of technologies. Demand for energy services (e.g. VKT) is described based on B2 scenario (IIASA 2007), IEA/WBCSD (2004) and Turton (2006a))
6	various	–	–	fleet growth from IEA/WBCSD (2004) projections to 2050, then Turton (2006a) to 2100
6	various	–	–	global energy use is determined endogenously by the choice of technologies; but the 'useful' energy demand for all sectors other than car transport is taken from the B2 scenario of the IPCC's SRES (IIASA 2007).
6	various	–	–	Costs are in 2000 US $, and a 5% discount rate is used in the model

Appendix B

Selected input assumptions, technology descriptions, and heuristics

Selected Powertrain Cost	US 2010v	US 2035ren	EU 2010v	EU 2035ren
ICE engine ($/l)	755.00	755.00	755.00	755.00
LiON battery ($/kWh)[a]	2000.00	1050.00	2000.00	1050.00
NiMH battery ($/kWh)[a]	1384.00	400.00	1384.00	400.00
Fuel cell cost ($/kW)	108.00	52.00	108.00	52.00
	US 2010v		US 2035ren	
Well-to-tank Emissions	Hydrogen	Electricity	Hydrogen	Electricity
VOC (g/kg)	1.40	0.06	0	0
CO (g/kg)	3.33	0.19	0	0
NO_x (g/kg)	8.16	0.77	0	0.01
PM10 (g/kg)	4.32	0.98	0	0
PM2.5 (g/kg)	2.20	0.26	0	0
CH4 (g/kg)	40.95	0.98	0	0.01
N2O (g/kg)	0.06	0.01	0	0
CO2 (g/kg) / (g/kWh)	12456.55	727.02	0	3.60
Fuel Cost Assumptions	US 2010v	US 2035ren	EU 2010v	EU 2035ren
Gasoline ($/l) / (€/l)	0.52	1.15	1.20	1.55
Ethanol (E–85) ($/l) / (€/l)	0.54	1.20	1.05	1.36
Diesel ($/l) / (€/l)	0.41	0.91	0.84	1.09
Biodiesel ($/l) / (€/l)	0.16	0.34	1.01	1.31
C. Natural Gas ($/kg) / (€/kg)	0.79	1.77	0.53	1.18
Hydrogen ($/kg) / (€/kg)	0.40	0.90	0.80	1.81
Electricity ($/kWh) / (k Wh)	0.09	0.15	0.11	0.13

[a] *Note:* Costs are scaled by W/Wh. These values are for high-power packs.

Appendix C

Characteristics of the present and future vehicle designs

Shading indicates the 'best' vehicle architecture with the darkest shading and the 'worst' architecture with the lightest shading for each criterion. Please note that this appendix provides representative results that can be used to compare engine and hybridisation technologies; the MCDA algorithm may select vehicles that are not necessarily the best in this table because it has access to the full design set.

Chemical Energy Converter	Otto	Diesel	Otto	Diesel	Otto	Diesel	Fuel Cell	Otto	Diesel	EV

2010 US compact (1.1-1.9L) steel passenger vehicles

Hybridisation:	None	None	Mild	Mild	Series	Series	Series	Parallel	Parallel	EV
Chemical Energy Converter:	Otto	Diesel	Otto	Diesel	Otto	Diesel	Fuel Cell	Otto	Diesel	EV
Weight (kg)	1124	1147	1157	1179	1608	1644	1977	1622	1671	1929
Total power (kW)	99	89	102	92	157	157	100	168	178	140
Cost										
Purchase (2010 USD)	17468	19516	18686	20742	37251	39557	48479	36482	39019	45432
Operating/Maintenance (2010 USD)	1529	1379	1484	1350	1454	1426	1306	1601	1542	1436
Total cost of ownership (2010 USD)	28489	29460	29441	30535	48048	50139	58249	48231	50297	56171
Performance										
0-100 Acceleration time (s)	8.8	10.0	8.8	9.9	7.9	8.1	15.3	7.5	7.4	10.7
80-120 Acceleration time (s)	3.5	4.0	3.5	4.0	3.1	3.2	6.1	3.0	2.9	4.2
Top Speed (kph)	218	210	221	213	255	255	219	260	265	245
Utililty										
Range (km)	705	956	767	1037	1357	1595	545	939	1180	327
All-electric range (km)	0	0	0	0	125	124	122	40	55	327
Passenger volume (m^3)	2.61	2.61	2.61	2.61	2.56	2.56	2.35	2.56	2.56	2.39
Towing capacity (kg)	1511	1211	1558	1258	2580	2544	685	2844	3072	1798
Environment										
Gasoline Equivalent cons. (L/100km)	8.5	7.2	7.8	6.6	4.9	4.7	3.7	6.7	6.3	2.2
Life cycle CO_2 (g/km)	259	216	240	202	184	176	174	236	220	215
Life cycle NO_x (g/km)	0.18	0.15	0.17	0.14	0.16	0.15	0.13	0.19	0.17	0.23
Life cycle PM10 (g/km)	0.14	0.07	0.13	0.06	0.11	0.05	0.02	0.12	0.06	0.02

2035 (renewable fuel) US compact (1.1-1.9L) steel passenger vehicles

Hybridization:	None	None	Mild	Mild	Series	Series	Series	Parallel	Parallel	EV
Chemical Energy Converter:	Otto	Diesel	Otto	Diesel	Otto	Diesel	Fuel Cell	Otto	Diesel	EV
Weight (kg)	1124	1147	1157	1179	1608	1644	1977	1622	1671	1929
Total power (kW)	99	89	102	92	157	157	100	168	178	140
Cost										
Purchase (2006 USD)	17614	19546	18281	20226	22268	24525	26547	21531	23955	21322
Operating/Maintenance (2006 USD)	2794	2345	2638	2234	1994	1872	1285	2411	2202	1432
Total cost of ownership (2006 USD)	43298	41180	42570	40869	40793	41943	38614	43787	44324	34770
Performance										
0-100 Acceleration time (s)	8.8	10.0	8.8	9.9	7.9	8.1	15.3	7.5	7.4	10.7
80-120 Acceleration time (s)	3.5	4.0	3.5	4.0	3.1	3.2	6.1	3.0	2.9	4.2
Top Speed (kph)	218	210	221	213	255	255	219	260	265	245
Utility										
Range (km)	705	956	767	1037	1357	1595	545	939	1180	327
All-electric range (km)	0	0	0	0	125	124	122	40	55	327
Passenger volume (m^3)	2.61	2.61	2.61	2.61	2.56	2.56	2.35	2.56	2.56	2.39
Towing capacity (kg)	1511	1211	1558	1258	2580	2544	685	2844	3072	1798
Environment										
Gasoline Equivalent cons. (L/100km)	8.5	7.2	7.8	6.6	4.9	4.7	3.7	6.7	6.3	2.2
Life cycle CO_2 (g/km)	260	220	242	206	185	178	41	238	224	74
Life cycle NO_x (g/km)	0.13	0.10	0.12	0.10	0.12	0.11	0.05	0.14	0.13	0.08
Life cycle PM10 (g/km)	0.09	0.03	0.08	0.03	0.06	0.02	0.00	0.07	0.03	0.01

Appendix D

Survey questionnaire and aggregated responses

1 How old are you?

16–21	21–28	28–35	35–50	50+
0	32	29	20	7

2 What is your gender?

male	female
60	28

3 Do you own a car?

yes	no
62	21

4 Daily travel by car (as driver or passenger, please answer in km):

Various (not listed) Average = 26.5 km

5 How much do you agree with each of the following statements?

	not at all	a little	yes and no	pretty close	completely right
a I like to be the first to pull away from a stop light	36	18	14	12	3
b I am very concerned about the future	0	9	23	30	21
c I take my career very seriously	4	8	18	29	24
e I commute using a car daily	30	7	8	10	28
f I care very much about costs	1	8	28	28	18
g I have a high level of education	0	3	8	21	51
h I have a strong feeling of social responsibility	1	2	22	37	21
i I am very aware of environmental issues	0	1	18	35	29
j I welcome change	0	2	32	32	17
k I always make rational choices	0	12	32	30	9
l I often do detailed financial analysis and planning	8	20	24	23	8

6 How important are these COST criteria to you?

	not at all	very little	moderately	quite a bit	extremely
a Purchase cost	1	3	19	37	21
b Operating cost (fuel, oil etc)	1	1	26	35	18
c Maintenance cost (service and repairs)	1	3	25	35	17

7 How important are these PERFORMANCE criteria to you?

	not at all	very little	moderately	quite a bit	extremely
a Off-the-line acceleration (0–100 kph)	14	20	32	13	2
b Passing acceleration (80–110 kph)	4	18	31	25	3
c Top speed	19	36	20	6	0
d Braking	2	4	14	38	23
e Handling	3	3	10	40	25
f Reliability	0	2	3	21	41
g Refuelling time	11	14	18	14	10

8 How important are these ENVIRONMENTAL criteria to you?

	not at all	very little	moderately	quite a bit	extremely
a Global-warming causing emissions (CO_2, methane, etc)	1	3	17	41	18
b Health-related emissions (NO_x, Ozone, etc)	0	3	14	39	24
c Noise	0	9	33	25	13

9 How important are these UTILITY criteria to you?

	not at all	very little	moderately	quite a bit	extremely
a Range (autonomy)	3	10	17	38	12
b Passenger capacity	2	7	22	45	4
c Cargo capacity	3	9	30	33	5
d Towing capacity	38	21	13	6	2
e Multimedia features (on-board video, stereo, etc)	12	25	18	8	3
f Sufficient drink holders	20	13	20	7	6

10 How important are these SAFETY criteria to you?

	not at all	very little	moderately	quite a bit	extremely
a Fatality risk to drivers/passengers	0	3	9	19	49
b Fatality risk to pedestrians	1	9	19	21	30
c Presence of active safety technology (anti-lock brakes, traction control etc)	0	0	20	26	34
d Presence of passive safety technology (seat belts, air bags, crumple zones)	0	0	8	21	51

Appendix E

Assumptions and inputs driving fleet dynamics simulation

The development of the vehicle fleet in the selected countries was estimated regarding their demographic and economic perspectives. Recent predictions by Alho *et al.* (2006) are that the population size in the European Economic Area (EEA) including Switzerland will increase slightly from the current level of 392 million to 427 million inhabitants by 2050. This median scenario by Alho *et al.* (2006) was chosen as the basis for the estimation of the fleet sizes for the first half of the twenty-first century, followed by a slight decrease from 2050 to 2100. This has been complemented by data from the Council of Europe (2005) on the population increase between 1995 and 2004, where Germany, Italy, Great Britain, Sweden, and Poland show a 0% increase, France and Switzerland between 2% and 5%, and Spain between 6% and 10%.

The economic aspect of the development is represented by the GDP, which shows the increase in value of all final goods and services produced within a nation in a given year (not taking into account purchasing power parity but taking into account inflation). It is used as a measure of economic development. The values are taken from the World Factbook (CIA, 2008). The computed fleet sizes are a result of data-deduced growth curves and take the values shown in Figure E.1.

In Table E.1, the most important fleet model input data for Germany and the year 2015 are given. The data are mostly taken from literature as described in the text.

Table E.1. *Illustrative presentation of reference input data for Germany in 2015.*

Data for 2015	PP f(t) Euro	Fuel con. l/100km	km/year/ vehicle	FP f(t) Euro	CO_2 emission g/l
Petrol ICE	15 700	7.4	13 100	1.32	2380
Petrol HEV	18 700	5.5	13 100	1.32	2380
Diesel ICE	17 000	6.2	15 600	1.24	2650
Diesel HEV	20 000	4.9	13 100	1.24	2650
NGV	17 100	6.5	13 100	0.72	1829
LPGV	17 600	6.2	13 100	0.72	2057
H_2	22 000	6.4	13 100	0.72	0
Car lifetime	12.2	Years			
uPP	–0.13	Utility constant for purchase price			
FiC Euro	260 000	Fuelling station investment cost			
	300 000	H_2-Fuelling station investment cost			
u FP	–0.0068	Utility constant for fuel price			
Inherent attractiveness f(t) (see Figure E.6)					
PP	Purchase price				
FP	Fuel price				

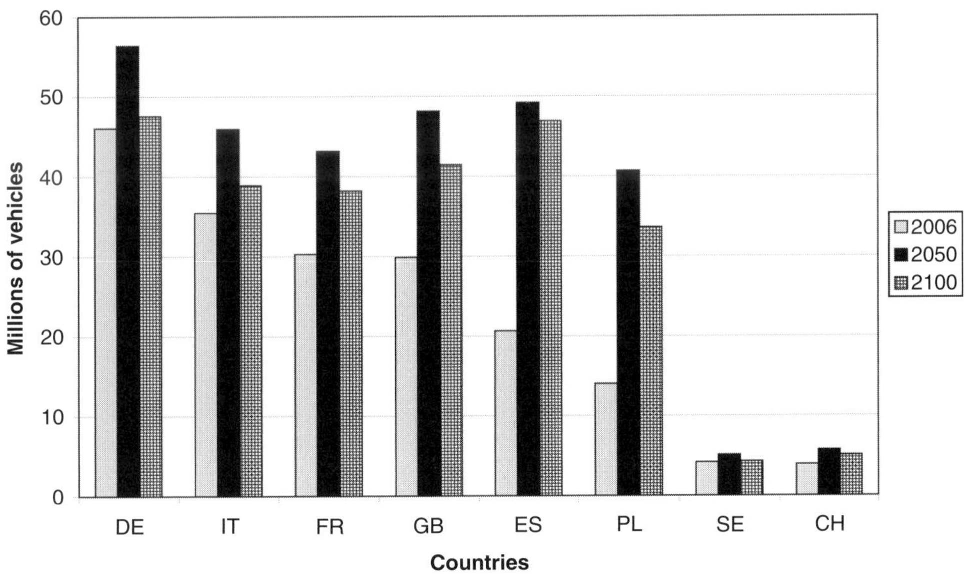

Figure E.1 Total passenger vehicle fleets for selected countries in 2006, 2050, and 2100 (vehicles are given in millions).

In the following the input values of the most important look–up functions are displayed in Figures E.2 to E.6. Most of the inputs are for the baseline BAU scenario, with the exception of the input fuel price for hydrogen shown in Figure E.3.

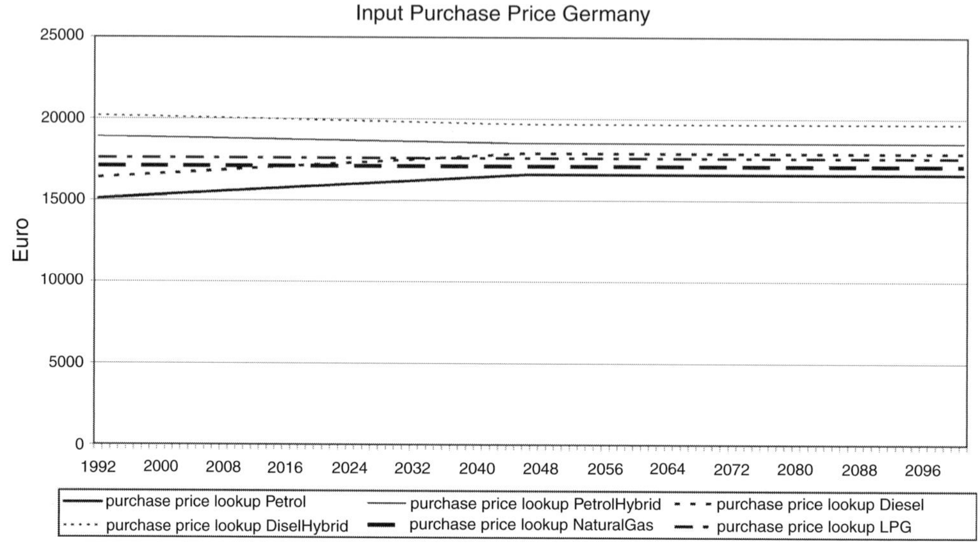

Figure E.2 Vehicle purchase price assumptions.

Input data for the baseline (BAU) scenario for Germany.

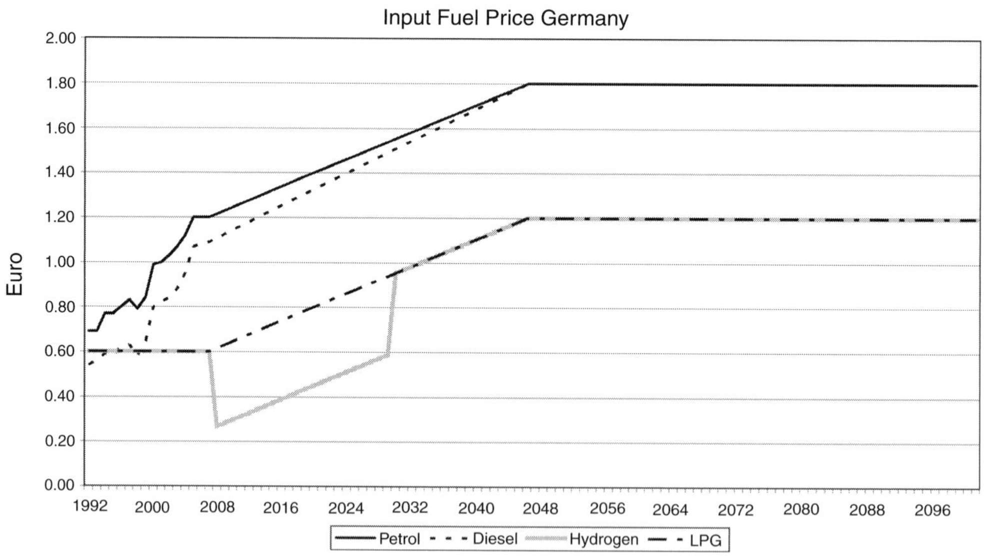

Figure E.3 Fuel price assumptions.

Input data for the baseline (strong promotion of hydrogen) scenario for Germany.

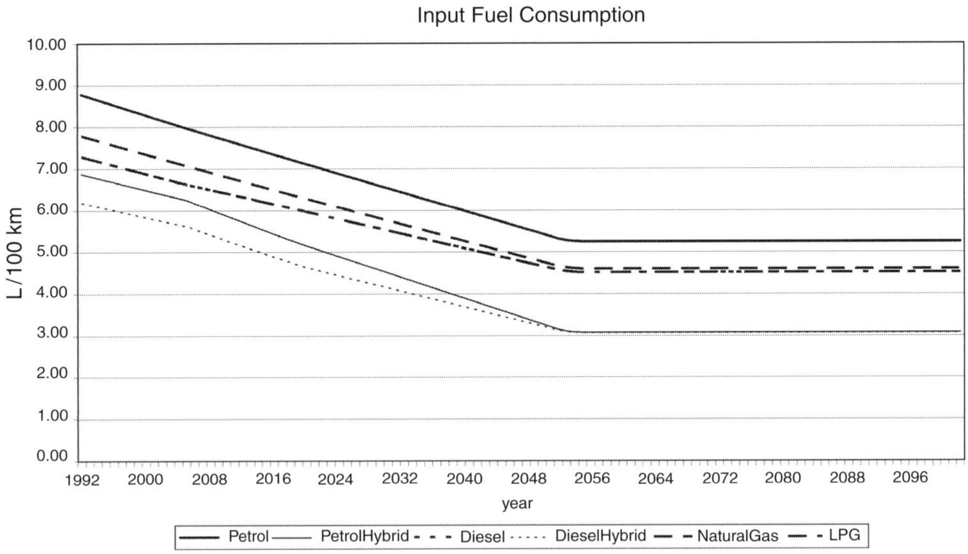

Figure E.4 Fuel consumption assumptions.

Input data for the assumed improvement in fuel consumption of the different drivetrain technologies. They apply equally to all selected countries.

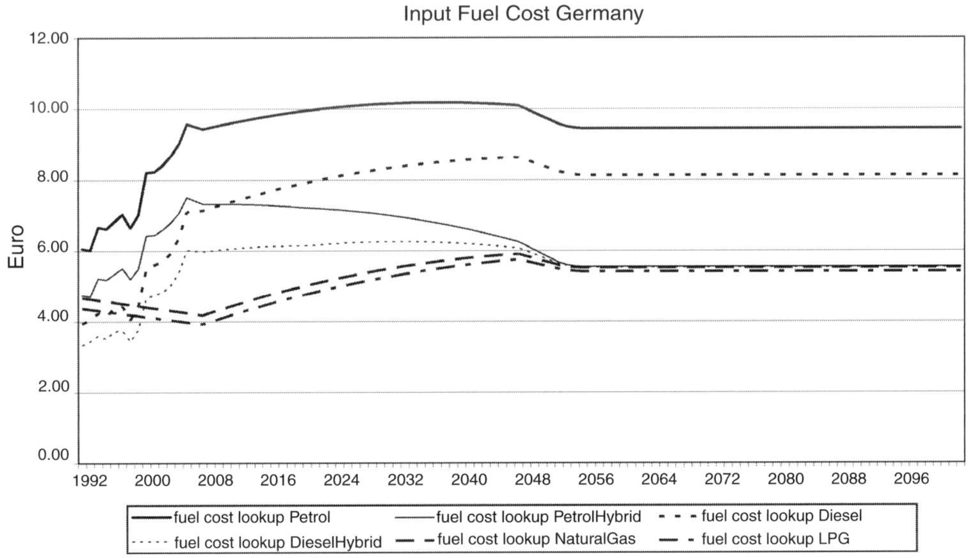

Figure E.5 Computed fuel costs.

Computed fuel costs development under the baseline (BAU) scenario for Germany.

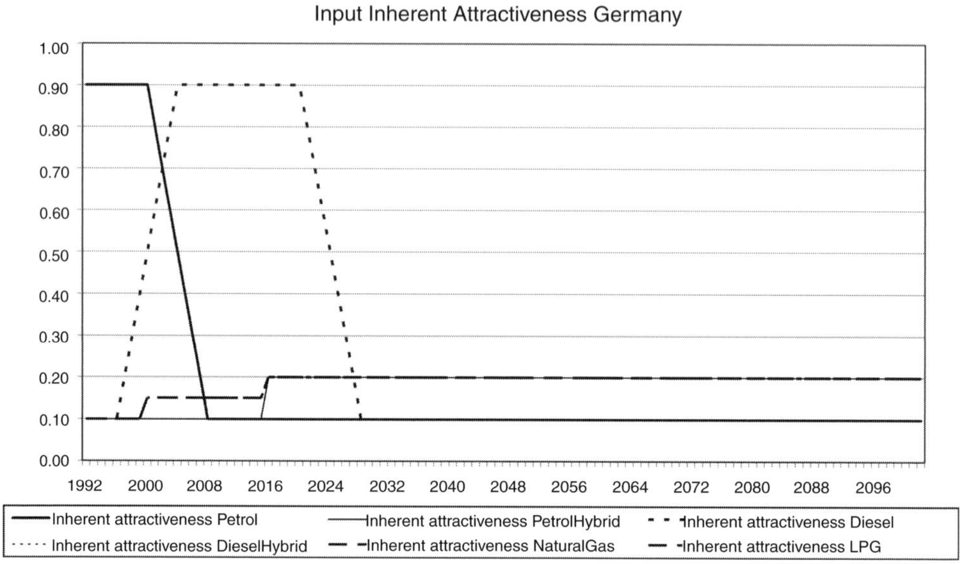

Figure E.6 Assumptions on the inherent attractiveness.

Input data for the assumed inherent attractiveness of the different drivetrain technologies under the baseline (BAU) scenario for Germany.

References

Aalto, T., Lallo, M., Hatakka, J., and Laurila, T. (2009). Atmospheric hydrogen variations and traffic emissions at an urban site in Finland. *Atmospheric Chemistry and Physics*, **9**(19), 7387–7396.

ACEA (2007). Sources of CO2 emissions in the EU: News of the ACEA. Association des Constructeurs Européen d'Automobile. Available online: http://www.acea.be/index.php/news/news_detail/sources_of_co2_emissions_in_the_eu, accessed 10 May 2010.

ACEM (2010). *The Motorcycle Industry in Europe*. Brussels: Association des Constructeurs Européens de Motocycles.

ADAC (2008). Der ADAC Autokatalog: Informationen zu über 7500 aktuellen Modellen. *Der ADAC-Autokatalog*. Available online: http://www1.adac.de/Auto_Motorrad/Autokatalog/default.asp, accessed 31 March 2010.

Ahluwalia, R.K. and Wang, X. (2007). Buildup of nitrogen in direct hydrogen polymer-electrolyte fuel cell stacks. *Journal of Power Sources*, **171**, 63–71.

Ahluwalia, R.K. and Wang, X. (2008). Effect of CO and CO_2 impurities on performance of direct hydrogen polymer–electrolyte fuel cells. *Journal of Power Sources*, **180**, 122–131.

Airliquide (2010). Unser Gas: Gasflaschen: Stahlflaschen (20–50 Litre). Accessed on 28.05.2010. Available online: http://www.airliquide.de/loesungen/produkte/gase/lieferformen/cylinder-versorgung.html, accessed 28 May 2010.

Aleklett, K., Höök, M., Jakobsson, K., Lardelli, M., Snowden, S., and Söderbergh, B. (2010). The peak of the oil age: analyzing the world oil production Reference Scenario in World Energy Outlook 2008. *Energy Policy*, **38**(3), 1398–1414.

Alho, J., Alders, M., Cruisen, H., Keilman, N., Nikander, T., and Pham, D.Q. (2006). New forecast: population decline postponed in Europe. *Statistical Journal of the United Nations*, **23**, 1–10.

Alvarez, R., Weilenmann, M., and Favez, J.-Y. (2009). Assessing the real-world performance of modern pollutant abatement systems on motorcycles. *Atmospheric Environment*, **43**, 1503–1509.

An, F., Gordon, D., He, H., Kodjak, D., and Rutherford, D. (2007). *Passenger Vehicle Greenhouse Gas and Fuel Economy Standards: A Global Update*. Washington, DC and San Francisco: The International Council on Clean Transportation.

ANL (2009). Argonne GREET Model 1.8c.0 & 2.7. *GREET Model*. Available online: http://www.transportation.anl.gov/modeling_simulation/GREET/index.html, accessed 7 December 2009.

Auckenthaler, T.S. (2005). Modelling and control of three-way catalytic converters. Unpublished PhD thesis, Swiss Federal Institute of Technology, Zurich. Available online: http://e-collection.ethbib.ethz.ch, accessed 7 January 2011.

Auto-Reporter.net (2007). VDA: Deutscher Strassenverkehr hat CO_2 um 12 Prozent reduziert. Available online: http://www.kfz.net/autonews/vda-deutscher-strassenverkehr-hat-co2-um-12-prozent-reduziert, accessed 10 May 2010.

Azar, C., Lindgren, K., and Andersson, B.A. (2003). Global energy scenarios meeting stringent CO_2 constraints: cost-effective fuel choices in the transportation sector. *Energy Policy*, **31**(10), 961–976.

BAFU, B. f. U. (2007). Klimapolitik der Schweiz: CO_2-Gesetz. Available online: http://www.bafu.admin.ch/klima/00493/00494/00496/index.html?lang=de, accessed 30 April 2010.

Bandivadekar, A. and Heywood, J. (2004). *Coordinated Policy Measures for Reducing the Fuel Consumption of the U.S. Light-Duty Vehicle Fleet*. Cambridge, MA: Laboratory for Energy and the Environment, Massachusetts Institute of Technology.

Bandivadekar, A., Bodek, K., Cheah, L., *et al.* (2008a). *On the Road in 2035: Reducing Transportation's Petroleum Consumption and GHG Emissions*. Cambridge, MA: Laboratory for Energy and the Environment: Massachusetts Institute of Technology. Available online: http://web.mit.edu/sloan-auto-lab/research/beforeh2/otr2035/, accessed 10 January 2011.

Bandivadekar, A., Cheah, L., Evans, *et al.* (2008b). Reducing the fuel use and greenhouse gas emissions of the US vehicle fleet. *Energy Policy*, **36**, 2754–2760.

Barnes, D.H., Wofsy, S.C., Fehlau, B.P., *et al.* (2003). Hydrogen in the atmosphere: Observations above a forest canopy in a polluted environment. *Journal of Geophysical Research*, **108**(D6), 4197.

Barreto, L. (2001). Technological Learning in Energy Optimisation Models and Deployment of Emerging Technologies. PhD thesis, ETH Zurich, Switzerland.

Barreto, L. and Kypreos, S. (2004). Emissions trading and technology deployment in an energy-systems 'bottom-up' model with technology learning. *European Journal of Operational Research* **158**(1), 243–261.

Barreto, L. and Kypreos, S. (2006). Extensions to the energy-system GMM model: an overview. PSI Report No. 06–07, Villigen, Switzerland: Paul Scherrer Institute.

Bauer, C. (2007). Holzenergie. In Dones, R. *et al.* (eds.), Sachbilanzen von Energiesystemen: Grundlagen für den ökologischen Vergleich von Energiesystemen und den Einbezug von Energiesystemen in Ökobilanzen für die Schweiz. Final report ecoinvent No. 6-IX, Paul Scherrer Institut Villigen, Swiss Centre for Life Cycle Inventories, Dübendorf, CH.

Bauer, C., Bolliger, R., Tuchschmid, M., and Faist-Emmenegger, M. (2007). Wasserkraft. In Dones, R. *et al.* (eds.), Sachbilanzen von Energiesystemen: Grundlagen für den ökologischen Vergleich von Energiesystemen und den Einbezug von Energiesystemen in Ökobilanzen für die Schweiz. Final report ecoinvent No. 6-VIII, Paul Scherrer Institut Villigen, Swiss Centre for Life Cycle Inventories, Dübendorf, CH.

Bauer, C., Dones, R., Heck, T., and Hirschberg, S. (2008a). Environmental assessment of current and future Swiss electricity supply options. In *Proceedings of The International Conference on the Physics of Reactors 'Nuclear Power: A Sustainable Resource'*, Interlaken, Switzerland, September 14–19, 2008.

Bauer, C., Heck, T., Dones, R., Mayer-Spohn, O., and Blesl, M. (2008b). Deliverable 7.2 RS 1a: Final report on technical data, costs, and life cycle inventories of advanced fossil power generation systems. New Energy Externalities Developments for Sustainability (NEEDS) Integrated Project, EU 6th Framework Programme, Brussels.

Bedel, L. and Junker, M. (2006) Natural gas pipelines for hydrogen transport. World Hydrogen Energy Conference, 13–16 June.

Bertsekas, D.P. (2005). *Dynamic Programming and Optimal Control* 3rd edn., Nashua, NH: Athena Scientific.

BFS, B. f. S. (2004). *Motorfahrzeuge in der Schweiz*. Neuenburg: Bundesamt für Statistik.

Bjelkengren, C. (2008). *The impact of mass decompounding on assessing the value of vehicle lightweighting*. Massachusetts Institute of Technology. Available online: http://dspace.mit.edu/handle/1721.1/44384, accessed 22 September 2009.

Bolhàr-Nordenkampf, M., Rauch, R., Bosch, K., Aichernig, C., Hofbauer, H. (2002). Biomass CHP plant Güssing: using gasification for power generation. In *Proceedings, International Conference on Biomass Utilisation*, Thailand.

Bond, S.W., Alvarez, R., Vollmer, M.K., Steinbacher, M. *et al.* (2010). Molecular hydrogen (H_2) emissions from gasoline and diesel vehicles. *Science of the Total Environment*, **408**(17), 3596–3606.

Bond, S.W., Gül, T., Reimann, S., Buchmann, B., and Wokaun, A. (2011a). Sources of anthropogenic hydrogen (H_2) to the atmosphere during the potential transition to an increasingly H_2-intensive economy. *International Journal of Hydrogen Energy*, doi:10.1016/j.ijhydene.2010.10.016.

Bond, S.W., Vollmer, M.K., Steinbacher, M., Henne, S., and Reimann, S. (2011b). Atmospheric molecular hydrogen (H_2): Observations at the high-altitude site, Jungfraujoch, Switzerland. *Tellus Series B*, **36**(1), 1122–1135.

Bossel, U., Eliasson, B., and Taylor, G. (2003). The future of the hydrogen economy: bright or bleak? *Distributed Generation and Alternative Energy Journal*, **18**(3), 29–70.

Brownstone, D., Bunch, D., and Train, K. (2000). Joint mixed logit models of stated and revealed preferences for alternative-fuel vehicles. *Transportation Research B: Methodological*, **34**, 315–338.

Burger, B. and Bauer, C. (2007) Windkraft. In Dones, R. *et al.* (eds.), Sachbilanzen von Energiesystemen: Grundlagen für den ökologischen Vergleich von Energiesystemen und den Einbezug von Energiesystemen in Ökobilanzen für die Schweiz. Final report ecoinvent No. 6-XIII, Paul Scherrer Institut Villigen, Swiss Centre for Life Cycle Inventories, Dübendorf, CH.

Burns, L.D., McCormick, J.B., and Borroni-Bird, C.E. (2002). Vehicle of change. *Scientific American*, (October), 40–49.

CEH (2007). *Marketing Research Report: Hydrogen*. Menlo Park, CA: Chemical Economics Handbook (CEH) – SRI Consulting.

CIA (2008). The World Factbook. Washington DC: CIA. Available online: https://www.cia.gov/library/publications/the-world-factbook/, accessed 10 May 2010.

CONCAWE, EUCAR, and European Commission (2007). Well-to-wheels analysis of future automotive fuels and powertrains in the European context (version 2c). Available online: http://ies.jrc.ec.europa.eu/uploads/media/WTW_Report_010307.pdf /, accessed 10 May 2010.

Conrad, R. and Seiler, W. (1981). Decomposition of atmospheric hydrogen by soil microorganisms and soil enzymes. *Soil Biology & Biochemistry*, **13**, 43–49.

Conrad, R. and Seiler, W. (1985). Influence of temperature, moisture, and organic carbon on the flux of H_2 and CO between soil and atmosphere: field studies in subtropical regions. *Journal of Geophysical Research*, **90**(D3), 5699–5709.

Council of Europe (2005). *Recent demographic developments in Europe*. Cedex, France: Council of Europe Publishing.

D'Alleva, B.A. and Lovell, W.G. (1936). Relation of exhaust gas composition to air–fuel ratio. *Society of Automotive Engineers Journal (Transactions)*, **38**(3), 90–98.

Daniele, S., Jansohn, P., and Boulouchos, K. (2009) Flame front characteristic and turbulent flame speed of lean premixed syngas combustion at gas turbine relevant conditions. *Proceedings of the ASME Turbo Expo*, June 8–12, 2009 Orlando, Florida, GT2009-59477.

Davis, S.C. and Diegel, S.W. (2007). *Transportation Energy Data Book*. Oak Ridge, TN: Oak Ridge National Laboratory.

Derwent, R.G., Collins, W.J., Johnson, C.E., and Stevenson, D.S. (2001). Transient behaviour of tropospheric ozone precursors in a global 3-D CTM and their indirect greenhouse effects. *Climatic Change*, **49**(4), 463–487.

Dones, R. (2007) Kernenergie. In Dones, R. *et al.* (eds.), Sachbilanzen von Energiesystemen: Grundlagen für den ökologischen Vergleich von Energiesystemen und den Einbezug von Energiesystemen in Ökobilanzen für die Schweiz. Final report ecoinvent No. 6-VII, Paul Scherrer Institut Villigen, Swiss Centre for Life Cycle Inventories, Dübendorf, CH.

Dones, R., Bauer, C., and Röder, A. (2007). Kohle. In Dones, R. *et al.* (eds.), Sachbilanzen von Energiesystemen: Grundlagen für den ökologischen Vergleich von Energiesystemen und den Einbezug von Energiesystemen in Ökobilanzen für die Schweiz. Final report ecoinvent No. 6-VI, Paul Scherrer Institut Villigen, Swiss Centre for Life Cycle Inventories, Dübendorf, CH.

Dufour, J., Serrano, D.P., Galvez, J.L., Moreno, J., and Garcia, C. (2009). Life cycle assessment of processes for hydrogen production. Environmental feasibility and reduction of greenhouse gas emissions. *International Journal of Hydrogen Energy*, **34**, 1370–1376.

Duret, A., Friedli, C., and Maréchal, F. (2005). Process design of synthetic natural gas (SNG) production using wood gasification. *Journal of Cleaner Production*, **13**, 1434–1446.

DVV (2006). *Verkehr in Zahlen*. Hamburg, Germany: Verkehrs-Verlag.

ecoinvent Centre (2007). The ecoinvent LCA database: ecoinvent data v2.0. Swiss Centre for Life Cycle Inventories, Dübendorf, Switzerland. Available online: www.ecoinvent.org, accessed 7 January 2011.

ecoinvent (2009). Ecoinvent Database. Available online: http://www.ecoinvent.org/database/, accessed 11 December 2009.

Ehhalt, D.H. and Rohrer, F. (2009). The tropospheric cycle of H2: a critical review. *Tellus B*, **61**(3), 500–535.

EIA (2007a). *International Energy Outlook 2007*, Pittsburgh, PA: US Energy Information Administration.

EIA (2007b). Estimated Natural Gas Pipeline Mileage in the Lower 48 States. Energy Information Administration 2007. Available online: http://www.eia.doe.gov/pub/oil_gas/natural_gas/analysis_publications/ngpipeline/mileage.html, 1 December 2007.

EIA (2008a). Highlights from EIA energy use by sector. In *International Energy Outlook 2008*. Washington DC: Energy Information Administration.

EIA (2008b). *The Impact of Increased Use of Hydrogen on Petroleum Consumption and Carbon Dioxide Emissions*. Washington, DC: Energy Information Administration.

EIA (2010a). Global Share of Total Energy Demand. Available at: http://www.eia.doe.gov/oiaf/ieo/highlights.html, accessed 5 March, 2010.

EIA (2010b). US Share of Total Energy Demand. Available at: http://www.eia.doe.gov/emeu/aer/pdf/pages/sec2_4.pdf, accessed 5 March, 2010.

EIA (2010c). US Transportation Petroleum Energy Use. Available at: http://www.eia.doe.gov/pub/oil_gas/petroleum/analysis_publications/oil_market_basics/demand_text.htm, accessed 12 March, 2010.

Epstein, M., Olalde, G., Santén, S., Steinfeld, A., and Wieckert, C. (2008). Towards the industrial carbothermal production of zinc. *Journal of Solar Energy Engineering*, **130**, 014505–1–014505–4.

European Commission (1995). Towards fair and efficient pricing in transport (Green paper). Available online: http://eur-lex.europa.eu/LexUriServ/LexUriServ.do?uri=COM:1995:0691:FIN:EN:PDF, accessed 10 May 2010.

European Commission (2007). Directive 2009/30/EC of the European Parliament and of the Council of 23 April 2009. Available online: http://eur-lex.europa.eu/smartapi/cgi/sga_doc?smartapi!celexplus!prod!CELEXnumdoc&lg=EN&numdoc=32009L0030, accessed 10 May 2010.

European Commission (2008). 20 20 by 2020: Europe's climate change opportunity. Communication from the Commission to the European Parliament, the Council, the European Economic and Social Committee and the Committee of the Regions. COM(2008) 30 final, Brussels, Belgium, 23 January 2008. Available online: http://ec.europa.eu/energy/climate_actions/doc/com_2008_030_en.pdf, accessed 7 January 2011.

European Commission (2009). Regulation (EC) No 443/2009 of the European Parliament and of the Council. *Official Journal of the European Union*: 1–14. Available online: http://eur-lex.europa.eu/LexUriServ/LexUriServ.do?uri=OJ:L:2009:140:0001:0015:EN:PDF, accessed 10 May 2010.

Eurostat (2005). Panorama of Transport. Available online: http://epp.eurostat.ec.europa.eu/cache/ITY_OFFPUB/KS-DA-09-001/EN/KS-DA-09-001-EN.PDF, accessed 10 May, 2010.

Eurostat (2008). Air pollutants and greenhouse gas. Available online: http://epp.eurostat.ec.europa.eu/statistics_explained/index.php/Air_pollution_statistics#Further_Eurostat_information, accessed 10 May 2010.

Evans, C., Cheah, L.. Bandivadekar, A., and Heywood, J. (2007). Getting more miles per gallon. *Issues in Science and Technology*, **25**(2), 71–80.

Faist Emmenegger, M., Heck, T., Jungbluth, N., and Tuchschmid, M. (2007). Erdgas. In Dones, R. *et al.* (eds.), Sachbilanzen von Energiesystemen: Grundlagen für den ökologischen Vergleich von Energiesystemen und den Einbezug von Energiesystemen in Ökobilanzen für die Schweiz. Final report ecoinvent No. 6-V, Paul Scherrer Institut Villigen, Swiss Centre for Life Cycle Inventories, Dübendorf, CH.

FAO (Food and Agriculture Organization of the United Nations) (2006). FAOSTAT database. Rome: FAO Statistics Division. Available online: http://faostat.fao.org/, accessed 10 October 2006.

Favez, J.-Y., Weilenmann, M., and Stilli, J. (2009). Cold start extra emissions as a function of engine stop time: evolution over the last 10 years. *Atmospheric Environment*, **43**, 996–1007.

Felder, R. (2007) Well-to-wheel analysis of renewable transport fuels: synthetic natural gas from wood gasification and hydrogen from concentrated solar energy. PhD Thesis No. 17437, ETH Zurich, CH.

Fishbone, L.G. and Abilock, A. (1981). MARKAL: a linear-programming model for energy system analysis: technical description of the BNL Version. *International Journal of Energy Resources*, **5**(4), 353–375.

Fishbone, L.G., Giesen, G., Goldstein, G., *et al.* (1983). *User's guide for MARKAL (BNL/KFA Version 2.0), A multi-period, linear programming model for energy system analysis, BNL 51701*. Jülich: Department of Applied Science, Brookhaven National Laboratory, NY and Kernforschungsanlage Julich, Julich, Germany.

Forrester, J.W. (1969). *Urban Dynamics*. Cambridge, MA: MIT Press.

Forster, P., Ramaswamy, V., Artaxo, P., *et al.* (2007). *Changes in Atmospheric Constituents and in Radiative Forcing*. Cambridge, UK: Cambridge University Press.

Frischknecht, R., Faist Emmenegger, M., Steiner, R., Tuchschmid, M., and Gärtner, S. (2008) Deliverable 15.1: LCA of background processes. New energy externalities developments for sustainability (NEEDS) integrated project, EU 6th Framework Programme, Brussels.

Frischknecht, R., Jungbluth, N., Althaus, H.-J., *et al.* (2007a). Overview and methodology: ecoinvent report No. 1. Dübendorf, Switzerland: Swiss Centre for Life Cycle Inventories.

Frischknecht, R., Jungbluth, N., Althaus, H.-J., Bauer, C., Doka, G., Dones, R., Hischier, R., Hellweg, S., Humbert, S., Köllner, T., Loerincik, Y., Margni, M., and Nemecek, T. (2007b). Implementation of life cycle impact assessment methods. ecoinvent report No. 3, v2.0. Dübendorf, Switzerland: *Swiss Centre for Life Cycle Inventories.*

Frischknecht, R., Tuchschmid, M., Faist Emmeneger, M., Bauer, C., and Dones, R. (2007c): Strommix und Stromnetz. In Dones, R. *et al.* (eds.), Sachbilanzen von Energiesystemen: Grundlagen für den ökologischen Vergleich von Energiesystemen und den Einbezug von Energiesystemen in Ökobilanzen für die Schweiz. ecoinvent report No. 6, v2.0. Dübendorf, Switzerland: *Swiss Centre for Life Cycle Inventories.*

Fulton, L. and Eads, G. (2004). *IEA/SMP* model documentation and reference case projection. Auxiliary material to: mobility 2030: meeting the challenges to sustainability; the sustainable mobility project. IEA/CRA. Available online: http://www.wbcsd.org/web/publications/mobility/smp-model-document.pdf, accessed 7 January 2011.

Gao, Y. and Ehsani, M. (2010). Design and Control Methodology of Plug-in Hybrid Electric Vehicles. *IEEE Transactions on Industrial Electronics*, **57**(2), 633–640.

General Motors (2001). *Well-to-Wheel Energy Use and Greenhouse Gas Emissions of Advanced Fuel/Vehicle Systems: North American Analysis/European Analysis*, Chicago, IL: Argonne National Laboratory. Available online: http://www.transportation.anl.gov/pdfs/TA/339.pdf.

Geodkoop, M., De Schryver, A., and Oele, M. (2008). Introduction to LCA with SimaPro 7. Amersfoort, NL: PRé Consultants B.V.

Gigerenzer, G. and Todd, P. (1999). *Simple Heuristics That Make Us Smart*, Berlin: Oxford University Press.

Goldman Sachs (2004). The BRICS and global markets: crude, cars and capital. *CEO Confidential*, **2004/09** (September).

Gomez, A., Azzaro-Pantel, C., Pibouleau, L., *et al.* (2007). Optimization of electricity / hydrogen cogeneration from generation IV nuclear energy systems. *Computer Aided Chemical Engineering*, **24**, 1109–1114.

Grahn, M., Azar, C., Lindgren, K., Berndes, G., and Gielen, D. (2007). Biomass for heat or as transportation fuel? A comparison between two model-based studies. *Biomass and Bioenergy*, **31**, 747–758.

Granovski, M., Dincer, I., and Rosen, M.A. (2006). Environmental and economic aspects of hydrogen production and utilization in fuel cell vehicles. *Journal of Power Sources*, **157**, 411–421.

Grant, A., Witham, C.S., Simmonds, P.G., Manning, A.J., and O'Doherty, S. (2010). A 15-year record of high-frequency, in situ measurements of hydrogen at Mace Head, Ireland. *Atmospheric Chemistry and Physics*, **10**(3), 1203–1214.

Greene, D. (2001). *TAFV Alternative Fuels and Vehicles Choice Model Description.* Springfield, VA: Oak Ridge National Laboratory.

Greene, D. L., Leiby, P.N., and Bowman, D. (2007). *Integrated Analysis of Market Transformation Scenarios with HyTrans*. Springfield, VA: Oak Ridge National Laboratory.

Griebel, P., Boschek, E., and Jansohn, P. (2007). Lean blowout limits and NOx emissions of turbulent, lean premixed, hydrogen-enriched methane/air flames at high pressure. *Journal of Engineering For Gas Turbines and Power-Transactions of the ASME*, **129**, 404–410.

Gucciardi Garcez, C. A. and de Souza Vianna, J. N. (2009). Brazilian Biodiesel Policy: Social and environmental considerations of sustainability. *Energy*, **34**(5), 645–654.

Gül, T. (2008). An energy–economic scenario analysis of alternative fuels for transport. PhD thesis No. 17888, ETH Zurich, Switzerland. Available online: http://e-collection.ethbib.ethz.ch.

Guzzella, L. and Sciarretta, A. (2007). *Vehicle Propulsion Systems: Introduction to Modeling and Optimization* 2nd edn., Berlin, Germany: Springer Verlag.

H2A (2006a). H2A spreadsheet models. Unpublished data from the H2A Project at the U.S. Department of Energy's Hydrogen Program and the National Renewable Energy Laboratory, H2A model-version 2006. Available online: http://www.hydrogen.energy.gov/h2a_production.html, accessed 22 July 2007.

H2A (2006b). H2A components model. National Renewable Energy Laboratory, U.S. Department of Energy. Version 10 July 2006. Available online: http://www.hydrogen.energy.gov/h2a_delivery.html, accessed 22 July 2007.

de Haan, P., Mueller, M.G., and Scholz, R.W. (2009). How much do incentives affect car purchase? Agent-based microsimulation of consumer choice of new cars. Part II: forecasting effects of feebates based on energy-efficiency. *Energy Policy*, **37**(3), 1083–1094.

Haeseldonckx, D. and D'haeseleer, W. (2007). The use of the natural-gas pipeline infrastructure for hydrogen transport in a changing market structure. *International Journal of Hydrogen Energy*, **32**, 1381–1386.

Hammer, S., Vogel, F., Kaul, M., and Levin, I. (2009). The H_2/CO ratio of emissions from combustion sources: comparison of top-down with bottom-up measurements in southwest Germany. *Tellus*, **61B** (3), 547–555.

Han, H. (1994). The competitive position of alternative automotive materials. Master's Thesis. Cambridge, MA USA: Massachusetts Institute of Technology.

Hankey, S. and Marshall, J.D. (2009). Impacts of urban form on future US passenger-vehicle greenhouse gas emissions. *Energy Policy*, **38/9**, 4880–4887.

Hauglustaine, D. A. and Ehhalt, D. H. (2002). A three-dimensional model of molecular hydrogen in the troposphere. *Journal of Geophysical Research*, **107**(D17), 4330, doi:10.1029/2001JD001156.

Heeb, N. V., Saxer, C. J., Forss, A., and Brühlmann, S. (2006). Correlation of hydrogen, ammonia and nitrogen monoxide (nitric oxide) emissions of gasoline-fuelled Euro-3 passenger cars at transient driving. *Atmospheric Environment*, **40**, 3750–3763.

Hertel, T. (2009). Analyzing the poverty impacts of biofuel mandates. Presented at the 12th Annual Conference on Global Economic Analysis, Santiago, Chile.

Heywood, J.B. (1988). *Internal Combustion Engine Fundamentals*. New York: McGraw-Hill Publishing Company.

Hirschberg, S. (2004). Hopes on hydrogen: no quick fixes. Energie-Spiegel N° 12. Villigen, Switzerland: Paul Scherrer Institute.

Holladay, J.D., Hu, J., King, D.L., and Wang, Y. (2009). An overview of hydrogen production technologies. *Catalysis Today*, **139**, 244–260.

Honda Motor Co. (2010). Specifications of the Honda FCX-Clarity. Available online: http://automobiles.honda.com/fcx-clarity/specifications.aspx, accessed 29 June 2010.

Hoogwijk, M., Faaij, A., van den Broek, R., *et al.* (2003). Exploration of the ranges of the global potential of biomass for energy. *Biomass and Bioenergy*, **25**(2), 119–133.

Hoogwijk, M., van Vuuren, D., de Vries, B., and Turkenburg, W. (2007). Exploring the impact on cost and electricity production of high penetration levels of intermittent electricity in OECD Europe and the USA, results for wind energy. *Energy*, **32**, 1381–1402.

Hydrogen Analysis Research Centre (HARC) (2006). Lower and higher heating values of hydrogen and fuels. Washington DC: US Department of Energy. Available online: http://hydrogen.pnl.gov/cocoon/morf/hydrogen/site_specific/fuel_heating_calculator?canprint=false, accessed 28 May 2010.

IEA (1999). *Automotive Fuels for the Future, the Search for Alternatives*. Paris: International Energy Agency IEA/OECD.

IEA (2002a). Energy Balances of OECD Countries 1971–2000. (Database on CD-Rom), Paris: International Energy Agency IEA/OECD.

IEA (2002b). Energy Balances of Non-OECD Countries 1971–2000. (Database on CD-Rom), Paris: International Energy Agency IEA/OECD.

IEA (2003). *World Energy Investment Outlook* Paris: International Energy Agency IEA/OECD.

IEA (2006). Energy Technology Perspectives Scenarios and Strategies to 2050. Available online: www.iea.org/speech/2006/ramsay/etp_beijing.pdf, accessed 1 June 2006.

IEA (2007). *World Energy Outlook 2007: China and India Insights*. Paris: International Energy Agency IEA/OECD.

IEA (2008a). Energy Statistics Division. Paris, France: Paris: International Energy Agency IEA/OECD. Available online: http://www.iea.org, accessed 9 January 2011.

IEA (2008b), *World Energy Outlook, OECD/IEA, 2008*. Paris, France. Available online: http://www.worldenergyoutlook.org/docs/weo2008/WEO2008.pdf, accessed 9 January 2011.

IEA/WBCSD (International Energy Agency/World Business Council for Sustainable Development) (2004). IEA/SMP Transport Spreadsheet Model. Available online: http://www.wbcsd.org/plugins/DocSearch/details.asp?type=DocDet&ObjectId=MTE0Njc, accessed 2 May 2007.

IIASA (2007). GGI Scenario Database. International Institute for Applied System Analysis (IIASA) GGI Scenario Database 2007. Available online: http://www.iiasa.ac.at/Research/GGI/DB/, accessed 18 September 2007.

International Organization for Standardization (ISO) (2006). European standard EN ISO 14044: Environmental Management – Life Cycle Assessment – Requirements and Guidelines. Geneva:ISO.

IPCC (2000). *Special Report on Emissions Scenarios*. Cambridge, UK: Cambridge University Press.

IPCC (2007a). Chapter 5: Transport and its infrastructure. In Mitigation of climate change, Working Group III Report, 4th Assessment Report, Intergovernmental Panel on Climate Change.

IPCC (2007b). Climate change 2007: Synthesis Report, 4th Assessment Report, Intergovernmental Panel on Climate Change.

IPPC (2007c). Summary for policymakers, in climate change 2007: the physical science basis. In Solomon, S., Qin, D., Manning, M., *et al.* (eds.) Contribution of Working Group I to the Fourth Assessment Report of the Intergovernmental Panel on Climate Change. Cambridge, UK: Cambridge University Press.

Ivy, J. (2004). Summary of Electrolytic Hydrogen Production: Milestone Completion Report. Golden, CO: National Renewable Energy Laboratory.

Jackson, M.C. (2009). Fifty years of systems thinking for management. *Journal of the Operational Research Society*, **60**, 24–32.

Janssen, A. (2004). Modeling the market penetration of passenger cars with new drive train technologies. Dissertation ETH No. 15855, Zurich, Swiss Federal Institute of Technology Zurich.

Janssen, A., Lienin, S.F., Gassmann, F., and Wokaun, A. (2006). Model aided policy development for the market penetration of natural gas vehicles in Switzerland. *Transportation Research A*, **40**, 316–333.

Joint Research Centre (2007). *Well-to-Wheels Analysis of Future Automotive Fuels and Powertrains in the European Context*, Brussels: EU JRC. Available online: http://www.co2star.eu/publications/Well_to_Tank_Report_EU.pdf, accessed 22 November 2010.

Jungbluth, N. (2007). Erdöl. In: Sachbilanzen von Energiesystemen: Grundlagen für den ökologischen Vergleich von Energiesystemen und den Einbezug von Energiesystemen in Ökobilanzen für die Schweiz In Dones, R. (ed.), ecoinvent report No. 6-IV, Swiss Centre for Life Cycle Inventories, Duebendorf, CH.

Jungbluth, N. and Tuchschmid, M. (2007). Photovoltaics. In Dones, R. *et al.* (eds.), Sachbilanzen von Energiesystemen: Grundlagen für den ökologischen Vergleich von Energiesystemen und den Einbezug von Energiesystemen in Ökobilanzen für die Schweiz. ecoinvent report No. 6-XII, Swiss Centre for Life Cycle Inventories, Dübendorf, CH, 2007.

Kammen, D.M. and Lipman, T. E. (2003). Letters: assessing the future hydrogen economy. *Science*, **302**, 226–229.

Kang, P. (1998). *A Technical and Economic Analysis of Structural Composite Use in Automotive Body-in-white Applications*. Cambridge, MA: Massachusetts Institute of Technology.

Kasseris, E.P. (2006). Comparative analysis of automotive powertrain choices for the near to mid-term future. MASc Mechanical Engineering. Massachusetts Institute of Technology.

Kasseris, E.P. and Heywood, J. (2007). Comparative analysis of automotive powertrain choices for the next 25 years. SAE paper 2007–01–1605, Transactions, V. 116, Journal of Fuels & Lubricants, Section 4, SAE World Congress, Detroit, MI, April 16–19, 2007.

Koroneos, C., Dompros, A., and Roumbas, G. (2008). Hydrogen production via biomass gasification: a life cycle assessment approach. *Chemical Engineering and Processing*, **47**, 1261–1268.

Kothari, R., Buddhi, D., and Sawhney, R.L. (2008). Comparison of environmental and economic aspects of various hydrogen production methods. *Renewable and Sustainable Energy Reviews*, **12**, 553–563.

Kromer, M.A. and Heywood, J. (2007). Electric powertrains: opportunities and challenges in the US light-duty vehicle fleet, publication No. LFEE 2007–03 RP. Cambridge, MA: MIT.

Kwon, J. *et al.* (2008). Impact of drive cycles on PHEV: component requirements. In SAE World Congress. Detroit, Michigan: SAE International.

Liu, J. (2007). *Modeling, configuration and control optimization of power-split hybrid vehicles*. PhD Thesis, University of Michigan.

Lomborg, B. (2009) Copenhagen consensus website. Available online: http://copenhagenconsensus.com/CCC%20Home%20Page.aspx, accessed 1 April 2010.

Loulou, R., Goldstein, G., and Noble, K. (2004). Documentation for the MARKAL family of models. Energy Systems Technology Analysis Programme (ETSAP), International Energy Agency (IEA), Paris, France: International Energy Agency.

Lovins, A.B. (2003). Letters: assessing the future hydrogen economy. *Science*, **302**, 226–229.

Maack, M. (2008). Deliverable 8.5, RS 1a: Generation of the energy carrier Hydrogen in context with electricity buffering generation through fuel cells. New Energy Externalities Developments for Sustainability (NEEDS) Integrated Project, EU 6th Framework Programme, Brussels.

Maass, P. (2005) The breaking point. *New York Times*, 31 August. Available online: http://www.nytimes.com/2005/08/21/magazine/21OIL.html, accessed 1 April 2010.

MacLean, H.L. and Lave, L.B. (2003). Evaluating automobile fuel/propulsion system technologies. *Progress in Energy and Combustion Science*, **29**(1), 1–69.

Maisonnier, G., Perrin, J., Steinberger-Wilckens, R., and Trümper, S.C. (2007). 'European Hydrogen Infrastructure Atlas' and 'Industrial Excess Hydrogen Analysis' Part II: Industrial surplus hydrogen and markets and production. Deliverable 2.1 and 2.1a, Brussels: Roads2HyCom, European Commission.

Makowski, M., Granat, J., and Ogryczak, W. (2009). *Multiple Criteria Analysis of Discrete Alternatives with a Simple Preference Specification: Pairwise-outperformance Based Approaches*. Laxenburg Austria: International Institute for Applied Systems Analysis.

Malen, D.E. and Reddy, K. (2007). *Preliminary Vehicle Mass Estimation Using Empirical Subsystem Influence Coefficients*. Ann Arbour, MI: University of Michigan.

Manne, A., Mendelsohn, R., and Richels, R. (1995). MERGE: A model for evaluating regional and global effects of GHG reduction policies. *Energy Policy*, **23**(1), 17–34.

Marques J.G. (2010). Evolution of nuclear fission reactors: third generation and beyond. *Energy Conversion and Management*, **51**(9), 1774–1780.

McCollum, D. and Yang, C. (2009). Achieving deep reductions in US transport greenhouse gas emissons: Scenario analysis and policy implications. *Energy Policy*, **37**, 5580–5596.

McDonald, A. and Schrattenholzer, L. (2001). Learning rates for energy technologies. *Energy Policy*, **29**(4), 255–261.

McKinsey & Company (2009). *Roads Toward a Low-carbon Future: Reducing CO_2 Emissions from Passenger Vehicles in the Global Road Transportation System*. New York: McKinsey & Company.

McNutt, B. and Rodgers, D. (2004). Lessons learned from 15 years of alternative fuels experience 1988 to 2003. In Sperling, D. and Cannon, J.S. (eds.), *The Hydrogen Energy Transition: Moving Toward the Post Petroleum Age in Transportation*, Amsterdam: Elsevier.

Melaina, M.W. (2007). Turn of the century refueling: a review of innovations in early gasoline refueling methods and analogies for hydrogen. *Energy Policy*, **35**(10), 4919–4934.

Messner, S. (1997). Endogenised technological learning in an energy systems model. *Journal of Evolutionary Economics*, **7**, 291–313.

Meyer, I. and Wessely, S. (2009). Fuel efficiency of the Austrian passenger vehicle fleet: analysis of trends in the technological profile and related impacts on CO_2 emissions. *Energy Policy*, **37**(10), 3779–3789.

Molburg, J.C. and Doctor, R.D. (2003). Hydrogen from steam-methane reforming with CO_2 capture. In *Proceedings of 20th Annual International Pittsburgh Coal Conference*, Sept. 15–19, Pittsburgh, PA.

Morrow, K., Karner, D., and Francfort, J. (2008). *Plug-in Hybrid Electric Vehicle Charging Infrastructure Review*, Idaho Falls, ID: Idaho National Laboratory.

Mueller, T. (2006). Your Move. In *The Best American Science Writing 2006*. New York: Harper Collins Publishing, pp. 1–15.

Mueller-Langer, F., Tzimas, E., Kaltschmitt, M., and Peteves, S. (2007). Techno-economic assessment of hydrogen production processes for the hydrogen economy for the short and medium term. *International Journal of Hydrogen Energy*, **32**(16) 3797–3810.

Nelson, D. R., Adger, W.N., and Brown, K. (2007). Adaptation to environmental change: contributions of a resilience framework. *Annual Review Environment and Resources*, **32**, 395–419.

Novelli, P.C., Lang, P.M., Masarie, K.A., *et al.* (1999). Molecular hydrogen in the troposphere: global distribution and budget. *Journal of Geophysical Research*, **104**(D23), 30427–30444.

Ogden, J.M. (2004). *Hydrogen as an energy carrier: Outlook for 2010, 2030 and 2050.* The 10–50 solution: technologies and policies for a low-carbon future. Available online: http://www.its.ucdavis.edu/publications/2004/UCD-ITS-RP-04-24.pdf, accessed 7 January 2011.

Owejan, J.P., Gagliardo, J.J., Falta, S.R., and Trabold, T.A. (2009a). Accumulation and removal of liquid water in proton exchange membrane fuel cells. *Journal of The Electrochemical Society*, **156**(12), B1475-B1483.

Owejan, J.P., Gagliardo, J.J., Sergi, J.M., Kandlikar, S.G., and Trabold, T.A. (2009b). Water management studies in PEM fuel cells, part I: Fuel cell design and in situ water distributions. *International Journal of Hydrogen Energy*, **34**(8), 3436–3444.

Pagerit, S., Sharer, P., and Rousseau, A. (2006). Fuel economy sensitivity to vehicle mass for advanced vehicle powertrains. In *Advanced Hybrid Vehicle Powertrains 2006. SAE Technical Paper Series*. Argonne, IL: Argonne National Laboratory.

Paneth, F.A. (1937). The chemical composition of the atmosphere. *Quarterly Journal of the Royal Meteorological Society*, **63**(271), 433–443.

Pock, M. (2007). *Gasoline and Diesel Demand in Europe: New Insights*. Vienna: Institute for Advanced Studies.

Prather, M.J. (2003). An environmental experiment with H_2? *Science*, **302**, 581–582.

Price, H., Jaeglé, L., Rice, A., *et al.* (2007). Global budget of molecular hydrogen and its deuterium content: Constraints from ground station, cruise, and aircraft observations. *Journal of Geophysical Research*, **112**(D22108), doi:10.1029/2006JD008152.

Rafaj, P. (2005). Analysis of policies contributing to sustainability of the global energy system using the global multi-regional MARKAL model (GMM). PhD thesis, ETH Zurich, Switzerland.

Rafaj, P. and Kypreos, S. (2005). Internalization of external cost in the power generation sector: analysis with the global multi-regional MARKAL model. *Energy Policy*, **35**(2), 828–843.

Rafaj, P., Kypreos, S., and Barreto, L. (2005). Flexible carbon mitigation policies: analysis with a global multi-regional MARKAL model. In Haurie, A. and Viguier, L. (eds.), *Coupling Climate and Economic Dynamics*, Dordrecht: Kluwer Academic Publishers, pp. 237–266.

Ragettli, M. (2007). Cost outlook for the production of biofuels. Master thesis, Paul Scherrer Institut, ETH Zurich, Villigen PSI, Switzerland.

Rahn, T., Eiler, J.M., Boering, K.A., *et al.* (2003). Extreme deuterium enrichment in stratospheric hydrogen and the global atmospheric budget of H_2. *Nature*, **424**, 918–921.

Rauch, R., Hofbauer, H., Bosch, K., *et al.* (2004). Steam gasification of biomass at CHP plant Güssing: status of the demonstration plant. *Proceedings of the 2nd World Conference and Technology Exhibition on Biomass for Energy, Industry and Climate Protection*, Rome: ETA.

Rhee, T.S., Brenninkmeijer, C.A.M., and Röckmann, T. (2006). The overwhelming role of soils in the global atmospheric hydrogen cycle. *Atmospheric Chemistry and Physics*, **6**, 1611–1625.

Ribeiro, S.K., Kobayashi, S., Beuthe, M., *et al.* (2007). Transport and its infrastructure. Climate Change 2007: Mitigation. Contribution of Working Group III to the Fourth

Assessment Report of the Intergovernmental Panel on Climate Change. Cambridge, UK: Cambridge University Press.

Roads2HyCom (2009) Roads2HyCom Final Report Available online: http://www.road-s2hy.com/r2h_downloads/Roads2HyCom%20R2H8500PUv6%20-%20Final%20Report.pdf, accessed 30 June 2009.

Rogers, E.M. (2003 5th Ed.). *Diffusion of Innovations*. New York: Free Press.

Rogner, H.-H. (1997) An assessment of world hydrocarbon resources. *Annual Review of Energy and the Environment*, **22**, 217–62.

Rohs, S., Schiller, C., Riese, M., *et al.* (2006). Long-term changes of methane and hydrogen in the stratosphere in the period 1978–2003 and their impact on the abundance of stratospheric water vapor. *Journal of Geophysical Research*, **111**(D14315), doi:10.1029/2005JD006877.

Ross, M. and Wenzel, T. (2001). *Losing Weight to Save Lives: A Review of the Role of Automobile Weight and Size in Traffic Fatalities*, Ann Arbour, MI: US Department of Energy.

Roth, S., Hirschberg, S., Bauer, C. *et al.* (2009). Sustainability of electricity supply technology portfolio. *Annals of Nuclear Energy*, **36**, 409–416.

Rousseau, A., Pagerit, S., and Gao, D. (2007). Plug-in Hybrid Electric Vehicle Control Strategy Parameter Optimization. *23rd International Electric Vehicle Symposium*. Anaheim, California: Argonne National Laboratory.

Sanderson, M.G., Collins, W.J., Derwent, R.G., and Johnson, C.E. (2003). Simulation of global hydrogen levels using a Lagrangian three-dimensional model. *Journal of Atmospheric Chemistry*, **46**, 15–28.

Schlapbach, L. and Züttel, A. (2001). Hydrogen-storage materials for mobile applications. *Nature*, **414**, 353–358.

Schmidt, U. (1974). Molecular hydrogen in the atmosphere. *Tellus*, **26**, 78–90.

Schneider, M.T., Schade, B., and Grupp, H. (2004). Innovation process 'fuel cell vehicle': What strategy promises to be most successful? *Technological Analysis and Strategic Management*, **16**, 147–172.

Schultz, M.G., Diehl, T., Brasseur, G.P., and Zittel, W. (2003). Air pollution and climate-forcing impacts of a global hydrogen economy. *Science*, **302**, 624–627.

Schweizerische-Eidgenossenschaft (2000). Bundesgesetz über die Reduktion der CO_2-Emissionen (CO2-Gesetzt) SR 641.71. Available online: http://www.admin.ch/ch/d/sr/6/641.71.de.pdf, accessed 10 May 2010.

Shaw, J. and Roth, R. (2002). Achieving an affordable low emission steel vehicle; an economic assessment of the Ulsab-Avc program design. *SAE World Congress & Exhibition*. Detroit, MI: SAE International.

Sherif, S.A., Zeytotinoglu, N. and Veziroglu, T.N. (2007). Liquid hydrogen: potential, problems, and a proposed research program. *International Journal of Hydrogen Energy*, **22**(7), 683–688.

Simbeck, D. and Chang, E. (2002). *Hydrogen Supply: Cost Estimate for Hydrogen Pathways – Scoping Analysis*. Golden, CO: National Renewable Energy Laboratory.

Simbeck, D.R. (2004). CO_2 capture and storage: the essential bridge to the hydrogen economy. *Energy*, **29**, 1633–1641.

Simbeck, D.R. and Chang, E. (2002). Hydrogen supply: cost estimate for hydrogen pathways: scoping analysis. Subcontractor Report NREL/SR-540–32525, Golden, CO: National Renewable Energy Laboratory.

Smit, R., Weeda, M., and de Groot, A. (2007). Hydrogen infrastructure developments in the Netherlands. *International Journal of Hydrogen Energy*, **32**, 1387–1395.

Sokolov, A.P., Stone, P.H., Prinn, R., *et al.* (2009). Probabilistic forecast for 21st century climate based on uncertainties in emission (without policy) and climate parameters.

Report Series of the MIT Joint Program on the Science and Policy of Global Change, January 2009.

Solomon, S., Qin, D., Manning, M., Alley, R.B., *et al.* (2007). Technical Summary. Climate Change 2007: The Physical Science Basis. Contribution of Working Group I to the Fourth Assessment Report of the Intergovernmental Panel on Climate Change. Cambridge, UK: Cambridge University Press.

Sorrell, S., Dimitropoulos, J., and Sommerville, M. (2009). Empirical estimates of the direct rebound effect: a review. *Energy Policy* **37**(4), 1356–1371.

Spath, P.L. and Mann, M.K. (2001). *Life Cycle Assessment of Hydrogen Production via Natural Gas Steam Reforming*. Golden, CO: National Renewable Energy Laboratory.

Spielmann, M., Bauer, C., Dones, R., and Tuchschmid, M. (2007). Transport Services. Ecoinvent report No. 14. Swiss Centre for Life Cycle Inventories, Dübendorf, Switzerland.

Steinbacher, M., Fischer, A., Vollmer, M.K., *et al.* (2007). Perennial observations of molecular hydrogen (H_2) at a suburban site in Switzerland. *Atmospheric Environment*, **41**, 2111–2124.

Stern, N. (2006). *The Economics of Climate Change*, Cambridge, UK: Cambridge University Press.

Sterner, T. (2003). *Policy Instruments for Environmental and Natural Resource Management*. Washington DC: Resources for the Future.

Stiegel, G.J. and Ramezan, M. (2006). Hydrogen from coal gasification: an economical pathway to a sustainable energy future. *International Journal of Coal Geology*, **65**, 173–190.

Stremler, B. (2008). The Rising Price of Fuel: The Toyota Perspective. Available online: www.pollutionprobe.org/Happening/pdfs/risingprice/Stremler.pdf, accessed 10 January 2011.

Struben, J. (2004). Technology transition: identifying challenges for hydrogen vehicles. *Proceedings of the 22nd International Conference of ths System Dynamics Society*, July 25–29, Oxford, England.

Struben, J. and Sterman, J.D. (2006). Transition challenges for alternative fuel vehicle and transportation systems. MIT Sloan Research Paper **4587–06**.

Sundström, O. and Stefanopoulou, A. (2007). Optimum battery size for fuel cell hybrid electric vehicle with transient loading consideration: part II. *Journal of Fuel Cell Science and Technology*, **4**, 176–184.

Sundström, O., Ambühl, D., and Guzzella, L. (2008). On implementation of dynamic programming for optimal control problems with final state constraints. *Les Rencontres Scientifiques de l'IFP: Advances in Hybrid Powertrains*. Advances in Hybrid Powertrains. Rueil-Malmaison, France: IFP.

Swissinfo (2007). UN rapporteur calls for biofuel moratorium. Available online: http://www.swissinfo.org/, accessed 11 October 2007.

Thomas, S. (2009). Transportation options in a carbon-constrained world: hybrids, plug-in hybrids, biofuels, fuel cell electric vehicles, and battery electric vehicles. *International Journal of Hydrogen Energy*, **34**, 9279–9296.

Thurnheer, T., Soltic, P., and Dimopoulos Eggenschwiler, P. (2009). S.I. engine fuelled with gasoline, methane, and methane/hydrogen blends: heat release and loss analysis. *International Journal of Hydrogen Energy*, **34**(5), 2494–2503.

Tian, A. (2010). China sells 13.6 million vehicles in 2009, *China Automotive Review*, 13 January

Tol, R.S. (2005). The marginal damage costs of carbon dioxide emissions: an assessment of the uncertainties. *Energy Policy*, **33**, 2064–2074.

Tromp, T., Shia, R., Allen, M., Eiler, J.M., and Yung, Y.L. (2003). Potential environmental impact of a hydrogen economy on the stratosphere. *Science*, **300**, 1740–1742.

Turton, H. (2006a). Scenario and policy analysis of sustainable energy systems and automobile transportation. PhD thesis No. 16886, ETH Zurich, Switzerland.

Turton, H. (2006b). Sustainable global automobile transport in the 21st century: An integrated scenario analysis. *Technological Forecasting and Social Change*, **73**(6), 607–629.

Ulli-Beer, S., Gassmann, F., Bosshardt, M., and Wokaun A. (2010). Generic structure to simulate acceptance dynamics. *System Dynamics Review*, **26**(2), 89–116.

Ulli-Beer, S., Bosshardt, M., Gassmann, F., and Wokaun, A. (2008). Citizens' choice: modeling long term technology transition in the automobile industry. *Proceedings of the 26th International Conference of the System Dynamics Society*, July 20–24, Athens, Greece.

UNFCCC (1999). The Kyoto Protocol to the Convention on Climate Change, UNFCCC Climate Change Secretariat and UNEP, UNEP/IUC/99/10, Geneva, Switzerland: United Nations Framework Convention on Climate Change.

Vasic, A.-M. and Weilenmann, M. (2006). Comparison of real-world emissions from two-wheelers and passenger cars. *Environmental Science and Technology*, **40**(1), 149–154.

Viebahn, P., Kronshage, S., Trieb, F., and Lechon, Y. (2008). Deliverable 12.2, RS Ia: Final report on technical data, costs, and life cycle inventories of solar thermal power plants. New Energy Externalities Developments for Sustainability (NEEDS) Integrated Project, EU 6th Framework Programme, Brussels.

Vinot, E., Scordia, J., Trigui, R., Jeanneret, B., and Badin, F. (2008). Model simulation, validation and case study of the 2004 THS of Toyota Prius. *International Journal of Vehicle Systems Modelling and Testing*, **3**(3), 139–167.

Vitart, X., Carles, P., and Anzieu, P. (2008). A general survey of the potential and the main issues associated with the sulfur–iodine thermochemical cycle for hydrogen production using nuclear heat. *Progress in Nuclear Energy*, **50**(2–6), 402–410.

Vollmer, M.K., Juergens, N., Steinbacher, M., Reimann, S., Weilenmann, M. and Buchmann, B. (2007). Road vehicle emissions of molecular hydrogen (H_2) from a tunnel study. *Atmospheric Environment*, **41**(37), 8355–8369.

Vollmer, M.K., Walter, S., Bond, S.W., Soltic, P., and Röckmann, T. (2010). Molecular hydrogen (H_2) emissions and their isotopic signatures (H/D) from a motor vehicle: implications on atmospheric H_2. *Atmospheric Chemistry and Physics*, **10**, 5707–5718.

Wallner, T., Lohse-Busch, H., Gurski, S., *et al.* (2008). Fuel economy and emissions evaluation of BMW Hydrogen 7 Mono-Fuel demonstration vehicles. *International Journal of Hydrogen Energy*, **33**, 7607–7618.

Warwick, N.J., Bekki, S., Nisbet, E.G., and Pyle, J.A. (2004). Impact of a hydrogen economy on the stratosphere and troposphere studied in a 2-D model. *Geophysical Research Letters*, **31**(L05107), doi:10.1029/2003GL019224.

WBCSD (2004). Mobility 2030: Meeting the challenges of sustainability, World Business Council for Sustainable Development. The Sustainable Mobility Project, Full Report. Geneva: WBCSD.

Weaver, J., Muci-Küchler, K., and Kamali, S. (2005). Heuristics for Architecting Automobiles and Automotive Systems: Educating the Next Generation of Automotive System Architects. *Proceedings of the 2005 SAE World Congress*. Detroit, MI: SAE World Congress.

Weilenmann, M., Favez, J.-Y., and Alvarez, R. (2009). Cold-start emissions of modern passenger cars at different low ambient temperatures and their evolution over vehicle legislation categories. *Atmospheric Environment*, **43**(15), 2419–2429.

Weilenmann, M., Soltic, P., Saxer, C., Forss, A.-M., and Heeb, N. (2005). Regulated and nonregulated diesel and gasoline cold start emissions at different temperatures. *Atmospheric Environment*, **39**(13), 2433–2441.

Weiss, M.A. *et al.* (2000). *On the Road in 2020: A Life-cycle Analysis Of New Automobile Technologies*. Available online: http://web.mit.edu/energylab/www/pubs/el00-003.pdf, accessed 16 September 2009.

Wenzel, T. and Ross, M. (2008). The weight relationship between vehicle weight, size and safety. *Proceedings of Physics of Sustainable Energy: Using Energy Efficiently and Producing It Renewably*. Berkeley CA.

Wieckert, C., Frommherz, U., Kräupl, S., *et al.* (2007). A 300 kW solar chemical pilot plant for the carbothermic production of zinc. *Journal of Solar Energy Engineering*, **129**, 190–196.

Wigley, T.M.L. (2003). MAGICC/SCENGEN 4.1: Technical Manual, Boulder, CO:. National Center for Atmospheric Research.

Wigley, T.M.L. and Raper, S.C.B. (1997). Model for the assessment of greenhouse-gas induced climate change (MAGICC Version 2.3.). Norwich, UK: Climatic Research Unit, University of East Anglia.

de Wit, M.P. and Faaig, A.P.C. (2007). Impact of hydrogen onboard storage technologies on the performance of hydrogen vehicles: a techno-economic well-to-wheel assessment. *International Journal of Hydrogen Energy*, **32**, 4859–4870.

World Health Organization (2006). WHO air quality guidelines for particulate matter, ozone, nitrogen dioxide and sulfur dioxide: global update 2005; Summary of risk assessment. Geneva, Switzerland:WHO.

Xiao, X., Prinn, R.G., Simmonds, P.G., *et al.* (2007). Optimal estimation of the soil uptake rate of molecular hydrogen from the Advanced Global Atmospheric Gases Experiment and other measurements. *Journal of Geophysical Research*, **112**(d07303), doi:10.1029/2006JD007241.

Yang, C. and Ogden, J. (2007). Determining the lowest-cost hydrogen delivery mode. *International Journal of Hydrogen Energy*, **32**, 268–286.

Yu, X., Pingwen, M., Ming, H., Baolian, Y., and Shao, Z.-G. (2009). The critical pressure drop for the purge process in the anode of a fuel cell. *Journal of Power Sources*, **188**(1), 163–169.

Yver, C., Schmidt, M., Bousquet, P., Zahorowski, W., and Ramonet, M. (2009). Estimation of the molecular hydrogen soil uptake and traffic emissions at a suburban site near Paris through hydrogen, carbon monoxide and radon-222 semi-continuous measurements. *Journal of Geophysical Research*, **114**(D18304), doi:10.1029/2009JD012122.

Zachariadis, T., Samaras, Z., and Zierock, K.-H. (1995). Dynamic modeling of vehicle populations: an engineering approach for emissions calculations. *Technological Forecasting and Social Change*, **50**, 135–149.

Zittel, W. and Altmann, M.A. (1996). *Molecular hydrogen and water vapour emissions in a global hydrogen energy economy*. 11th World Hydrogen Energy Conference, Stuttgart, Germany.

Index